Computer-Mediated Communication:

Issues and Approaches in Education

Sigrid Kelsey
Louisiana State University, USA

Kirk St.Amant
East Carolina University, USA

Information Science
REFERENCE

Managing Director:	Lindsay Johnston
Book Production Manager:	Sean Woznicki
Development Manager:	Joel Gamon
Development Editor:	Hannah Abelbeck
Acquisitions Editor:	Erika Carter
Typesetters:	Mackenzie Snader, Milan Vracarich Jr.
Print Coordinator:	Jamie Snavely
Cover Design:	Nick Newcomer

Published in the United States of America by
Information Science Reference (an imprint of IGI Global)
701 E. Chocolate Avenue
Hershey PA 17033
Tel: 717-533-8845
Fax: 717-533-8661
E-mail: cust@igi-global.com
Web site: http://www.igi-global.com

Library of Congress Cataloging-in-Publication Data

Computer mediated communication : issues and approaches in education / Sigrid Kelsey and Kirk St. Amant, editors.
p. cm.
Includes bibliographical references and index.
Summary: "This book examines online interactions from different national, cultural, linguistic, legal, and economic perspectives, exploring how the increasingly international and intercultural Internet affects the ways users present ideas, exchange information, and conduct discussions online"--Provided by publisher.
ISBN 978-1-61350-077-4 (hardcover) -- ISBN 978-1-61350-078-1 (ebook) -- ISBN 978-1-61350-079-8 (print & perpetual access) 1. Communication of technical information. 2. Cyberspace. 3. Social media. 4. Distance education--Computer-assisted instruction. I. Kelsey, Sigrid. II. St. Amant, Kirk, 1970-
T10.5.C5929 2012
371.33'4--dc23
2011017913

British Cataloguing in Publication Data
A Cataloguing in Publication record for this book is available from the British Library.

Table of Contents

Section 3
Case Studies

Section 4
Online Collaboration / Online Communities

Foreword

THE CHANGING NATURE OF EDUCATION: EVERY EDUCATIONAL DELIVERY FORMAT CAN INCORPORATE NEW TECHNOLOGIES

Computer-mediated communication permeates our daily lives. We check and post status updates on Facebook, Tweet, text, check apps on mobiles, and still spend significant amounts of time on email. Social networking and the use of devices such as clickers are incorporated into many face-to-face and virtual classrooms. In addition to providing means to cover and explore course content, such technology-based communications provide options for individual work, as well as countless ways to acquire and strengthen project management skills for effective teamwork. As information professionals, we are active partners in the use of technologies to support and extend communication in educational settings, whether those settings are formal courses or individual interactions with users in their search for information. It is this environment that sets the context for modern educational practices.

Today, online education is meaningful for millions of Americans. The eighth annual national survey on online learning, *Class Differences: Online Education in the United States*, found that in the fall of 2009, almost 30 percent of students took one or more online courses. This number represents more than 5.6 million individuals — a million more students than reported in the previous year. These include courses the study categorizes as web facilitated, blended/hybrid, or entirely online (Allen & Seaman, 2010). Around half of the respondents to this survey (which included 2,500 higher education institutions) noted that the recession has contributed to increased interest in completing online courses and/or entire programs online. The forecast is that interest in and involvement with online education will continue to grow.

Online education has created opportunities for students around the world, for it permits them to enter into learning environments at their convenience. Whether or not librarians have completed online courses as part of their professional education, it behooves them, at a minimum, to become aware of the options of online communication and education. Regardless of whether library faculty members in graduate programs accredited by the American Library Association (ALA) deliver online courses, they are called on to assist their students to be successful in these settings.

A glance at current job vacancy announcements illustrate these needs as applicants are called on to support, advance, and create college and university's online education initiatives. Prospective employees should demonstrate "knowledge of and interest in exploring emerging technologies" (Missouri State University, 2011), or, more specifically, they should lead "the creation of library instructional videos, screencasts and other multimedia presentations" (Houston Academy of Medicine, 2011). And, of course, applicants are still expected to have traditional skills including excellent written, oral, and interpersonal communication skills.

These responsibilities are obvious in the current accreditation standards for LIS programs, developed by ALA's Committee on Accreditation. Standards stipulate that each LIS program curriculum "integrates the theory, application, and use of technology" (American Library Association, 2008, p. 7). The role of the librarian is more specifically addressed in competency documents developed to delineate the skills, attributes, and knowledge needed for entry into specific information service areas. These changes in LIS education are most often seen in references to skills needed in employing technologies in teaching and other forms of communication. For example, in their teaching roles, music librarians "promote the effective use of all technologies" (Hunter, 2002, p. 5). Art librarians "have a broad understanding of information technology and are skilled in the implementation and utilization of technological tools, regardless of specific format, medium, or method of delivery . . . And help shape information technology products" (Ball & Harrington, pp. 8-9). The Reference and User Services Association (RUSA) has developed a number of excellent guidelines for information professionals, including the revised "Guidelines for Behavioral Performance of Reference and Information Service Providers" and the "Guidelines for Implementing and Maintaining Virtual Reference Services," approved in 2010 (RUSA, 2010).

These statements might be perceived as ironic, as I am speaking as a Professor in, perhaps, the only library and information science (LIS) program that proudly continues to deliver its coursework only in face-to-face format. Still, even our courses involve aspects of online education: faculty and students use Blackboard for email, discussion threads, document sharing, and e-grading. My colleagues have established presences in Second Life and supervised a doctoral dissertation on the results of a Second Life educational cooperative (Chavez, 2008). Faculty might serve as guest lecturers in other LIS programs and participate in conversations through web conferencing services such as Adobe Acrobat Connect, Clarity, and Elluminate.

We incorporate active learning experiences to support differing learning styles and to model to our students the strategies that they might use in their future careers. Students and faculty use Skype to continue personal and professional connections. Students participate in a mentoring program in the University of Texas at Austin University Libraries that may expose them to digital library services. Similarly, student Capstone projects, culminating efforts completed at the end of students' coursework, have included using social media to market library services and to establish new services including online book clubs. Thus, even within face-to-face educational settings, online technologies enhance learning and help present alternative educational options to students. Educators can and should embrace such methods to experience them now and to prepare for the next innovations.

Loriene Roy
The University of Texas at Austin, USA

***Loriene Roy** is Professor in the School of Information at the University of Texas at Austin. She serves on the boards for the American Indian Experience, Academic Solutions Database, Outreach Connections Native Health Information Portal, International Children's Digital Library, WebJunction.org, and TexShare. Roy is founder/director of "If I Can Read, I Can Do Anything," a reading club for Native students. She was the 2007-8 American Library Association President. Her awards include the 2009 Leadership Award, National Conference Tribal Archives, Libraries, and Museums; 2007 State of Texas Senate Proclamation; 2006 ALA Equality Award; and 2007 Library Journal "Mover & Shaker." She is a Trustee of the LeRoy C. Merritt Humanitarian Fund and Convener of the IFLA SIG on Indigenous Matters. She teaches reference, library instruction, reader's advisory, and public librarianship courses. She co-edited books on librarian education and service/civic engagement, has written numerous articles, and delivered over 500 presentations. She is Anishinabe (White Earth Reservation), a member of the Minnesota Chippewa Tribe.*

REFERENCES

Allen, I. E., & Seaman, J. (2010). *Class differences: Online education in the United States, 2010.* USA: Babson Survey Research Group. Retrieved from http://sloanconsortium.org/ publications/ survey/ pdf/ class_ differences.pdf

American Library Association. (2008). *Standards for accreditation of master's programs in library & information studies*. Chicago, IL: American Library Association. Retrieved on April 17, 2011, from http:// ala.org/ ala/ educationcareers/ education/ accreditedprograms/ standards/ standards_2008.pdf

Ball, H., & Harrington, S. (n.d.). *ARLIS/NA Core competencies for art information professionals*. Retrieved on April 17, 2011 from http://www.arlisna.org/ resources/ onlinepubs/ corecomps.pdf

Houston Academy of Medicine. (2011). *Instructional Technologies Librarian*. Retrieved on April 17, 2011 from http://www.ischool. utexas.edu/ jobweb/ JobDetails.php? JobID=37868

Hunter, D. (2002). *Core competencies and music librarians*. Retrieved on April 17, 2011, from http:// www.musiclibraryassoc.org/ uploadedFiles/ Employment_ and_ Education/ Music_ Librarianship/ Core_ Competencies.pdf?n=7658

Missouri State University. (2011). *Reference Librarian*. Retrieved on April 17, 2011, from http://www. ischool. utexas.edu/ jobweb/ JobDetails. php?JobID= 37842

Reference and User Services Association. (2004). *Guidelines for behavioral performance of reference and information service providers*. Chicago, IL: American Library Association. Retrieved from http:// www.ala.org/ ala/ mgrps/ divs/ rusa/ resources/ guidelines/ guidelinesbehavioral.cfm

Reference and User Services Association. (2010). *Guidelines for implementing and maintaining virtual reference services*. Chicago, IL: American Library Association. Retrieved on April 17, 2011, from http:// www.ala.org/ ala/ mgrps/ divs/ rusa/ resources/ guidelines/ virtual-reference-se.pdf

Sanchez, J. (2009). Virtual worlds: New spaces for service learning . In Roy, L., Jensen, K., & Meyers, A. H. (Eds.), *Service learning: Linking library education and practice* (pp. 169–178). Chicago, IL: American Library Association.

Preface

Approaches to education continuously evolve as educators harness the ever-growing number of available technologies in an effort to meet the changing needs of students. It is a process in which today's instructors are continually adjusting their teaching methods to accommodate new modes of communication available to their students. As quickly as new technologies arise, educators must design approaches and pedagogical techniques to take advantage of their potential for education. At the same time, they need to address issues and problems the technology may introduce. The result is a complex balancing act between that which is known and that which is new.

In today's classrooms and libraries, computer-mediated communication (CMC) is the norm. With rare exceptions, students are digital natives: individuals who have grown up immersed in online media and who are well-adapted to CMC. For this generation of students, communicating with smart phones, laptops, and other Internet devices is almost as natural as breathing, and as a result, they have mastered the use of applications such as SMS, social networking, e-mail, and video calling. Educators must therefore seek ways in which to enrich their teaching to keep pace with the daily lives of contemporary students. It is through interacting with students as digital peers that effective discussions and debates can take place and education can meet the needs and expectations of the 21st century.

New technologies, complemented by a greater amount of educational material available to a wider audience than ever before, introduce new avenues of communication. The Internet and its related technologies offer instructional designers, educators, and librarians opportunities to develop learning environments that are conducive to instruction for today's students. As Dumova puts forth in her chapter in this collection, new technologies "promise to create learning environments that could transform many aspects of the traditional relationships between instructors and their students, including the presentation of course content, learning activities, student-teacher and student-to-student interactions, as well as assessment." Computer-mediated communication is the current mode of delivery for many instructional activities, and the number of technologies available is often inspiring, and is, at times, overwhelming.

However, approached correctly, new technologies for computer-mediated communication have the potential to generate effective approaches to education by quickly integrating themselves into mainstream instruction in context where teachers and students alike swiftly adopt new communication techniques. The developments emerging from such situations bring about benefits and challenges, and these factors affect education in a myriad of ways. As an integral part of technology, CMC is a vital topic to research and explore as it relates to education, and each new technological development brings with it new facets of CMC to explore and research in relation to their potential use in educational contexts.

Computer-Mediated Communication: Issues and Approaches in Education highlights current concepts, issues, and emerging trends relating computer-mediated communication and its issues in, and approaches to, education. To this end, this volume contains eighteen thematically organized chapters that highlight best practices, strategies, approaches, technologies, case studies, testing, trends, information literacy, examples, and models for using computer-mediated communication technologies in different educational settings. In order to provide readers with an understanding how CMC affects issues of and approaches to education this book

1. Introduces readers to a variety of technologies that facilitate computer-mediated communication in educational settings
2. Provides case studies and research regarding these topics
3. Discusses theoretical frameworks for studying CMC

A large and important component to this collection is the introduction and review of communication technologies used to mediate and facilitate education. Blogs, mini-blogs, virtual worlds, video conferencing, ePortfolios, and course management systems are just a few of the ways in which educators are using computers to communicate with their students. Thus, the collection's first section, "Overview of Technologies and Current Trends / Fundamental Concepts" includes six chapters that introduce, review, and provide ideas for integrating current technologies into educational practices. To do so, the chapters in this first section provide readers with an in-depth look at various CMC technologies. In so doing, the authors of these chapters provide analyses and recommendations for the strategic uses of such technologies in a variety of educational settings. Now, more than ever, educators must teach about the numerous tools available online, and how students can find, use, and evaluate information. The entries in this section offer an initial set of ideas readers can use to achieve these objectives.

Jenna Ryan's chapter, "Meeting Them Halfway: Using Social Networking to Connect with Students," examines several popular technologies and their possibilities for use in higher education. From the simple-to-use and popular social networking tools, such as Facebook and Twitter, to the more complicated CMC technologies like virtual worlds, Ryan reviews the technologies and related literature. She also makes recommendations for the uses of such technologies in a way that would enhance the use of CMC in educational settings.

Ryan's chapter is followed by another discussing the use of virtual worlds in the educational setting. Specifically, Stoerger puts forth a case study of the use of Second Life in several courses, including the use of Voice over Internet Protocol (VoIP) with Second Life. Martin and Hesseldenz, in turn, discuss the approaches to online information taken in academic libraries. Dumova's chapter presents results of a much-needed survey that determined how students perceive the usability of online testing in Blackboard. These findings bring to light several ways in which educators can make better use of online testing by using teaching strategies that utilize different pedagogical tools than face to face environments for better assessment practices. Next, Rice's chapter suggests ways in which educators can employ ePortfolios, and Whitehair and Tally's chapter examines how web conferencing has brought about significant changes to the ways in which educators are able to communicate with their distant students. The overall section then concludes with Rusnak's chapter that discusses the implications of online setting for visual arts, a discipline traditionally set in the classroom.

Just as educators must adapt to the changing technological landscape, so must they prepare their students for working in a world where understanding CMC is critical to professional success. Moreover, today's student must be prepared to work in a global society where communicating with worldwide teams via online media is commonplace. "Social Implications" – the second section in this collection – contains chapters that address the new global nature of society created by the international proliferation of CMC.

This section uses case studies and models that can inform educational practices and help educators who strive to prepare their students for participation in the modern global economy.

The section begins with Henry and Lima's presentation of a critical instance case study that describes a collaborative project between grade schools in the rural United States and in Rio de Janeiro, Brazil. Students involved in the collaboration learn to communicate through blogs as well as learn about the differences and similarities in each other's cultures and daily life. Through this use of CMC, students develop a sense of global citizenship, a vital trait to teach in today's society. The following chapter by Boehm and Aniola-Jedrzejek describe a similar kind of collaboration among college-aged students. The chapter again emphasizes the importance of students to develop a global perspective in order to be successful in today's economy. Thus, the students are assigned to partake in a collaborative cross-global team effort. It is a context that requires them to work as part of a globally-distributed team in which fellow students from another country and of a different culture interact via CMC technologies.

In the section's third chapter, Picherit-Duthler describes international collaboration as well in terms of how students' perceive working on global virtual teams. Next, Liberati suggests a framework that would enable physically disparate team members to pool resources and complete projects: for example, scientific experiments, without traveling. The paradigm he presents could allow experiments to be executed by scientists and science educators working apart from each other.

Assessing and evaluating online communication at any level can be challenging. In an environment that often lacks the cues available in face-to-face communication, conveying meaning can sometimes be more difficult. Hart suggests using speech codes theory, which is grounded in the ethnography of communication, as a framework by which to evaluate and analyze computer-mediated communication. Specifically, she suggests that is has the potential to shed light on CMC in educational settings. Xia's chapter, in turn, examines student opinions of online vs. on-site classes and reveals what factors can affect student perceptions of what makes an online class more, as, or less successful than an on-site one.

Section three, entitled "Case Studies" presents case studies and models relating to CMC in educational settings. Gunn and Kraemer, for example, discuss several models to integrate information literacy across a university's curriculum. According to these authors, information is communicated in the classroom and online through numerous technologies. As a result, it is more important than ever for students to know how to find, evaluate, and use information. Information literacy is often a focus of a university library's mission, and Kraemer and Gunn present ways in which librarians can integrate information literacy into courses across the curriculum.

Next, Eckstein's chapter presents a course project in which the students are required to make a public service announcement (PSA) video and share it by using Web 2.0 technology. According to Eckstein, Web 2.0 technology has turned the web from a passive tool to a contributive environment, with students contributing to the web by posting on YouTube, Facebook, blogs, and other 2.0 platforms. Studying how students interact with information is thus a vital component to understand the issues of CMC in education. In examining this idea, Eckstien presents an approach for teaching students how to harness the possibilities of CMC to effect social change.

Section Four – "Online Collaboration / Online Communities" – contains chapters that analyze how online communities form and how CMC affects them. Salam's chapter, for example, points out how applying activity theory to analyze online discussion transcripts can assist educators in understanding participation and knowledge construction in an online environment. Similarly, Tanner and Amato's chapter examines online discussions and provides research-based suggestions to improve participation among students. Then, Valentino's chapter analyzes more than two hundred online discussions among students, and uses the results to deduce what types of instructor facilitation encourage the best discussions.

CMC is an ever-changing and growing phenomenon that will continue to shape and be shaped by educational practices. These technologies offer educators incredible potential for transforming how they communicate with and teach students. The chapters in the book offer educators a foundation to understand some of the issues related to CMC and education, and tools and frameworks to apply theories to teaching. The editors hope readers will build upon this foundation to create new pedagogical approaches and generate original research that can facilitate the understanding of and uses of such technologies in a range of educational situations.

Sigrid Kelsey
Louisiana State University, USA

Kirk St.Amant
East Carolina University, USA

Acknowledgment

I would like to thank my family for supporting me in everything I do, and graduate assistant Daniel Vance for his assistance with this book.

Sigrid Kelsey

For my daughters, Lily and Isabelle, who are a source of inspiration in all that I do, and for my mother, Joan Pelletier St.Amant, whose example taught me to appreciate both the beauty and the value of education in all settings.

Kirk St.Amant

Section 1

Overview of Technologies and Current Trends / Fundamental Concepts

Chapter 1
Meeting Them Halfway:
Using Social Networking to Connect with Students

Jenna Ryan
Boston University, USA

ABSTRACT

Computer-mediated communication (CMC), especially via Web 2.0 technologies like social networking and casting software, has become an essential part of the lives of the current generation of college students. This chapter will explore the possibilities for the use of these technologies by universities and colleges to connect with their students and to enhance instruction. Topics covered will include Facebook, Twitter, virtual worlds, casting software, and future trends.

INTRODUCTION

The development of the Internet and computer-mediated communication (CMC) may be one of the most significant influences on higher education in the last 30 years (Hendley, 2000). Everett Rogers, in his seminal work on the diffusion of innovation through society, coined the term "early adopter" to describe those individuals who, while not the creators of new technology, are among the first to adopt new innovations into their daily lives, and serve as role models for others (1983, pp. 248-249). The qualities Rogers assigns to his "early adopter" category are qualities often found in academic environments: the ability to deal with abstract ideas, rationality, intelligence, favorable attitudes toward education and science, higher achievement motivation and high aspirations (1983, pp. 257-258).

As institutions that attract the early adopter type, colleges and universities – and in particular their students – have been at the forefront of

DOI: 10.4018/978-1-61350-077-4.ch001

the information revolution. The incorporation of computer-based technologies into the college classroom and library is also nothing new. From the replacement of the card catalog with the Online Public Access Catalog (OPAC) to the submission of assignments through email, e-textbooks and e-journals, course management systems, and fully web-based distance education, the electronic world has been rapidly incorporating itself into the traditional educational arena for years.

The "Net Generation" (also called Generation Y, Millennials, Bridgers, and NextGens) – those with birthdates range from 1982 to 2002 (Jones, 2008, p. 64) – and the emergence of "Web 2.0," or the interactive web, have made a whole new world of electronic communication open up. It is a world that is central to the lives of today's college students, the majority of whom were raised in a world in which technology was a pre-established part of everyday life, the extent to which has only increased during their lifetime (AlemaÏn & Wartman, 2008, p. 1). A recent study found that 90.3% of undergraduates use social networking sites to communicate with friends and acquaintances (Caruso, Smith, & Salaway, 2009, p. 14). While our students use computer-mediated communication as much, possibly even more than traditional forms of communication, faculty members, many who remember a time before the Internet revolution, may find these tools less than intuitive.

New CMC tools develop rapidly and their impacts can be unpredictable – technologies and services that show great initial promise may either completely fail to catch on or be quickly toppled by competitors (such as Friendster, whose initial popularity was quickly eclipsed by MySpace and Facebook), while another product getting off to a slow start may become a mainstay of the digital arsenal (Twitter, for example). Well-meaning and dedicated educators, in an attempt to reach their students on the students' own home ground, can quickly lose their bearings in a maze of websites and services, where the learning curve for one not raised as part of the Net Generation can be high.

While keeping pace with new technologies can be challenging, universities that ignore social networking tools do so at their peril. Heather Mansfield, who regularly consults for colleges and universities on social media, warns "If colleges don't start profiles for their school on Facebook, MySpace, Twitter, etc., then an alum or current student will do it for them [...] they run the risk of losing complete control of their brand online (Gilroy, 2010, p. 21)."

In this chapter, I examine a number of computer-mediated communication software programs that are being used by the current generation of college and university students, and examine how these tools can be adapted to the educational arena in general and the classroom in particular. While I acknowledge the importance of email and instant messaging as methods of computer-mediated communication in education, I focus specifically on Facebook, Twitter, and virtual worlds, while also touching upon podcasting, videocasting, and screencasting, and noting possible future trends.

A BRIEF HISTORY OF WEB 2.0

While the Internet properly began with the development of ARPANET in 1969 and more importantly the development of the Internetwork Protocol (IP) in 1974, it really was not until the late 1980s and early 1990s that computers with Internet access became common household items (AlemaÏn & Wartman, 2008, pp. 14-15). With the early Internet, the general user was primarily a consumer of information, not a producer. The development of Internet Browsers (Mosaic in 1990, Netscape, Internet Explorer, Firefox and others soon after) allowed the non-specialist to load and view web pages relatively easily, but at this time, it still took specialized knowledge to actually construct a web page (p. 17).

The development of what has been alternately called "Web 2.0" and the "participatory web" has dramatically changed the nature of the Internet.

New technologies began to be developed that allowed those with little or even no programming knowledge to create and post their own content online. Email, the earliest version of computer-mediated communication, began to be supplemented by Web 2.0 technologies that not only allowed the uninitiated user to create content, but also provided a mechanism for users to comment upon others' content and to communicate with other site owners (Pankl & Ryan, 2008, p. 847).

The term "Web 2.0" was coined in 2004 during a brainstorming session between O'Reilly Media, a leading American media company, and Media Live International, a company that specializes in developing and organizing conferences and events in the technology fields. The term mirrors the tradition in computing software development to label each major release or edition of a program as a whole number, with incremental updates and revisions as decimals. O'Reilly had an idea for a conference based on what was recognized as a new trend in computer culture: the participatory web (2005). In essence, Web 2.0 is a "new edition" of the Internet. Web 2.0 technologies assume that every user is an author in his or her own right, and the technologies take online collaboration and community ownership for granted.

The Web 2.0 movement began in the late 1990s with a technology called weblogs, typically simplified to "blogs." Strictly speaking, a blog is defined by two criteria: it is a website that consists of dated entries, and it is organized in reverse chronological order. The true impact of blogs was their facility as a method of quickly disseminating information and ideas among Internet users. While the first bloggers were primarily computer programmers, software quickly surfaced that provided WYSIWYG (What You See Is What You Get) interfaces, allowing the general user to become not just a consumer of information on the Internet, but also a producer. Early programs such as OpenDiary and LiveJournal gave way to more robust services such as Blogger and Wordpress – all aimed at allowing the general public to create and publish their own little corner of the web. Most of these services provided a method for other users and readers to comment on each original blog post, thus creating conversations as bloggers read and commented and hyperlinked their way through one another's online worlds.

The concept of online dialog that was helped along by the development of blogs became the cornerstone of future Web 2.0 technologies. At their core, all Web 2.0 programs and services aim to facilitate communication, conversation, and collaboration among often-times geographically dispersed users. As such, they hold great potential for use in higher education, where communication is essential and collaboration is increasingly valued.

CMC TOOLS

The various forms of Web 2.0 technologies are limited only by the imaginations of those who create them. From sites like MySpace and Facebook, with which users create profiles presenting an online face to the world, to entire virtual worlds in which the user's digital self-representation, or "avatar," can interact with others in an entirely digital landscape, CMC technologies exist for users of nearly any level of technological savvy or online engagement.

Facebook

Facebook is both a boon and a bane in higher education. While some educators find it a useful tool, it can also be a tempting distraction for students (Selwyn, 2009). Begun in 2004 at Harvard University, Facebook was, from the beginning, intended exclusively for universities (boyd & Ellison, 2007, p. 218), but has since expanded its reach well beyond the world of higher education. As a social networking site, Facebook allows users to establish a "profile," providing information about themselves and their interests. Once registered, users can then designate other users as "friends," expanding their social network. Facebook also allows those who designate each other as friends to share pictures, chat, and post comments on their own or a friend's "wall," a whiteboard-like space for communication. Additionally, organizations including colleges and universities are able to create Facebook "pages," with many of the same functionalities as a personal profile and in addition, the ability to send announcements to those subscribed and to create events to be sent to subscriber's Facebook calendars.

Despite its close association with and origins in higher education, however, Facebook remains primarily an extra-curricular social network. A 2007 study found that students who use Facebook and other social network sites consistently prefer that these technologies "remain within the scope of their private lives" (Salaway & Caruso, 2007, p. 12).

Nonetheless, trends change rapidly, and more and more educators are using Facebook to various ends and purposes. For example, the Facebook Page Directory for Colleges & Universities on the Varsity Outreach website contains links to more than 600 official university Facebook pages in the United States (2010). Unless otherwise specified by the user, messages sent from any University pages are emailed to everyone who has signed up as a "fan" of that page. Further, by default, posts on the page's wall also show up in individual fans' news feeds, a section of a user's profile that allows them to view updates on their friends' accounts, and on those pages of which they are fans.

College and university admissions offices frequently use Facebook as a method of outreach to prospective students, and Alumni Associations use it as a way of keeping in touch with their members (AlemaÏn & Wartman, 2008, p. 105). A recent study by the Center for Marketing Research at the University of Massachusetts, Dartmouth, found that the number of university admissions offices using social networking sites increased from 29% in 2007 to 61% in 2008 alone (Barnes & Mattson, 2009). University libraries have also been taking advantage of the outreach capabilities of Facebook, using their Facebook profiles to provide access to their online catalogs, to suggest resources, and to keep their fans updated on various topics, such as library hours and events at the library and on campus. Faculty and instructors are increasingly reaching out to students on Facebook (Magolis, 2008; Schwartz, 2009) particularly in online distance education classrooms where a Facebook page can serve as a "virtual homeroom" (Harrington & Floyd, 2009, p. 3). Moreover, a 2009 study determined that a Facebook presence and high levels of self-disclosure on the part of instructors actually increases student perceptions of an instructor's credibility (Mazer, Murphy, & Simonds, 2009).

Educators who choose to have a presence on Facebook must exercise caution, however, particularly if they intend to use a single profile for communicating with students as well as with their family and friends. While a certain level of self-disclosure can be beneficial to student-teacher relations, it is unwise for an instructor to allow students to see pictures and posts that would compromise an educator's integrity, or appear unprofessional. Students and faculty alike must keep in mind that the information they publish on Facebook, or any other social networking software site, leaves the user's control once it is posted or uploaded. Once online, any opinion or piece of information may be difficult or even impossible to retract. How we conduct ourselves online has an impact on the professional image we portray as educators (Hegna & Johnson, 2010). This is particularly relevant when using social networking to discuss student assignments. For example, it may seem like a simple proposition to respond to a student's Facebook post about frustration with an assignment. However, providing guidance through the Facebook interface means that the comments will be archived indefinitely and perhaps viewable to many, possibly violating teacher-student confidentiality ("Facebook: Some Implications for Academic Freedom, Intellectual Property, and Education.," 2009, p. 6).

Used responsibly, however, Facebook can be a useful tool for educators to connect with their students. Professor Harriet Schwartz reflects, "Facebook, instant messaging, and the like keep my metaphorical office door open" (2009, p. B13).

Twitter

Another CMC technology that has been infiltrating the realm of higher education is the Twitter service. Credited with popularizing the concept of "microblogging," Twitter allows users to post pieces of information or news, called "tweets," each containing 140 characters or less, similar to mobile text messaging. Users can also become "followers" of other Twitter users, in order to receive their friends' tweets on their own Twitter page. Twitter can be used for anything from detailing the daily minutia of one's life to reporting news as it happens, and everything in between. Services that provide multimedia hosting with URL shorteners such as TinyURL and bit.ly began to be developed in response to the 140 character limit of tweets and some SMS messages, with the shortened URSs allowing users to share links, pictures, and video without exceeding the character limit.

As with Facebook, many universities are establishing Twitter accounts for various purposes. Many universities now have one or more Twitter feeds, updating their followers in short bursts of information with related news, campus services, events, and sports scores. University libraries have also developed Twitter feeds, which can be used to announce changes in library hours, services available through outreach programs, and library events, while also promoting new books or database purchases, exhibits, and unique collections (DeVoe, 2009).

In addition to official university Twitter feeds providing information to the university community, some instructors have begun to integrate the use of Twitter into their curriculum. These faculty members are using Twitter to post information about extra credit opportunities, or to respond to questions from their students (E., 2009). Purdue University professor Sugato Chakravarty, for example, has set up a Twitter account for his large lecture classes, allowing students to comment or ask questions in real-time during class (Young, 2010). Twitter can also be used in the classroom to perform quick polls of student opinions on the topic being discussed, without the need for easily lost or damaged devices like clickers (Thames, 2009).

Some instructors, like Professor David Parry at the University of Texas at Dallas, have used Twitter to foster a sense of community within their classes. Professor Parry's students follow him and each other and post a few tweets a week as part of a writing assignment. Mr. Parry has said that using Twitter in his courses has helped to break down the barrier between in-class and out-of-class (Young, 2008). Proponents of integrating Twitter into the classroom maintain that it helps to re-imagine the professors' role from "sage on the stage" to "guide on the side" (Young, 2010, p. 10).

Twitter also has great potential as a communication tool in computer-based distance education. Twitter has a feature that allows users to create a "list" of other users, to be grouped together in a separate feed. The respective list may then be subscribed to by other users. If the instructor creates a list of the accounts of students in a particular class, and has the students subscribe to it, the list is able to become a forum for sharing and discussion among class members. The Twitter list can be used by the instructor to post course-relevant links and news stories, assignment reminders, and to propose discussion questions. Students can use the list to ask questions of each other and of the instructor, to share insights into course material, and to connect intellectually with other class members. The "reply" function on Twitter allows tweets to be sent to specific users as well as to the list as a whole.

Pedagogical studies have found that informal, personal interaction with instructors and with classmates enhances the student' intellectual commitment to the material and provokes critical thought about personal values and goals (Dunlap & Lowenthal, 2009, p. 130). Twitter facilitates an informal, immediate connection among instructors and students that is analogous to bumping into one another in the hall, or chatting before class. A primary benefit of Twitter is its sense of immediacy – in order to fully take advantage of this it is important for the instructor to be reachable via Twitter more than just occasionally. For example, an instructor choosing to integrate Twitter into a course should make an effort to have the program to running in the background, whenever he or she is in the office, as well as checking it as often as possible on one's home computer or smart phone. Students appreciate prompt responses to their questions and comments, and being available enhances their belief that the instructor cares about them personally.

Not everyone is enthusiastic about the potential use of Twitter in the educational arena. "It's not a research-based tool," Daniel T. Willingham, professor at the University of Virginia in Charlottesville, warns. "The most important thing to remember is that we have no idea what impact these tools will have on learning, and it will take a decade to answer that question (Manzo, 2009)." Privacy is perhaps the most serious concern when using Twitter as an educational tool (Ritchie, 2009). By default, all Twitter tweets are public, which may lead to students, or even teachers, sharing information that would more appropriately be kept within the closed environment of a class or university community. Thus, it is important that educators employing tools such as Twitter, educate themselves and their students about potential risks of using CMC technologies. As with any social networking service, identity theft and cyber-bullying, in which CMC tools are used to communicate offensive and sometimes threatening messages, are dangers that students must be made aware of when using those services, whether for personal or educational use.

In addition to privacy concerns already mentioned, many educators fear that Twitter represents a dangerous distraction from the academic material that should engage the students' attention. Studies have shown that students are not nearly as adept at multitasking as they believe they are (Manzo, 2009). In a world where college students can often be found surfing the web, texting on their phones, playing online games, or doing just about anything except paying attention while in class, many instructors fear that encouraging the use of Twitter and other Web 2.0 technologies in the classroom equates to tacit approval of "fooling around" in class. Twitter can also be used by students to share answers on tests and assignments and such cheating is harder to visually detect in multimedia classrooms than pre-technology cheating methods like passing notes.

Virtual Worlds

Another Web 2.0 technology that has sparked interest in the realm of higher education is the virtual world, also known as an immersive environment, or MUVE (Multi-User Virtual Environment.) Virtual worlds, of which Second Life is the most well-known, evolved from MMORPGs (Massive Multi-user Online Role Playing Games) such as EverQuest and World of Warcraft. The primary difference between the two is that virtual worlds do not have a storyline, or any sort of quests that users or participants must accomplish. The virtual world exists simply as an immersive environment in which users, represented by characters that have a 3D-like quality. These likenesses, called "avatars," are able to interact with other users in a digital universe that attempts to replicate physical space.

Virtual world avatars can walk, run, and in some cases, fly around a computer-generated landscape.

Users, through their avatars, can interact with virtual objects such as buildings, furniture, or even storefronts. An avatar is also able to interact with other users via chat, direct messaging, and often VoIP – "Voice over Internet Protocol," a protocol used for transmitting voice over the Internet using computer microphones and speakers. The very immersive quality of the virtual world, in which physical actions, body language, and voice can all be transmitted across distances, creates a sense of intimacy and personal connection that some feel is lacking in many other types of computer-mediated communication.

Most virtual world programs allow individuals or organizations, like colleges and universities, to lease 'virtual property' for a fee, at which point the lessee may build and shape their property as they wish. Some universities, such as the University of New Orleans (http://virtualcampus.uno.edu), have chosen to create entire "virtual campuses" using Second Life, constructing buildings and landscaping that mirror the real life campus, thus allowing the university to conduct "virtual tours" for prospective students. In addition to touring, virtual classrooms may be used for seminars or distance education classes, creating a sense of community by providing a visual representation of geographically distant participants.

University libraries, frequently early adopters of technology, were some of the first to explore virtual worlds for educational purposes. For example, the Alliance Library System, a multiregional library system in Illinois, established a library in Second Life in April 2006 (Bell, Peters, & Pope, 2007). The effectiveness of library services in virtual worlds is still in question, however. Two of the primary services that are offered by libraries: reference and collection access, are both ill-suited to virtual worlds. Providing reference services in the virtual world (sometimes referred to as 'in-world') requires significant training on the part of the library staff, and would reach fewer patrons than an easier to use chat reference service using an instant messenger client. Like Facebook and Twitter, there is also the issue of privacy – unless the virtual patron has 'friended' the virtual reference librarian and sent a private message, chat questions may be seen ('heard') by all avatars in the area, creating a public and often-times confusing level of chatter. Another other primary library service, collection access, is complicated by vendor agreements that would make access to electronic collections from in-world difficult if not impossible (Ryan, Miller, & Porter, in press).

Despite these problems, several library-like virtual communities exist, in Second Life and other virtual worlds, deriving their success primarily from displays and the distribution of themed 'notecards' on which information on a particular topic, including links to relevant online material, may be picked up automatically by an avatar and saved in the user's inventory for future reference.

Where virtual worlds truly excel in terms of their potential educational impact is in the creation of three dimensional learning modules or exhibits, generally called "sims" (short for simulation) in-world (Ryan, et al., in press). University libraries, for example, are able showcase digital images of their rare materials in virtual rooms that are designed to tie in with the subject matter of the collection. Science disciplines can create interactive models of complex reactions, microscopic organisms, or physical processes. Architecture students can use plans to see a building before it is built. In the fields of literature and history, scenes from fiction or historical eras can be recreated in amazing detail. Thus, sims offer a creative and new way in which to demonstrate learning material.

An existing module that exemplifies this ability is the Alliance Library System's "Land of Lincoln" sim in Second Life. This area features a historically accurate small-town street in which one's avatar can don period clothing provided at the entrance, and in which one can view PDF files of "The Liberator" – an abolitionist newspaper – at the newspaper office, browse the library for books popular in Lincoln's, time accessible in full text through Project Gutenberg, and view biographies of important figures by clicking on the headstones in the virtual graveyard. The sim also includes a Civil War army camp and a fully navigable 3-D version of the White House as it existed when Lincoln lived there, complete with portraits on the walls, period furniture, and even coal in the boiler rooms (Pasteur, 2007b). Another example of such a virtual setting is Eloise Pasteur's Dante's *Inferno* sim, created for a DeSales University class, which allows the user to descend through the various levels of Hell, complete with falling ash, hellish scenery, audible screams of the damned, and plenty of information, quotes, and multimedia pertaining to the famous work of literature (2007a).

However useful and engaging these CMC tools are, the use of virtual worlds for this purpose does have significant disadvantages. To begin with, it can be costly – as of the time of this writing, the cost in Second Life for a private region of the type most universities use is a $1000 initial purchase price with a $295 monthly maintenance fee (Second Life, 2010). In addition, building within most virtual worlds requires specialized knowledge of that world's programming language forcing, universities to either spend time and money training employees, or hiring outside assistance.

From the students' perspective, the learning curve for participating in virtual worlds is high. One university that holds distance classes in Second Life requires its students to attend two one-hour online sessions just to learn how to use the software (Rockinson-Szapkiw & Walker, 2009). The technical requirements of the software are a concern as well. Second Life in particular experiences high levels of lag in loading graphics and video even on a high-end gaming computer (Ryan, et al., in press). Students using lower-end or even average systems may well find navigating the virtual world to be more trouble than it is worth.

Podcasting, Videocasting, and Screencasting

In addition to providing new methods of instruction using a variety of media, Web 2.0 technologies are also able to make more traditional instructional formats more accessible and portable for students. For example, by using podcasting, videocasting, or screencasting, an educator can take a traditional college lecture or PowerPoint presentation and turn it into something that can be listened to or viewed on a home computer, smartphone, or an iPod-like device.

Podcasting, named for the frequent use of Apple's iPod mobile music device as the medium of distribution, involves the creation of an audio file, which may be downloaded to a computer or portable device and listened to, at a time convenient to a student's personal schedule. Podcasting software comes in many varieties, with some free options and others that are commercial, and some based entirely online, with others requiring software to be downloaded. What most of them have in common is the ability to record audio from a microphone as an mp3 file, and then edit and upload the file to a server with an RSS (Real Simple Syndication) feed. Some podcasting software allows the files to be hosted on the software company's servers.

An instructor using podcasting to distribute lectures or other class materials, provides the students with the URL for the podcast, and the students download the files to listen to on their own time. The student may either manually download the specific files desired or 'subscribe' to a podcast RSS, so that any new 'episodes' posted to the podcast site are automatically downloaded to the student's computer or portable device.

The ability to record lectures, post them online, and have the students listen at a time convenient for them is particularly helpful for distance education and non-traditional students, as an alternative to rigid and sometimes inconvenient traditional on-campus classes. As well, it provides a way for traditional students to catch up on missed classes. Software and hardware companies, presumably sensing an emerging market in educational multimedia, are going out of their way to support the educational use of their products. For example, Apple's powerful mobile device manager iTunes has a special section, iTunes U, which serves as a directory of education-related podcasts, as well as a subscription and download manager for its popular iPod and iPhone devices.

Videocasting, as the name suggests, is much like podcasting, only with the added ability to record video as well as audio. Videocasting is sometimes called vidcasting, vodcasting, or video podcasting. The popularity of the videocasting site YouTube, established in 2005 (boyd & Ellison, 2007, p. 212) has led to an increased use of videocasting in higher education. As with podcasting, instructors can record their lectures, including visual aids such as a white-board or physical demonstration, and post them online, where students may view the lectures at their convenience. With a video-enabled portable device, students can even take the lectures with them and attend class anywhere, from on the train to in a coffee shop.

Again, as with podcasting, the various media companies are aligning themselves with educators. YouTube has established YouTube EDU, to host education-related videos from verified educators. In order to post on YouTube EDU, qualifying two-year and four-year institutions must submit an application (Gilroy, 2010, p. 18). YouTube EDU has competitors as well, including BigThink (http://bigthink.com), Academic Earth (http://academicearth.org), and FOR A.tv (http://fora.tv), all in the business of hosting videocasts by and for educators.

The final 'casting' technology, screencasting, combines many of the strengths of the other two. Screencasting technology allows an instructor to record literally anything that can be displayed on a computer screen, and edit the final footage into a cohesive instructional video. Screencasts can include PowerPoint-like slides, video, audio, and on-screen demonstrations. They can be particularly useful in teaching research methods, where instructors can record their own online actions as they search the web or an academic database, demonstrating which menus to select and buttons to click while using audio or superimposed text to narrate the steps.

Screencasting can be used to create tutorials to search students how to use particular software or database, or to demonstrate computer-generated modeling or digital animation. Simple screencasting software, such as CamStudio or Jing, is available free, while more powerful commercial applications like Adobe Captivate offer advanced features, such as interactivity, which allows the students to 'try it themselves' from within the screencast. Screencasting, unlike podcasting or videocasting, is difficult to use on a portable device, but the advantages in most cases outweigh the disadvantages, in comparison, because it is the most versatile method of casting, allowing an instructor to present over the web most aspects of traditional classroom teaching, from the audible lecture and video demonstrations to interactive quizzes or activities.

A study by librarians teaching research skills at SUNY Oswego in 2000 concluded that students learned just as well from online tutorials as they did from traditional in-class instruction (Lo & Dale, 2009, p. 150). With screencasting software able to duplicate the classroom environment with much more authenticity than the software available in 2000, one would expect the same result in increasingly complex areas of study as technology improves.

One of the largest advantages to casting technologies is their ability to provide access to higher education to a significantly greater portion of the population than the traditional campus model. In addition to sites like YouTube EDU and iTunesU, universities like Massachusetts Institute of Technology with their OpenCourseware project (http://ocw.mit.edu), and University of California Berkeley with their Webcast Berkeley site (http://

webcast.berkeley.edu), have begun to provide course-related multimedia free to the public. Now, anyone with access to a computer and the Internet (or to a public library that provides such) can attend a course at MIT, UC Berkeley, or any of a number of other colleges and universities that are providing lectures and other course materials free online.

FUTURE TRENDS

In addition to the technologies discussed above, new technologies are always on the horizon. Some that may find their way into the classroom are discussed below.

Text Messaging

Similar in function to Twitter and to instant messaging programs such as AIM, text messaging technology utilizes SMS (Short Message Service) protocol to allow people to use their mobile phones to send short (usually 200 characters or less) messages to another mobile phone. In fact, Twitter's 140-character limit was designed, according to Twitter's webpage, so that tweets could be sent to or from mobile phones with 160-character limits. "Minus 20 characters for author attribution," the site says, "that gives users just enough room (http://twitter.com/about.)"

SMS has only recently begun to be used in the classroom, due to the fact that it previously required the instructor to have a phone with a text messaging plan. SMS gateway clients (programs that allowed the sending and receiving of text messages from devices other than a mobile phone) have been available for some time, but most, such as Skype, charged a fee per text message. More recent services, such as Yahoo! Mobile and Google Voice have begun to allow users to send and receive SMS text messages from their computers for free. In non-computer equipped classrooms, students are more likely to have mobile phones than laptops (Day & Kumar, 2010, p. 130). As with Twitter, SMS text messaging would allow the students to send questions and comments directly to the instructor and to receive a response, all using their mobile phones.

Mobile Applications

In a similar vein, mobile applications are becoming more relevant to higher education with the release and extreme popularity of "smart phones" such as the BlackBerry, the iPhone, and the Android. Far more than just phones, these mobile devices connect wirelessly to the Internet and can run any number of software applications specifically designed for the mobile platform.

Many colleges and universities have already designed relevant applications (apps) to communicate with their students and stakeholders. For example, students at Northwestern University can download free mobile applications provided by the university that allow them to search the library catalog and campus directory, and to view a campus map with their position indicated using the phone's internal GPS (Engebretson, 2010, p. 36). Students at Duke University are able to use their mobile devices to pull up menus from the university's dining facilities, manage their campus cash card accounts, view an events calendar, search the library catalog, and access course descriptions by using the university's free mobile applications suite (Duke University, 2010).

As these smart phones become increasingly ubiquitous on college campuses, institution-specific mobile applications will become an important campus resource, although at present, this is hampered by the need to create separate applications for use on various phone platforms (an iPhone application will not run on Android phones, and vice versa).

CONCLUSION

Of course, we are not yet at a point where Web 2.0 technology and computer-mediated communication can entirely replace face-to-face instruction, or the need for physical campus facilities and traditional communication. Indeed, we may never get to that point. But used creatively and appropriately, these technologies have the potential to both enhance traditional instruction and to open up the world of higher education to students who might be otherwise unable to participate in the traditional on-campus model of teaching.

Technologies like Facebook, Twitter, SMS, and mobile applications can improve communication among students, faculty, and administrators, as well as serving as recruitment tools. The social nature of these services fosters a sense of community and allows educators to meet the students on their own ground, promoting the students' perception that the faculty and staff are interested in them and in their educational progress.

Technologies like virtual worlds, podcasting, videocasting, and screencasting can enhance student learning by providing alternate pathways to instruction. Virtual worlds can be used to afford students the opportunity to interact virtually with learning environments that they could never experience in a classroom, such as wandering through a historical period or viewing an atom from within. Podcasting, videocasting, and screencasting are able to present information in a variety of formats and can be particularly helpful for students who are auditory and visual learners (Rockinson-Szapkiw & Walker, 2009, p. 180). For distance students in particular, these technologies allow for a greater engagement with the material than the text-based communication on which most distance courses are built.

The advances of the recent few decades have reshaped higher education into a more fluid, interactive, and responsive system. Barriers are breaking down between the physical and the virtual; between students and teachers. The focus has turned from one-way instruction to two-way communication. Computer-mediated communication and Web 2.0 technologies are providing more opportunities for students to become active participants in their own education, as opposed to passive recipients of knowledge.

These technologies are not some sort of panacea for success, however. Rather, they are tools to be wielded by competent and dedicated educators – an enhancement of, not a replacement for, traditional pedagogy. Although the majority of today's college students have grown up using these and other forms of social media, the technologies themselves may be a barrier to others, for example those with visually impairments, or students who do not own their own computers. CMC tools are also to be used with caution, since, as with most tools, they can be misused. Instructors should always keep in mind that they are the face behind the technology and that the ways in which they make use of the technology will be reflected back upon them. In addition, the electronic medium can be used to blur the distinction between acceptable collaboration among students and academic dishonesty. For an example, witness the 2008 controversy over a Ryerson University student's Facebook "study group" (Goodall, 2008).

We are living and teaching in interesting times. The software and services discussed in this chapter afford the contemporary instructor with unprecedented levels of flexibility and creativity in course design. Multiple platforms and media types allow for greater accommodation of different learning styles as well. Although it is too early to truly evaluate the impact these technologies will have on student success, computer-mediated communication has given us the tools needed to create not just a learning experience but also a learning environment.

REFERENCES

Alemań. A. M. M. N., & Wartman, K. L. (2008). *Online social networking on campus: Understanding what matters in student culture* (1st ed.). New York, NY: Routledge.

Barnes, N. G., & Mattson, E. (2009). *Social media and college admissions: Higher-ed beats business in adoption of new tools for third year*. Dartmouth, MA: Center for Marketing Research - University of Massachusetts Dartmouth.

Bell, L., Peters, T., & Pope, K. (2007). Get a (second) life! *Computers in Libraries, 27*(1), 10–15.

Boyd, D. M., & Ellison, N. B. (2007). Social network sites: Definition, history, and scholarship. *Journal of Computer-Mediated Communication, 13*(1), 210–230. doi:10.1111/j.1083-6101.2007.00393.x

Caruso, J. B., Smith, S. D., & Salaway, G. (2009). *The ECAR study of undergraudate students and Information Technology*. 2009: EDUCAUSE Center for Applied Research.

Day, J. M., & Kumar, M. (2010). Using SMS text messaging to create individualized and interactive experiences in large classes: A beer game example. *Decision Sciences Journal of Innovative Education, 8*(1), 129–136. doi:10.1111/j.1540-4609.2009.00247.x

DeVoe, K. M. (2009). Bursts of information: Microblogging. *The Reference Librarian, 50*(2), 212–214. doi:10.1080/02763870902762086

Duke University. (2010). *Mobile Web at Duke University*. Retrieved May 28, 2010, from http://m.duke.edu/

Dunlap, J. C., & Lowenthal, P. R. (2009). Tweeting the night away: Using Twitter to enhance social presence. *Journal of Information Systems Education, 20*(2), 129–135.

E., M. (2009). Higher ed: A-Twitter or not? [Article]. *University Business, 12*(9), 14-14.

Engebretson, J. (2010). Universities log on to hand-held mobile apps. *Connected Planet*, 36-38.

Facebook: Some Implications for Academic Freedom, Intellectual Property, and Education. (2009). *Distance Education Report, 13*(24), 4-6.

Gilroy, M. (2010). Higher education migrates to YouTube and social networks. [Article]. *Education Digest, 75*(7), 18–22.

Goodall, B. (2008, March 19). *Ryerson U. Won't expel student over Facebook study group*. Retrieved from http://chronicle.com/ blogPost/ Ryerson-U-Won-t-Expel/3771

Harrington, S. J., & Floyd, K. S. (2009). *Enhancing engagement and the value of the course to the student through course organization and active learning* [Article]. (pp. 3–5). Online Classroom.

Hegna, J., & Johnson, D. (2010). Guidelines for educators using social and educational networking sites. [Article]. *Library Media Connection, 28*(5), 50–51.

Hendley, V. (2000). 30 years of higher education. *AAHE Bulletin, 52*(7), 3–8.

Jones, A. R. (2008). Riding the technological rapids with the millennials. [Article]. *Christian Librarian, 51*(2), 64–67.

Lo, L. S., & Dale, J. M. (2009). Information literacy "learning" via online tutorials: A collaboration between subject specialist and instructional design librarian. [Article]. *Journal of Library & Information Services in Distance Learning, 3*(3/4), 148–158. doi:10.1080/15332900903375325

Magolis, D. (2008). Connecting with online library research methods course students in Facebook. [Article]. *Pennsylvania Library Association Bulletin, 63*(6), 8–10.

Manzo, K. K. (2009). Twitter lessons in 140 characters or less. (Cover story). [Article]. *Education Week*, *29*(8), 1–14.

Mazer, J. P., Murphy, R. E., & Simonds, C. J. (2009). The effects of teacher self-disclosure via Facebook on teacher credibility. [Article]. *Learning, Media and Technology*, *34*(2), 175–183. doi:10.1080/17439880902923655

O'Reilly, T. (2005). *What is Web 2.0: Design patterns and business models for the next generation of software.* Retrieved May 11, 2010, from http://oreilly.com/web2/ archive/ what-is-web-20.html

Pankl, E., & Ryan, J. (2008). Information commons and Web 2.0 technologies: Creating rhetorical situations and enacting Habermasian ideals in the academic library. In Kelsey, S., & Amant, K. S. (Eds.), *Handbook of research on computer-mediated communication* (*Vol. II*, pp. 845–854). Hershey, PA: Information Science Reference. doi:10.4018/978-1-59904-863-5.ch060

Pasteur, E. (2007a, July 12, 2007). *The inferno comes to Second Life*. Retrieved from http://www.secondlifeinsider.com /2007/07/12/ the-inferno-comes -to-second-life/

Pasteur, E. (2007b, October 24). *Land of Lincoln to come to SL*. Retrieved from http://www.secondlifeinsider.com/2007/10/24/land-of-lincoln-to-come-to-sl/

Ritchie, M. (2009). Chirping about Twitter. [Article]. *Times Educational Supplement*, *4836*, 18–21.

Rockinson-Szapkiw, A. J., & Walker, V. L. (2009). Web 2.0 technologies: Facilitating interaction in an online human services counseling skills course. [Article]. *Journal of Technology in Human Services*, *27*(3), 175–193. doi:10.1080/15228830903093031

Rogers, E. M. (1983). *Diffusion of innovations* (3rd ed.). New York, NY: The Free Press.

Ryan, J., Miller, R., & Porter, M. (in press). Library services in virtual worlds: An examination of the potential for library services in immersive environments. *Journal of Information Technology Education.*

Salaway, G., & Caruso, J. B. (2007). *The ECAR study of undergraduate students and Information Technology*. 2007: EDUCAUSE Center for Applied Research.

Schwartz, H. L. (2009). Facebook: The new classroom commons? [Article]. *The Chronicle of Higher Education*, *56*(6), B12–B13.

Second Life. (2010). *Private region pricing.* Retrieved May 21, 2010, from http://secondlife.com/land /privatepricing.php

Selwyn, N. (2009). Faceworking: Exploring students' education-related use of Facebook. *Learning, Media and Technology*, *34*(2), 157–174. doi:10.1080/17439880902923622

Thames, G. (2009). *Twitter as an educational tool* (*Vol. 22*, pp. 235–235). Wiley-Blackwell.

Twitter. (2010). *Twitter is the best way to discover what's new in your world.* Retreived November 23, 2010, from http://twitter.com/about

Varsity Outreach. (2010). *Facebook page directory for colleges & universities.* Retrieved May 14, 2010, from http://www.varsityoutreach.com/pagedirectory.aspx

Young, J. R. (2008). Forget e-mail: New messaging service has students and professors atwitter. [Article]. *The Chronicle of Higher Education*, *54*(25), A15–A15.

Young, J. R. (2010). Teaching with Twitter. [Article]. *Education Digest*, *75*(7), 9–12.

KEY TERMS AND DEFINITIONS

Facebook: A social networking service that allows users to create a personal profile and share photos, notes, and to communicate with a network of friends made up of other users.

Instruction: The dissemination of information from educators to students.

Mobile Applications: Software specifically designed for smart phones.

Podcast: A series of audio files recorded by a user and made available for others to download and listen to on their computers or mobile devices.

Screencast: A multimedia presentation constructed by recording the user's computer screen output, allowing the inclusion of audio as well as anything that may be displayed on a computer.

Text Messaging: The transmission of short messages using SMS protocol, usually between mobile phone users.

Twitter: A social networking service that allows users to broadcast short, 140-character messages online.

Videocast: A series of video files recorded by a user and made available for others to download and watch on their computers or video-enabled mobile devices.

Virtual Worlds: Any of a number of programs that allow users to create a visual representation of themselves and to interact with others in an immersive three dimensional like virtual environment without the constraints of plot associated with similar gaming technologies.

Web 2.0: Any Internet-based service or technology that allows the average user to create and disseminate content, while interacting with fellow content-creators.

Chapter 2
Transmedia Communication: The Virtual Classroom Experience

Sharon Stoerger
University of Wisconsin-Milwaukee, USA

ABSTRACT

Virtual worlds have the potential to foster new forms of educational communication among students and their instructors. These digital exchanges in virtual worlds are facilitated by computer-mediated communication (CMC) tools such as text-based media and Voice over Internet Protocol (VoIP). This chapter will investigate the media that were used to support student-instructor interactions in three continuing education courses situated in Second Life (SL). Based on these observations, text chat was more effective than VoIP at supporting educational discussions during these class sessions.

INTRODUCTION

Three-dimensional virtual worlds are visually and textually rich spaces that have, since their inception, facilitated new forms of communication (New Media Consortium, 2007). *Second Life* (SL), for one, is a highly complex, multimodal environment that is made up of text, audio, and images. Unlike massively multiplayer online games (MMOGs) that are set in virtual worlds, there is no pre-scripted back story that guides the SL activities. Registered users (i.e., residents) imagine and create the content for the virtual world. In SL, residents also have control over the appearance and exploration activities of their avatar, or digital character, which can be a powerful learning experience alone. As this suggests, being a resident in SL requires active participation.

Many technologies such as SL have flattened the world (Friedman, 2005) in ways that enable individuals from around the globe to come together virtually to interact and participate in shared experiences. These technologies allow users to take part in the creation and exchange of information in

DOI: 10.4018/978-1-61350-077-4.ch002

dynamic ways (Dede, 2008). Moreover, the sense of "being there" is heightened when individuals actively participate and interact with others in the virtual world (i.e., in-world) (Merchant, 2009). Because SL is robust, educators can create a wide variety of in-world activities, including ones that may be difficult or impossible for students to experience in the physical world.

For example, instructors can situate the virtual classroom in a replica of 19th century England, and discuss Jane Austen's novel *Pride and Prejudice* while dressed as Elizabeth Bennet. Instructors can also take students on a tour of Virtual Harlem to learn about Harlem, New York during the 1920s, or integrate sound into the environment to enhance an in-world dance performance by Ballet Pixelle. Some early adopters include programmers of innovative simulations such as the Virtual Stomach Museum where students can experience the look and feel of conditions such as ulcers and stomach cancer. SL is also a space where geographically dispersed individuals can virtually come together to participate in professional activities such as educational conferences and research initiatives. Because of the learning potential associated with SL, some scholars are predicting that all universities will have an in-world presence within 5 years (Salt, Atkins, & Blackall, 2008).

Text-based and VoIP options are available in many digital environments. Yet, the availability of specific communication options and the stability of these media vary from one virtual world to the next. Also, it is not uncommon for new modes of communication to be integrated into existing virtual worlds. In SL, for example, synchronous text chat and instant messaging (IM) were the initial options for in-world communication. VoIP, which is a common feature in many MMOGs, was introduced to the SL environment in 2007 (Lewis PR, 2009). To support in-world activities, synchronous and asynchronous out-of-world (that is, outside of the virtual world) options may be utilized. Discussion forums (e.g., Google Groups, Moodle), VoIP (e.g., Skype), microblogging technologies (e.g., Twitter), and email are a few examples of external tools used to supplement in-world interactions.

It is well-documented that SL has a steep learning curve (e.g., Fitzgibbon, Oldham, & Johnston, 2008; Luo & Kemp, 2008). In order for residents to thrive in SL, they must have a certain degree of skill, an understanding of the environment, and possession of virtual world related knowledge. The acquisition of expertise in SL requires socialization, learning through doing, and the development of domain knowledge. Further, the choice of communication media by individuals who frequent virtual worlds can be a complex process. The availability and the affordances of the technology impact these decisions and influence the use of these tools. The social norms of these spaces play a role in these choices, as well, and can change as new communication options become available to the in-world group. Yet, groups that express a preference for a particular communication mode may be reluctant to adopt a new one.

Merchant (2009) posits that virtual worlds foster new forms of communicative relationships among teachers and students. With the exception of Steinkuehler (2004), who examined the discourse among players in the MMOG, *Lineage*, there is scant research that investigates the characteristics of the in-world interactions (Mon, 2009). While much of the literature touts the benefits of virtual worlds, scholars such as Salmon (2009) stress that more experimentation and further examination of the pedagogical benefits of these digital environments is needed. This study examines the educational and social interactions that took place during three SL-based continuing education courses and the media used to facilitate this process. The findings illustrate the patterns that emerged from the SL course interactions, as well as the ways in which these patterns changed as students engaged in active learning tasks. In addition, the factors that impacted the computer-mediated communication (CMC) tools used by the students and the instructors will be described.

BACKGROUND

Computer-Mediated Communication and Virtual Worlds

Text-Based Communication Media

Until 2007, the primary forms of communication in SL were text chat and real-time communication using instant messaging (IM). Unlike IM, text chat in SL can be seen by users in the area. Text chat and IM are viewed as "media poor" forms of communication (Short, Williams, & Christie, 1976) due to their lack of cues (e.g., voice tone, inflections, facial expressions, body language). But some advocates consider text-based options to be highly versatile and expressive media that support social connections (Nardi, Whittaker, & Bradner, 2000). Text chat, for one, is easier to set up than audio; it requires low bandwidth, and no extra equipment such as microphones, headsets, or speakers (Table 1). Individuals who can login to the system can communicate via text. Further, text messages are prefixed with an avatar ID tag, which identifies the speaker. Chat is also less intrusive; the text message is displayed on the computer screen, which is considered less disruptive than an audio message (Bradner, Kellogg, & Erickson, 1999; Scholl, McCarthy, & Harr, 2006). By using the chat feature, it is possible to arrange private conversations with in-world participants, as well.

Because the conversations are persistent, the text chat has asynchronous qualities. The program is also designed to foster a sense of presence in the digital environment; once residents login to the system, their avatars appears on the screen

Table 1. SL communication modes: Strengths and weaknesses

Communication Mode	Strengths	Weaknesses
Text-based (IM, text chat)	Eliminates disruptions to other individuals in the physical world	Does not eliminate grammatical errors, poor spelling, or typographical errors
	Levels the communication hierarchy (i.e., all class participants can communicate at any time)	Favors fast typists
	Provides a chat history window for review	Hinders communication for non-English speakers and individuals with learning disabilities (e.g., dyslexia)
	Records a transcript of the discussion	
	Requires no extra equipment	
VoIP	Allows the transmission of extra information such as gender, tone of VoIP, inflection, accent, etc.	Adds technical difficulties
	Reduces communication difficulties for individuals with learning disabilities	Disturbs individuals in the physical world
	Transmits physical world talking in the virtual world. Reduces communication difficulties for individuals with learning disabilities	Lacks communication cues that support turn taking
		May be difficult for ESL and speakers with strong accents to be understood
		Requires additional equipment (e.g., headset, microphone, speakers)
		Reveals physical world characteristics of the speaker

and remain visible until they logout. While the avatar may be present, these individuals may be physically and/or mentally AFK (away from the keyboard) (Boellstorff, 2008). However, a resident who has been AFK and missed part of the discussion can scroll back through the history of the text to catch up. Residents can also wait to respond or ignore messages and come back to them later. Some chat advocates also note that text chat allows users to correct and/or think about what they write prior to sending the message (Scholl et al., 2006). The research suggests that it is easier to communicate with a large group via chat than with VoIP, as well (Scholl et al., 2006).

The CMC literature notes that real or perceived anonymity can create different types of interactions. These can be positive in that individuals feel freer to express their ideas (Finholt & Sproull, 1990; Garton & Wellman, 1995). In addition, anyone can contribute to the chat discussion at any time, which creates a more egalitarian communication environment (Fanderclai, 1996; Jenkins, 2004) while challenging the notion of authority (Kress, 2005). With text chat, hearing impaired and deaf individuals can also participate in SL discussions (Peters & Bell, 2007). Conversely, anonymity can encourage negative communication traits, including the use of abusive language or "flaming" (Lea, O'Shea, Fung, & Spears, 1992; Sproull & Kiesler, 1991). It may also take longer for groups to make a decision (Garton, Haythornthwaite, & Wellman, 1999). Because the communication pace is rapid and the discussion threads are multi-braided, the norms of turn-taking and conventional spelling are often violated. As Merchant (2009) notes, however, these chaotic chat conversations, combined with unrestricted movement around the virtual environment can create an exhilarating atmosphere.

When individuals are coming together from geographically dispersed locations, language proficiency may be a communication factor to consider, as well. Non-native English speakers may have difficulty contributing in text-based communication in an English language setting, due to the rapid pace of discussions, and support strategies, such as rephrasing or summarizing earlier posts may be needed (e.g., Park & Bonk, 2008). Moreover, not everyone is an accomplished typist. With chat, fast typists are privileged and may monopolize discussions. Individuals with good computing skills also have an advantage in the initial chat discussions, which may enable them to more actively participate in the discussions (Harmon & Jones, 2001; Ingram, Hathorn, & Evans, 2000).

VoIP Communication Media

Until recently, online communication in virtual worlds has occurred largely through text and images (Halloran, Rogers, & Fitzpatrick, 2003; Williams, Caplan, & Xiong, 2007). According to the CMC literature, the availability of VoIP offers the potential of a rich communication medium to support virtual interactions (e.g., Daft & Lengel, 1984). Compared to text-based communication modes, which have fewer cues, VoIP is a rich media (Table 1). Because of the presence of additional cues, the communication is less anonymous – gender, personality characteristics, and emotions may be revealed (Nass & Gong, 2000). VoIP is also closer to face-to-face communication than text, which may enhance the communication experience and foster collaboration (Jensen, Farnham, Drucker, & Kollock, 2000). Moreover, some research suggests that individuals communicate more via VoIP than with text (Sallnäs, 2005).

Since the introduction of VoIP, which transmits computer-mediated VoIP communication over the Internet, the use of VoIP-mediated communication is growing in SL (Linden Research, 2009). The integration of VoIP into virtual worlds enables individuals to verbally talk to each other. Some scholars suggest that this makes text communication unnecessary. For example, some argue that VoIP creates greater immersion in the virtual world and helps co-located individuals get to know each other better (Halloran et al., 2003). VoIP has also been asserted to emulate a "real" experience (Scholl et al., 2006). But individuals such as virtual world researcher Richard Bartle believe that VoIP as a medium is too rich. Bartle (2003) argues that VoIP does not enhance the feeling of immersion, but rather distracts from that experience. Nonetheless, some users prefer VoIP because it is easier to use (e.g., no spelling, faster than typing) and is thought to be a more natural and expressive medium (Scholl et al., 2006).

At first glance, it would appear that the integration of VoIP communication in virtual worlds would be useful and support residents' interactions and conversations. Research shows that it is easier to discuss complex topics using media with multiple cues such as VoIP, rather than through less rich media such as synchronous text chat modes (Teng & Taveras, 2004-2005; Walther, 1999). Therefore, VoIP has the potential to support and enhance more advanced discussions. Further, the addition of VoIP communication in SL has been beneficial to and supported interactions with visually impaired residents – individuals who could not participate in SL via text chat (Mon, 2009).

But while VoIP is viewed as being more aligned with face-to-face communication, it may create difficulties for individuals engaged in online interactions. First, it can make it difficult to recognize in-world communicators (Halloran, et al., 2003; Wadley, Gibbs, Hew, & Graham, 2003). As research by Wadley et al. (2003) found, VoIP lacked sufficient contextual cues to help users identify who was talking, as well as where they were situated in the virtual environment. Stated another way, the talk is not marked with an avatar ID tag like the text communication. There is also no visual indication that the VoIP transmittal was sent or heard successfully (Hew, Gibbs, & Wadley, 2004). This is a valid concern given that sometimes there are technical difficulties that distort the sound of the VoIP or cause it to break up.

These technical difficulties, combined with the lack of contextual cues, may degrade the usefulness of online VoIP communication (Wadley et al., 2003). Further, VoIP can be distracting to individuals around the user in the physical world (Bartle, 2003), and the conversations lack the sense of privacy that text affords (Scholl et al., 2006). The presence of non-conversation sounds such as ambient noises and the ability to overhear the conversations of other in-world residents is problematic, as well (Hew et al., 2004). Many users expect VoIP to be superior to text. In practice, however, VoIP may be difficult to employ and in some cases, it has been rejected as a mode of communication by users (Hew et al., 2004).

Beyond Text and VoIP

While many SL residents used text chat or VoIP to communicate in-world, they also have the option of using both methods simultaneously. With the assistance of in-world devices such as a head up display (HUD), for example, residents can broadcast text and VoIP simultaneously. The HUD works like a teleprompter. In the courses utilizing this method, the instructor copied-and-pasted the text from a pre-scripted lecture into the HUD. This allowed her to read the lecture from the screen without students seeing the HUD device. Some users believe that this hybrid approach (text plus VoIP) creates the best of both worlds. VoIP provides them with an alternative communication channel. In addition, participants have access to the chat history window, as well as the SL system text transcript that is recorded by the system. An added bonus is that reading the text discussion in real time is not required. There are also findings that suggest that the use of VoIP plus text in virtual worlds is superior to text alone (Williams et al., 2007). Because research examining in-world communication is in a nascent stage, it is important to examine the media options that are available to virtual world users and the ways in which they are employed.

The interactions that take place in-world, such as chatting and talking, are not restricted to the virtual world environment. As Steinkuehler, Black, and Clinton (2005) discovered, some individuals continue the conversation in areas beyond the boundaries of these worlds. Discussion forums, websites, course management systems, and email all support out-of-world communication. Novices, for example, may use these digital spaces to access information and gain wisdom from more expert residents (Schrader & Lawless, 2009). For some individuals, these interactions extend beyond online environments to include offline interactions via telephone and even face-to-face meetings.

Communication Media Skills

Virtual worlds such as SL are complex virtual spaces that require users to acquire skills that will enable them to be literate across multiple media. For example, this may involve making sense of a chat message, articulating their thoughts via VoIP, and examining an external website during their tenure in this digital space. In SL, information may be represented through auditory (e.g., VoIP chat, soundscapes, ambient noise) and visual (e.g., text chat, IM, notecards) modes of communication. Some of these tools typically require additional equipment in order to work effectively. For example, a headset, microphone, and speakers may be needed to interact in SL via VoIP. While VoIP and text chat tools are typically at the forefront for communication, there are other available methods of communication. Examples of these include gifts, SL educational spaces, SL environmental tools, and wearable information objects, just to name a few.

While there are a variety of ways to communicate in SL, information management can nonetheless be difficult in this digital space. Users who receive numerous SL inventory items, for example, may find that the options to organize this information in-world are lacking. Thus, external media are sometimes used to support the in-world communication experience. Email, Google Groups, and the Moodle course management system are examples of media used to supplement the in-world communication resources. Potential uses of these media include posting session transcripts, disseminating additional reading materials, sharing URLs, and continuing in-world conversations. Users typically come to SL with some computer skills, yet not everyone is familiar with or comfortable using many of these types of media. To express themselves effectively in-world, these users need to learn the strengths and weaknesses of the SL communication options (Table 1). At the same time, they must also acquire the ability to express themselves through creations, interactions, and words.

THE RESEARCH

Prior to the current study presented here, the researcher conducted a pilot study to investigate educational activities and interactions taking place in virtual worlds. The researcher's rez date (i.e., the day the SL avatar is created) is August 2006, but a more systematic exploration of SL began in the summer of 2007. These pilot investigations of SL-based courses, which were part of a professional development series, helped the researcher gain insider's knowledge into the continuing education activities conducted in this digital environment. It also formed the foundation for building a level of in-world expertise and vocabulary necessary to understand the instructors and students who enrolled in these courses (see Stoerger, 2010). The research described in this section began in May 2008 and continued through the summer of 2009.

The Courses

The SL-based courses included in this study are the joint effort of a research university's continuing professional development program of study and a library consortium, both located in the Midwest. From the beginning, the coordinators of the series offered courses covering a variety of topics ranging from instructional tools to scripting to machinima (i.e., three-dimensional, real-time computer animation). Unlike some courses available in SL, individuals interested in this continuing education series had to pay a fee to enroll. Typically, the fee ranged from $125 to $300.

The courses examined for the present study were designed for librarians, museum professionals, educators, and others interested in learning and teaching in virtual worlds. Courses at the beginner, intermediate, and advanced level were identified and selected for this study. Each selected course included four sessions. This process of identifying the course level was based on prerequisites noted in the course descriptions along with the assistance of the course coordinator. Institutional Review Board (IRB) approval to access these sites was granted on February 23, 2009.

SL Course Enrollment

Interest in the educational uses of virtual worlds such as SL has grown over the years (e.g., de Freitas, 2008). However, the courses under investigation were small, and the number of students continued to decline in 2009. The summer 2009 beginner course, for example, had to be canceled and rescheduled twice due to low enrollment figures. The number of students participating in this summer course eventually reached a total of only four. In general, enrollment figures for the intermediate course were higher at 11 students. Enrollment was also an issue for the advanced course, which was scheduled to begin during the first week of August 2009. Due to the lack of interest, the course was cancelled. At the end of July 2009, an announcement was made that this continuing education series would be discontinued in September 2009.

Not only did the number of students vary by course level, but so did the number of instructors. For the beginner course, one individual conducted the four sessions. The intermediate course, on the other hand, was conducted by two instructors, and three individuals were responsible for teaching the advanced course.

Research Questions

To understand more fully the communication that takes place in SL courses, the following questions served to guide the research:

- What participation patterns emerge from the student-instructor interactions in SL?
- How do these participation patterns differ at the three course levels?
- What factors play a role in determining which media are used to support these interactions?

Data Collection

For the three selected courses, session transcripts from each of the 12 sessions were collected for analysis. Once the data were cleaned for extraneous system-related characters, each transcript ranged from approximately 440 to slightly more than 1,000 lines of text chat. Further, observational field notes were collected from the beginner and intermediate courses. Because the advanced course that was originally scheduled for August 2009 was canceled, transcripts were obtained from the May 2008 sessions. Thus, observations for the advanced course were made based on the transcripts alone rather than in combination with field notes collected during the class period.

The students and instructors from the SL courses were also invited to participate in email interviews. Given that the classes were small,

all students and instructors for each course were invited to participate in the study. Individuals who agreed to participate were sent a list of questions to answer and return to the researcher. In some cases, the researcher had follow-up questions for the participants. These interactions took place via email or through text chat in SL.

Analytic Methods

Discourse analysis methods and ethnographic techniques were used to assess the student-instructor interactions that took place during the course sessions under investigation. The first part of the analysis assessed the participation characteristics of the chat discussions such as the number of words, the number of characters, and the participation levels (i.e., the number of messages posted to the discussion) of the participants. Herring's (2004) Computer-Mediated Discourse Analysis (CMDA) approach was used to direct this portion of the research. The results of this analysis were supported or refuted by ethnographic methods including participant observations, collected field notes and unstructured interviews.

Previous work by the researcher suggests that the SL courses are teacher-centered (Stoerger, 2010). Therefore, attention was paid to the SL class sessions investigated in this current study that appeared to be more student-centered. Transcripts and field notes were reviewed to determine what activities were employed, and the types of interactions that were taking place during the student-centered sessions. Further, the ways in which these activities impacted the student-centered nature of the sessions were examined, as well. To gain additional insight into the emerging patterns, a student version of Atlas.ti, a qualitative analysis tool, was used to code data and build conceptual frameworks. Overall, the qualitative data were reviewed, coded, examined, and analyzed for additional evidence to answer each research question. This process took place as the data were collected throughout the research.

FINDINGS AND DISCUSSION

On the whole, the SL instructors contributed more to the class discussions than the students. However, there were times when the participation level gap between the instructors and students became less pronounced. This was true for the participants in the beginner course, where the difference between the instructors and students decreased, particularly in session three (Figure 1). In session three, students created an in-world poster rather than "listening" to a two-hour lecture.

Overall, however, the chart shows that the instructor and student participation percentages are fairly consistent and flat. Because there is a large amount of information to learn, and the students are unfamiliar with SL, it logically follows that the instructor would participate at a higher level; therefore, more teacher lead instruction is needed in a beginner course.

The chart for the intermediate course illustrates a different picture. While the instructors posted more messages than the students in sessions one and two, the participation levels of the students came closer to those of the instructors in sessions three and four (Figure 2). Sessions one and two were structured as lectures. Unlike the beginner and advanced courses, the first hour of both sessions was also conducted using a hybrid communication approach. In contrast, sessions three and four included more active learning tasks, such as dancing and manipulating scripts.

Unlike their beginner and intermediate peers' participation levels, the participation levels of the students attending session two of the advanced course exceeded those of their instructors (Figure 3). Students attending session two participated in a field trip to view educational uses of small SL spaces. They also met the designers of those spaces and were given the opportunity to ask those individuals questions. The other three sessions were structured around lectures.

Interestingly, the higher the course level, the greater the number of instructors present for those

Figure 1. Beginner student-instructor participation levels by session

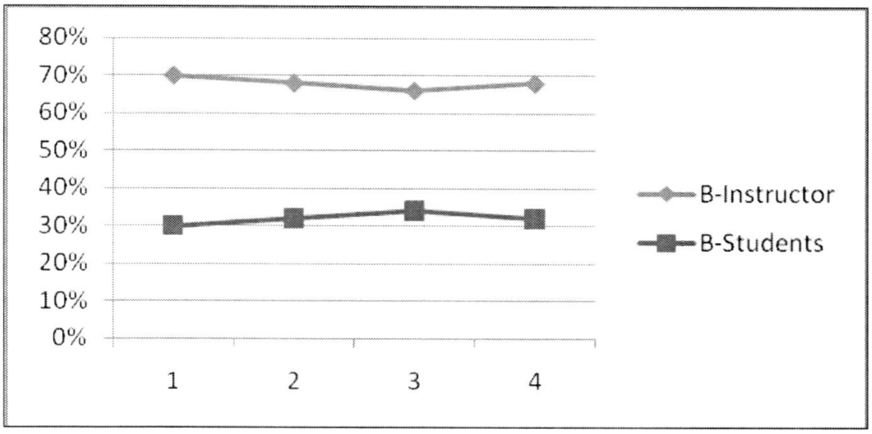

Figure 2. Intermediate student-instructor participation levels by session

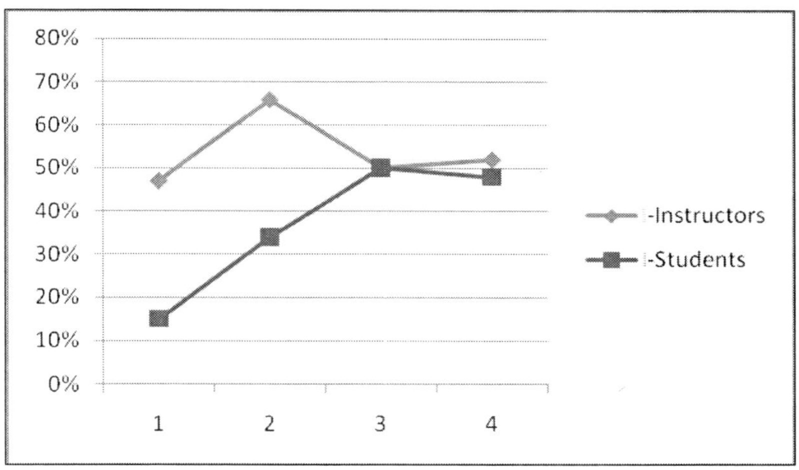

sessions. For instance, the beginner course had one instructor, the intermediate course had two, and the advanced course had three. Yet, it appears that an increase in the number of instructors present per session did not squelch the students' participation levels when active learning tasks were involved.

The structural characteristics (e.g., word length and message length) of the messages posted during the three courses followed similar patterns. Overall, the instructors posted longer messages to the class discussions and used longer words than the students (Table 2). However, there were instances when the complexity of the students' words and messages became more complex. Many of these sessions involved virtual field trips and hands-on activities. For example, the fourth session of the beginner course involved a field trip to educational sites in SL where they experienced a virtual tsunami, for example. The average message and word length for the students participating in session four was 5.9 and 4.1, in that order. In contrast, the message and word length for the instructor was 5.8 and 3.9.

Figure 3. Advanced student-instructor participation levels by session

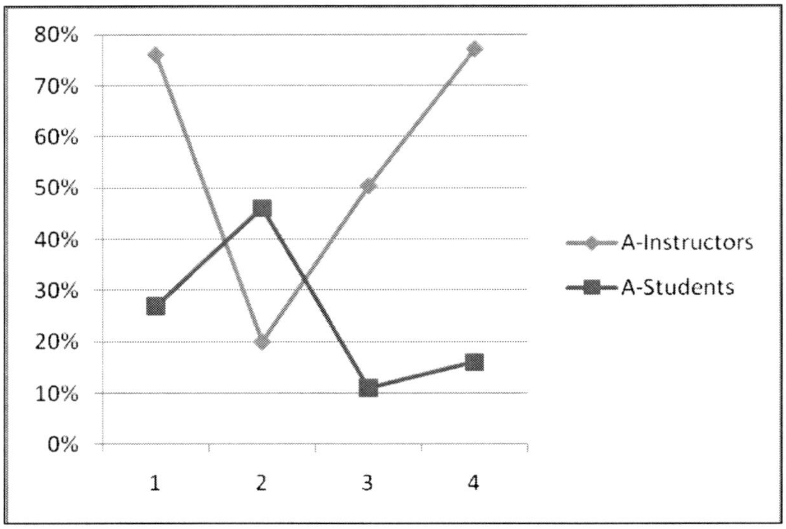

Interestingly, Table 2 also shows an upward pattern that emerged for students when they moved from the beginner to the intermediate to the advanced levels. The students in the beginner course took shorter turns and used simpler words in comparison to students attending the other levels. Indeed, students in the advanced class posted on average almost three more words per message more than the beginner students. The advanced students also used longer words in their messages than their peers in the beginner and intermediate courses. In general, the messages contributed by the students participating in these SL sessions were longer and more complex as the course level increased.

While VoIP has been an option in SL since 2007 (Second Life Wiki, n.d.), it was still an unstable communication medium for the courses described in this chapter. Thus, the instructors in this study relied primarily on text chat to communicate with students. Instant messaging was used on occasion for private conversations between students and instructors. When asked about their communication media choices, many

Table 2. Average message and word length by course level, session, and role

Course Level	Role	Average Message Length (words/ messages)	Average Message Length Range	Average Word Length (characters/words)	Average Word Length Range
Beginner Ave. Total	B-Instructor	6.4	5.8 – 7.1	3.9	3.9 – 3.9
Beginner Ave. Total	B-Student	5.0	4.2 – 5.9	4.1	4.0 – 4.1
Intermediate Ave. Total	I-Instructors	10.0	9.0 – 11.3	4.5	4.2 – 4.7
Intermediate Ave. Total	I-Students	6.7	4.5 – 7.9	4.3	4.0 – 4.4
Advanced Ave. Total	A-Instructors	9.7	7.4 – 12.2	4.4	4.1 – 4.7
Advanced Ave. Total	A-Students	7.3	4.6 – 8.8	4.4	4.2 – 4.5

of these instructors stated that text chat created a more egalitarian environment where no single voice could dominate (see also Jenkins, 2004). They also noted that text chat is more accessible for individuals who are deaf or who have hearing impairments (see also Peters & Bell, 2007). No students in these courses had hearing problems, so the instructors did attempt to use VoIP. In fact, there were several unsuccessful attempts to use VoIP in sessions at the three course levels. Due to technical difficulties associated with the SL VoIP option, the instructors typically reverted back to text chat alone. In their examination of online VoIP communication, Wadley et al. (2003) argue that problems such as these could be addressed through better design of VoIP chat.

One instructor at the intermediate level was able to use VoIP without many difficulties. This instructor utilized a HUD that enabled her to easily transmit VoIP and text chat simultaneously. While several comments were made about the effectiveness of this approach, the data imply that it may have had a negative impact on student participation levels. When the hybrid approach was employed, the use of multiple communication modes gave students options. At the same time, however, it reduced their levels of participation. For instance, in session two of the intermediate course, the instructors used a hybrid approach to disseminate the lecture material to the students. During that portion of the session, the instructors posted 75% of the messages, whereas students contributed 25% of the postings. In other intermediate sessions when the hybrid approach was not undertaken, the student and instructor participation levels were more aligned and almost equal.

Communication Obstacles

Technical difficulties were common during the SL class sessions. For example, during the first beginner session, two of the students' computers crashed several times in the first 15 minutes of class. It was at the end of that session that the instructor seemed to disappear and was unable to return to class. Crashing was not the only SL system difficulty encountered during these courses. While much of the session time did not involve moving around the SL space, lag was evident when the group attempted to perform activities beyond sitting and listening to a lecture. Because the images rendered in real time, it was not unusual for several minutes to pass before the images completely appeared on the screen, as well (see also Warburton, 2009). This resulted in avatars that initially emerged without clothes and mere outlines of buildings to be visible, for example. These problems caused class time to be spent concentrating on overcoming technical problems, rather than on discussing course material.

While the intermediate and advanced courses had enrollment prerequisites, students came to the courses with a wide range of SL and technology-related skills. In general, the students were comfortable with computers but were relatively new to SL. This was evident primarily in the beginner courses. However, one student in the intermediate course had spent 3 years in-world, but she had problems using the external communication tools selected by the instructors. In the second intermediate course session, a student asked what to do with the "http thing" the instructor posted to the group, indicating that she had little experience with the Internet. These difficulties surprised the instructors who assumed students would be familiar with email and accessing websites, for example.

Despite the technological challenges faced by some of the SL students, attempts were made by the beginner, intermediate, and advanced course instructors to use VoIP during their sessions. In fact, much class time was spent by the instructors to get it to work. It was not uncommon for the instructors to repeatedly state in VoIP and rebroadcast via text chat, "Can you hear me now?" The typical student response in text chat to this question was "No." In several sessions, one or more students could not hear the instructors or guest speakers. Instructors had problems with the sound reverberating and static, as well. Of the three levels, the advanced course instructors attempted to use VoIP the least. VoIP was only used in the first part of one session by the guest speaker; the other members of the group remained in text chat. During that portion of the session, however, several students could not hear the speaker, the sound stopped for those who could, and the instructors had to summarize the guest speaker's comments while participating in the session.

Unlike text chat, VoIP users must have equipment such as a microphone and headphones or speakers. In addition to having the proper equipment, there are other concerns to consider in regard to using VoIP. For one, VoIP users must find a location where their SL conversation will not disturb individuals in the physical world. Even in SL, VoIP carries farther than other in-world communication modes. Thus, a VoIP conversation may disrupt other virtual residents. Given that there was no immediate solution to the difficulties of VoIP communication, it was simply easier for the instructors to interact with the students via text-based media during these SL sessions. In other words, the text-based options were the communication method of choice for the courses at the beginner, intermediate, and advanced levels. As social information processing theory (SIP) suggests, CMC users will adapt to a medium over time and will work with it to develop strategies which compensate for its lack of richness (Walther, 1994; Walther & Parks, 2002).

Interestingly, the intermediate instructor had the most success using VoIP, and did so using a hybrid approach. When asked about communication options used in these courses, the SL instructors and students claimed that the hybrid communication was a valuable communication style. However, as was noted earlier, the actual data suggest that it discouraged interactions between and among students and instructors. In other intermediate sessions when the hybrid approach was not undertaken, the student and instructor participation levels were more aligned and almost equal. The research on multimedia technology suggests that supplying information in more than one form can increase cognitive load and reduce understanding (Carr, 2010).

Hybrid communication is not unique to SL. In fact, broadcasting text and VoIP simultaneously has been used in other library-related online courses and programs (Jenkins, 2004). Because the SL courses were designed by librarians, the findings presented by Jenkins (2004) may have influenced the SL instructor's decision to use a hybrid approach when lecturing.

Communication Rules

One benefit that is often touted about communication in virtual worlds is that the lines of communication are open to everyone at any time. The assumption is that no one "voice" can dominate the discussion. Text-based communication media support this notion by enabling individuals to participate at any time (Jenkins 2004). This tends to disrupt the physical classroom hierarchy. Virtual world instructors are forced to give up some of the control they typically enjoy in the physical world classroom setting (Rodrigues & Sedo, 2008), and some educators are not comfortable with this approach.

For instance, the intermediate instructor noted the importance of outlining the rules of communication at the beginning of a session or event to address questions such as who will use VoIP, who will use text chat, will the instructors be the only participants allowed to use VoIP, and will the students be instructed to ask questions and contribute comments through text chat only. The instructors for the intermediate session clearly stated that students were to use text chat only. While many instructors believe that this approach helps them reduce the chaos and manage the riskier aspects of these SL sessions (e.g., Selfe & Selfe,

1994), it also served to reestablish the classroom hierarchy that the literature claims is dismantled when using text chat alone.

As already noted, the use of VoIP chat was attempted on several occasions, with little success. The SL courses that did employ VoIP, at least to some extent, restricted the students' communication options to one mode (i.e., text), whereas instructors had the opportunity to use two (i.e., text and/or VoIP). The SL instructors noted that this arrangement made the sessions less chaotic and easier to manage. With the exception of an "experiment" using VoIP during a discussion about instructional tools that occurred in an intermediate course session, students never used VoIP to interact with others. Had each course extended beyond the four, 2-hour sessions, the participants may have gained greater technological and SL skills. And as a result, the media they selected, as well as their use, may have changed over time (Contractor et al., 1998; Walther, 1995).

CONCLUSION

Previous research has shown that reading and writing are typically a central part of participating in virtual worlds (Steinkuehler, 2007). The SL courses examined in this study were no exception. Text-based media, particularly text chat, were heavily used by the course participants. VoIP is a relatively new communication medium available to SL participants. The assumption was that VoIP would enrich the virtual world experience – a step forward from text-based communication. While the instructors attempted to use VoIP, the students relied almost exclusively on text chat. In some courses, communication rules were established prior to the start of the session. The intermediate course is one example where the instructors specifically directed students to use text chat and not VoIP. Similar instructions were not given in the beginner and advanced courses. In those courses, the attempts to use VoIP were typically abandoned due to technical difficulties encountered by the instructors, as well as the students.

Several instructors wanted to use VoIP but were unsuccessful. Technological complications such as audio drops, echoes, and static were too prevalent to make it useable. Some of these VoIP problems were not consistent. There were times when some students could hear the VoIP discussions, whereas others during the same time period could not. Because the students were paying to attend these sessions, many instructors stopped using VoIP during the sessions, so that everyone could participate. On the whole, however, VoIP was not particularly usable during the courses.

The technological obstacles and the communication rules did not completely squelch the levels of participation. In sessions three and four, the participation gap narrowed between the instructor and the students in the beginner course (Figure 1). The participation levels of the students participating in sessions three and four in the intermediate course were comparable to that of the instructors (Figure 2). In the advanced course, the students participated more in the discussions than the instructors during session two (Figure 3). The common thread in these sessions is that active learning approaches employed by the instructors, approaches such as field trips and hands-on tasks also served to boost participation rates. Despite the fact that students indicated that they liked the hybrid approach used in the intermediate course, this study found that the SL students participated less when the hybrid approach was employed. This runs counter to research that suggests that VoIP plus text would be superior to text alone (Williams et al., 2007). Because the hybrid approach was used in only two of the SL sessions investigated in this study, more research is needed to examine its potential over time.

VoIP was not the only obstacle faced during the classes. When asked how SL does not work, two students mentioned the technical problems they encountered in-world such as lag, glitches, and slowness. The steep learning curve was recognized as an obstacle by the students and the instructors, as well. Regardless, text-based communication tools were able to support the educational activities that were carried out during these SL sessions. Because of technical difficulties, it is unclear the effect VoIP would have had on the student-instructor interactions in this virtual world. One thing is clear: technological challenges such as those associated with VoIP can serve to discourage the wider adoption of SL by educators (see also Warburton, 2009). Once these difficulties are addressed, then further investigations between the student-instructor interactions conducted via VoIP versus text chat could be made to determine their effectiveness.

REFERENCES

Bartle, R. (2003). Not yet you fools! *Game+Girl=Advance*. Retrieved from http://www.gamegirladvance.com/ archives/2003/07/28/ not_yet_you_fools.html

Boellstorff, T. (2008). *Coming of age in Second Life: An anthropologist explores the virtually human*. Princeton, NJ: Princeton University Press.

Bradner, E., Kellogg, W., & Erickson, T. (1999). The adoption and use of "Babble": A field study of chat in the workplace. *Proceedings of the European Conference on Computer Supported Cooperative Work (ECSCW '99), Copenhagen, Denmark*, (pp. 139-158). Retrieved from http:// www.ecscw.org/ 1999/08.pdf

Carr, N. (2010). *The shallows: What the internet is doing to our brains*. New York, NY: W. W. Norton & Company.

Contractor, N. S., Whitbred, R., Fonti, F., Hyatt, A., Jones, P., & O'Keefe, B. (1998). *Self-organizing communication networks in organizations: Validation of a computational model using exogenous and endogenous theoretical mechanisms.* Paper presented at the International Communication Association, Jerusalem, Israel. Retrieved October 18, 2009, from http://hyperion.math.upatras.gr/ commorg/nosh/comp.html

Daft, R., & Lengel, R. (1984). Information richness: A new approach to managerial behavior. In Straiv, B., & Cummings, L. L. (Eds.), *Research in organizational behavior* (pp. 191–223). Homewood, IL: JAI Press.

de Freitas, S. (2008). Emerging trends in serious games and virtual worlds. *Becta: Emerging Technologies for Learning, 3*, 57-72. Retrieved from http://partners.becta.org.uk/ upload-dir/ downloads/ page_documents/research /emerging_technologies08 _chapter4.pdf

Dede, C. (2008). New horizons: A seismic shift in epistemology. *EDUCAUSE Review, 43*(3), 80–81.

Fanderclai, T. L. (1996). Like magic, only real. In Cherny, L., & Weise, E. (Eds.), *Wired women: Gender and new realities in cyberspace* (pp. 224–241). Seattle, WA: Seal Press.

Finholt, T., & Sproull, L. (1990). Electronic groups at work. *Organization Science, 1*(1), 41–64. doi:10.1287/orsc.1.1.41

Fitzgibbon, A., Oldham, E., & Johnston, K. (2008). Are Irish student-teachers prepared to be agents of change in using IT in education? In K. McFerrin et al. (Eds.), *Proceedings of Society for Information Technology and Teacher Education International Conference 2008* (pp. 1397-1404). Chesapeake, VA: AACE.

Friedman, T. L. (2005). *The world is flat: A brief history of the twenty-first century*. New York, NY: Farrar, Straus & Giroux.

Garton, L., Haythornthwaite, C., & Wellman, B. (1999). Studying on-line social networks. In Jones, S. (Ed.), *Doing Internet research* (pp. 75–105). Thousand Oaks, CA: Sage.

Garton, L., & Wellman, B. (1995). Social impacts of electronic mail in organizations: A review of the research literature. *Communication Yearbook, 18*, 434–453.

Halloran, J., Rogers, Y., & Fitzpatrick, G. (2003). From text to talk: Multiplayer games and VoIP. *Proceedings of Level Up: 1st International Digital Games Research Conference,* (pp. 130-142).

Harmon, S. W., & Jones, M. G. (2001). An analysis of situated web-based instruction. *Educational Media International, 38*(4), 271–280. doi:10.1080/09523980110105123

Herring, S. C. (2004). Computer-mediated discourse analysis: An approach to researching online behavior. In Barab, S. A., Kling, R., & Grey, J. H. (Eds.), *Designing for virtual communities in the service of learning* (pp. 338–376). Cambridge, MA: Cambridge University Press.

Hew, K., Gibbs, M., & Wadley, G. (2004). *Usability and sociability of the Xbox Live VoIP channel.* Paper presented at the Australian Workshop on Interactive Entertainment, Sydney, Australia. Retrieved from http://www.ieconference.org/ie2004/proceedings/ 051%20hew.pdf

Ingram, A. L., Hathorn, L. G., & Evans, A. (2000). Beyond chat on the internet. *Computers & Education, 35*(1), 21–35. doi:10.1016/S0360-1315(00)00015-4

Jenkins, C. A. (2004). The virtual classroom as ludic space. In Haythornthwaite, C. A., & Kazmer, M. M. (Eds.), *Learning, culture and community in online education* (pp. 163–176). New York, NY: Peter Lang.

Jensen, C., Farnham, S., Drucker, S., & Kollock, P. (2000). The effect of communication modality on cooperation in online environments. *Proceedings of CHI '00* (pp. 470–477). The Hague, Amsterdam.

Kress, G. (2005). Gains and losses: New forms of texts, knowledge, and learning. *Computers and Composition, 22*, 5–22. doi:10.1016/j.compcom.2004.12.004

Lea, M., O'Shea, T., Fung, P., & Spears, R. (1992). "Flaming" in computer-mediated communication: Observations, explanations, implications. In M. Lea (Ed.), *Contexts of computer-mediated communication* (pp. 89-112). New York, NY: Wheatsheaf.

Lewis, P. R. (2009). *Over 15 billion minutes of VoIP have been delivered in Second Life.* Linden Lab. Retrieved from http://lindenlab.com/pressroom /releases/19_05_09

Linden Research, Inc. (2009). *Second Life wiki.* Retrieved from http://wiki.secondlife.com/ wiki/VoIP

Luo, L., & Kemp, J. (2008). Second Life: Exploring the immersive instructional venue for library and information science education. *Journal of Education for Library and Information Science, 49*(3), 147–166.

Merchant, G. (2009). Literacy in virtual worlds. *Journal of Research in Reading, 32*(1), 38–56. doi:10.1111/j.1467-9817.2008.01380.x

Mon, L. (2009). Questions and answers in a virtual world: Educators and librarians as information providers in Second Life. *Journal of Virtual Worlds Research, 2*(1). Retrieved from https://journals.tdl.org/jvwr /article/view/398/452.

Nardi, B. A., Whittaker, S., & Bradner, E. (2000). Interaction and outeraction: Instant messaging in action. *Proceedings of the ACM Conference on Computer Supported Cooperative Work (CSCW '00), Philadelphia, PA,* (pp. 79-88).

Nass, C., & Gong, L. (2000). Social aspects of speech interfaces from an evolutionary perspective: Experimental research and design implications. *Communications of the ACM*, *43*(9), 36–43. doi:10.1145/348941.348976

New Media Consortium. (2007). *The 2007 horizon report*. Retrieved from http://www.nmc.org/pdf/2007_Horizon_Report.pdf

Park, Y. J., & Bonk, C. J. (2007). Synchronous learning experiences: Distance and residential learners' perspectives in a blended graduate course. *Journal of Interactive Online Learning*, *6*(3), 245–264.

Peters, T., & Bell, L. (2007). MUVEing toward accessibility. *Computers in Libraries*, *27*(4), 34–36.

Rodrigues, D. B., & Sedo, D. R. (2008). Experiencing information literacy in Second Life. *Canadian Journal of Library and Information Practice and Research*, *3*(1). Retrieved from http://journal.lib.uoguelph.ca/ index.php/perj/article /view/426/861.

Sallnäs, E.-L. (2005). Effects of communication mode on social presence, virtual presence, and performance in collaborative virtual environments. *Presence (Cambridge, Mass.)*, *14*(4), 434–439. doi:10.1162/105474605774785253

Salmon, G. (2009). The future for (second) life and learning. *British Journal of Educational Technology*, *40*(3), 526–538. doi:10.1111/j.1467-8535.2009.00967.x

Salt, B., Atkins, C., & Blackall, L. (2008, October). *Engaging with Second Life: Real education in a virtual world*. Retrieved from http://slenz.files.wordpress.com/2008/12/slliteraturereviewa1.pdf

Scholl, J., McCarthy, J., & Harr, R. (2006). *A comparison of chat and audio in media rich environments*. Paper presented at the CSCW, Banff, Alberta, Canada.

Schrader, P. G., & Lawless, K. A. (2009). The hidden literacies of massively multiplayer online games. In Pullen, D. L., Gitsaki, C., & Baguley, M. (Eds.), *Technoliteracy, discourse and social practice: Frameworks and applications in the digital age* (pp. 200–219). Hershey, PA: IGI Global. doi:10.4018/978-1-60566-842-0.ch012

Selfe, C. L., & Selfe, R. J. Jr. (1994). The politics of the interface: Power and its exercise in electronic contact zones. *College Composition and Communication*, *45*(4), 480–504. doi:10.2307/358761

Short, J., Williams, E., & Christie, B. (1976). *The social psychology of telecommunications*. New York, NY: Wiley.

Sproull, L., & Kiesler, S. (1991). *Connections: New ways of working in the networked organization*. Cambridge, MA: MIT Press.

Steinkuehler, C. (2007). Massively multiplayer online game as a constellation of literacy practices. *E-learning*, *4*(3). Retrieved from http://website.education.wisc.edu /steinkuehler/papers/ Steinkuehler_eLearn.pdf. doi:10.2304/elea.2007.4.3.297

Steinkuehler, C. A. (2004). Learning in massively multiplayer online games. In Y. B. Kafai, W. A. Sandoval, N. Enyedy, A. S. Nixon & F. Herrera (Eds.), *Proceedings of the Sixth International Conference of the Learning Sciences: Embracing Diversity in the Learning Sciences* (pp. 521-528). Mahwah, NJ: Lawrence Erlbaum Associates. Retrieved from http://website.education.wisc.edu/steinkuehler/papers/ SteinkuehlerICLS2004.pdf

Steinkuehler, C. A., Black, R. W., & Clinton, K. A. (2005). Researching literacy as a tool, place, and way of being. *Reading Research Quarterly*, *40*(1), 95–100.

Stoerger, S. (2010). In a virtual classroom, who has a "voice"? A discourse analysis of student-instructor interactions in two Second Life-based courses. In Park, J., & Abels, E. (Eds.), *Interpersonal relations and social patterns in communication technologies: Discourse norms, language structures and cultural variable* (pp. 237–255). Hershey, PA: IGI Global. doi:10.4018/978-1-61520-827-2.ch013

Teng, T. L., & Taveras, M. (2004-2005). Combining live video and audio broadcasting, synchronous chat, and asynchronous open forum discussions in distance education. *Journal of Educational Technology Systems*, *33*(2), 121–129. doi:10.2190/XNPJ-5MQ6-WETU-D18D

Wadley, G., Gibbs, M., Hew, K., & Graham, C. (2003, November). Computer supported cooperative play, "third places" and online videogames. *Proceedings of the Thirteenth Australian Conference on Computer Human Interaction (OzChi 03), Brisbane, Australia*, (pp. 238-241).

Walther, J. (1994). Anticipated ongoing interaction versus channel effects on relational communication in computer-mediated interaction. *Human Communication Research*, *20*, 473–501. doi:10.1111/j.1468-2958.1994.tb00332.x

Walther, J. (1995). Relational aspects of computer-mediated communication: Experimental observations over time. *Organization Science*, *6*(2), 186–203. doi:10.1287/orsc.6.2.186

Walther, J. (1999). *Visual cues and computer mediated communication: Don't look before you leap*. Paper presented at the International Communication Association, San Francisco, CA.

Walther, J., & Parks, M. (2002). Cues filtered out, cues filtered in: Computer-mediated communication and relationships. In Knapp, M., & Daly, J. (Eds.), *Handbook of interpersonal communication* (3rd ed.). Thousand Oaks, CA: Sage.

Warburton, S. (2009). Second Life in higher education: Assessing the potential for and the barriers to deploying virtual worlds in learning and teaching. *British Journal of Educational Technology*, *40*(3), 414–426. doi:10.1111/j.1467-8535.2009.00952.x

Williams, D., Caplan, S., & Xiong, L. (2007). Can you hear me now? The impact of VoIP in an online gaming community. *Human Communication Research*, *33*, 427–449. doi:10.1111/j.1468-2958.2007.00306.x

KEY TERMS AND DEFINITIONS

Computer-Mediated Communication: Communication that occurs between two or more individuals who are connected to each other through separate, networked computers.

Cues: Elements that add to the richness of the communication between individuals and may impact how the receiver interprets the communicated message. Examples of cues include voice tone, inflections, facial expressions, body language, and gestures, just to name a few.

Head Up Display (HUD): A transparent device that conveys information to the user. One of its uses in Second Life, specifically, is as an information display. Users may also use a HUD as a control panel for automation overrides that replace standard animations (e.g.., walk, fly, stand) and scripted attachments (e.g., re-colorable clothes, hair).

Instant Messaging (IM): Text-based communication that occurs in real time between specified individuals who are connected to each other through networked computers.

Second Life (SL): A three-dimensional virtual world that was released to the public by Linden Lab in 2003. Basic membership is free, and individuals from around the world create SL accounts.

Student-Centered Learning: An instructional method that focuses on the educational needs of the student, where the student takes an active role in their learning process. This approach is typically contrasted with teacher-centered learning where students adopt a more passive educational role.

Synchronous Text Chat: Written online communication that takes place in real time between two or more individuals who are using networked computers.

Teacher-Centered Learning: As opposed to student-centered learning, an instructional method where the focus is on the teacher, and where students typically adopt a more passive role in the learning process. In addition, this approach emphasizes lower-order thinking skills such as memorization and recall.

Virtual Worlds: Online environments that are intended for users to inhabit and where they can interact with one another.

Voice over Internet Protocol (VoIP): The transmission of voice communication in real time over the Internet.

Chapter 3
Library Resources and Services in 21st Century Online Education

Heath Martin
University of Kentucky Libraries, USA

Peter Hesseldenz
University of Kentucky Libraries, USA

ABSTRACT

This chapter analyzes the roles of academic libraries in computer-mediated instruction through examination of past and current practices, existing opportunities and challenges, and emerging trends. By examining key concepts, activities, and scholarship related to library resources and services-- information and communication technology, access to resources, scholarly publishing, information literacy, and models of collaboration — the authors demonstrate the importance of those resources and services to online education and the need to work with other stakeholders to meet the opportunities and challenges that lie ahead.

INTRODUCTION

Computer-mediated learning has become commonplace in higher education, with most colleges and universities offering online classes for both on- and off-campus students (Gurney & Wilkes, 2008). Though academic libraries have long worked with distance education students, many adjustments have had to be made to successfully serve students in an online environment. However, academic librarians have found that, in most cases, the needs of online students and in-person students are merging. All students now make extensive use of electronic materials; in most cases, they expect and prefer them. Many students, both on campus and off, prefer to contact a reference librarian virtually rather than going to the physical library. And, while in-class library instruction sessions are still widely used, many librarians are making innovative use of online technology to deliver information literacy

DOI: 10.4018/978-1-61350-077-4.ch003

concepts to students. Therefore, while academic libraries continue to purchase many books and still work in buildings made of brick and mortar, many have adopted the practice of treating all students as potential online students. This chapter discusses the roles academic libraries play in educating online students, paying particular attention to library resources and library services.

As far as library resources are concerned, librarians have the same basic concerns for online users as they do for face-to-face users. Librarians provide students with resources they need in order to complete their academic pursuits, whether they be complex data sets or readings for a beginning English class. Librarians try to have materials in place when the patrons need them in a form that is easily accessible. If that is not possible, libraries have mechanisms to quickly acquire the materials. When the users are distant or online, the library acts as digital gateway to resources, with proxy connections designed to be as minimally disruptive as possible. Once on a library's website, clear, easy-to-follow descriptions can provide seamless access to materials such as e-books, e-journals, and online databases, as well as freely available scholarly material. When these practices are followed, online students can easily find library resources and put them to use.

Today's students have grown up digitally, which has allowed them to become, for the most part, savvy computer users. Often, students mistake technical proficiency for mastery of the research process, arriving at college thinking they are expert web searchers. In general, however, they are unaware of the untapped world of scholarly material available through the library's databases. A great challenge for academic librarians is to make students aware of the multitude of material available beyond Google and other search engines, and then to teach them how to use those materials effectively. For distance learners, sometimes far from the actual library, there is a temptation to continue with a comfortable routine of using Google or other search engines to perform research (Brophy and Bawden, 2005). Librarians work hard to anticipate this, and to make online students aware of the importance of appropriate materials and the methods for using them.

LIBRARY RESOURCES

The Legacy of Print Collections

Long before electronic resources became prevalent in academic library collections, print resources were employed in the service of distance students. During the later years of the twentieth century, circulation policies quickly adapted to the emergence of distance education, extending borrowing privileges to students in remote locations. At the same time, interlibrary loan services allowed academic libraries to make significant parts of their collections available to remote students affiliated with other institutions, as well as provide access to other libraries' holdings for their own distance education students. Other methods of providing traditional library resources and collections services to remote users included access to video resources, deployment of library personnel to satellite locations, and reciprocal resource-sharing agreements with other libraries and institutions (Snyder, Logue, and Preece, 1996).

Today, print resources continue to contribute to distance education through some of the same mechanisms and services. Online library catalogs persist as a primary way to search a library's holdings and determine what is available. Once identified, users accomplish access to material by requesting delivery of the items, which may be physical or electronic. While remote access to electronic databases satisfies many resource needs for distance students, ensuring access to and delivery of print and other tangible formats remains a fundamental component of library support for distance education.

Given this continued hybrid quality in library collections between print and electronic resources and collections services, it is no wonder collection management at academic libraries continues to include serious consideration of the universe of print publications. Even as electronic resources supplant many print holdings historically found in libraries, those same print materials serve as a basis of reference for strategic acquisition of electronic equivalents. Indeed, in many cases, as library user demand has negotiated the transition from print to electronic, some library collection managers have found it advisable to acquire the same resource in multiple formats. As budget constraints have been imposed on academic library budgets in recent years, however, more collection managers have eschewed this luxury by canceling the print component of dual holdings (Schonfeld, 2007). Nevertheless, the transitory correspondence of print and electronic collections has ensured a continuity between the formats, making it possible for libraries to provide immediate remote access to traditional print resources deemed essential by instructors teaching in distance education.

Electronic Resources and Online Education

This is not to say development of libraries' electronic collections has been driven solely by consulting the universe of print scholarly production. Collection management in the electronic age has also been influenced by emerging demands of online education courses and the teaching methods themselves. The responsiveness to computer-mediated instruction is evident in the next-generation library resources without direct print counterparts, such as aggregated databases, cataloged websites, unique digital collections of primary sources, online journals, and other born-digital resources including certain repository content and targeted collections of data sets. However, current and emerging models of collection management also reflect the important direct influence by an institution's distance education programs on the development of library collections. The liaison model, a common collection management model employed in academic libraries today, grew to prominence on the philosophy that librarians involved in collection management should communicate directly with teaching faculty in order to understand current teaching and research needs, and respond to them effectively. As this model continues to evolve today, librarians are placing more emphasis on interactivity with instructors and the online curriculum through increased collaboration with faculty on course planning, identification and delivery of scholarly resources, copyright clearance for electronic content, and assessment of collections-related aspects of course delivery.

The remote access necessary for effective delivery of electronic library resources has evolved over time in an effort by libraries, content owners, and resource vendors to meet demands of online use and instruction. Access to resources through usernames and passwords has largely been replaced by access rights determined by IP address, allowing libraries to create seamless linking to licensed content at the points of discovery. For remote users, this may not mean the absence of log-in requirements, since users accessing resources from personal computers or other devices not registered to the library's institution must be authenticated. Authentication usually takes place through a library's proxy server, and once users successfully authenticate, they are generally able to access multiple resources for the duration of their session.

Widespread adoption of link resolver software has also contributed to enhanced remote access to electronic library resources. The software uses OpenURL technology, first introduced in 2000, to permit context-sensitive linking among various library databases, including the online library catalog. This technology is invaluable to libraries in their efforts to provide article- and chapter-level access to electronic documents, as well as searching across multiple databases for full-text content.

In addition, as Lagace and Chisman (2007) have described, libraries have succeeded in employing link resolver software to enhance services such as interlibrary loan and local document delivery, which may also support online instruction. Used with IP authentication, link resolver software provides seamless redirection among discovery tools and disparate library resources, allowing online users and instructors to navigate their research and share classroom materials more efficiently and without specialized knowledge of specific databases and other content repositories.

While advances continue regarding the provision of remote access to library resources, some limitations remain. One limitation is the mixed results of efforts to provide federated searching of both print and electronic library resources. Most academic libraries would prefer to present a single interface to search for library resources in all formats. Online library catalogs are able to approximate this ideal, as batch-loadable records containing necessary metadata have become available for an increasing number of resources too numerous to be cataloged by local staff. However, such records generally represent content at the title level only, and for journal content, this means searching at the journal title level, but not at the level of individual articles. Similarly, for monographs, edited collections, and other book formats, users are able to search titles of works, but often not chapters or sections within those works. Since much e-book content is now indexed at the chapter or section levels as part of its electronic publication, this limitation requires solutions to provide more precise points of access in support of online instruction.

One significant way libraries provide access at the article, chapter, or section levels is through the link resolver technology already described. Once a piece of scholarship is discovered through open-URL linking, instructors and librarians assisting in the integration of resources into a course management shell often use a persistent URL to direct students to, for direct access to the desired article, chapter, or section. Instructors and course designers benefit by working closely with their liaison librarian when planning online courses and preparing online delivery of units, lessons, assignments, and research assistance. Librarian expertise concerning specific library databases, portals, books, and journals aids course planners in locating specific pieces of scholarship not readily apparent through searching the online library catalog.

Additional limitations confronted by libraries as they attempt to meet the demands of online instruction are the technological and license restrictions placed on the content itself. While content providers continually try to balance publishers' concerns over intellectual property rights with libraries' imperative to deliver manageable content to their users, the ability to access and manipulate scholarly sources can vary among various databases and repositories. These variations sometimes create frustration or uncertainty as instructors incorporate library electronic resources into course delivery and student assignments. Printing limitations imposed by digital rights management (DRM) technologies and unavailability of preferred download formats, for example, may have practical impacts on an instructor's ability to assign reading materials to students with varying ability or capacity to interact with course materials online.

In addition to technological restrictions placed on methods of use, licensing terms required by content providers may also affect users' ability to access and manipulate library resources. While limitations similar to those enforced by DRM technologies are sometimes stipulated in licenses negotiated between libraries and vendors, usually these contracts also determine the scope of users permitted access to the resources. Standard licensing terms have come to resemble the typical needs of library users, such as instructors requiring access to resources for students in online or hybrid classes. However, the dispersed nature of some academic library branch structures, as well as collaborative partnerships with other academic institutions or community organizations, can introduce obstacles to effective delivery of electronic resources for online instruction. For instance, some branch libraries might acquire resources independently of centralized library purchasing operations. In other cases, colleges and universities may enter into resource-sharing agreements or formulate inter-institutional curriculum offerings not supported by the terms of all contracts involving library electronic resources, creating confusion among users about which materials are available to remote populations.

While libraries continue to work with content providers to align technological constraints and licensing terms with evolving trends in institutional structures and inter-institutional collaboration, libraries and stakeholders in computer-mediated instruction may address existing incompatibilities more immediately through effective communication about license restrictions and the situational needs of online instruction. On the libraries' side, administrators, curriculum designers, instructors and students should note that many libraries have adopted (or are in the process of adopting) electronic resource management (ERM) software, allowing sharing of selected information concerning the terms of use attached to specific library resources. Additionally, librarians serving as liaisons to particular disciplines or programs are focusing their attention not only on the topics and subject matter of classes and courses, but on the methods of delivery informing library resource needs. Through these and other methods of communication, librarians can better inform stakeholders about the qualities of access and use available for their resources.

Finally, other institutional stakeholders in online curriculum and instruction can contribute to effective communication about resource availability by including librarians and library administrators in discussions concerning the conception, design, implementation, and use of library materials in those online educational endeavors. Through increased awareness of the composition and needs of online programs and offerings at each of these levels, library stakeholders can anticipate priorities when selecting resources and negotiating licensing terms, illuminate the possibilities and limitations associated with relevant library resources, and make those involved aware of library resources and services that may remove or ameliorate difficulties associated with particular licensing constraints.

An Expanding Universe of Available Resources

While our discussion of remote access to electronic library resources thus far has focused on access to existing resources, it is also crucial to recognize the impact of the expanding universe of electronic resources being added to academic library collections. Scholarly and popular journal content has long been available in electronic format and continues to increase as publishers and other organizations publish new issues in electronic format and, in many cases, expand the portion of backfiles available electronically through digitization initiatives. In addition, numerous electronic-only titles have been introduced that further add to the universe of electronic scholarly resources. These recently introduced electronic resources enhance libraries' abilities to support computer-mediated instruction.

The emergence of electronic books, or e-books, is perhaps the most recent significant development important to academic libraries' role in online instruction. E-book usage is growing significantly among the US population, with 51 percent of consumers indicating an increase in e-book purchases during 2009 (Milliot, 2010). This general trend has impacted academic libraries, as the number of titles available electronically increases and user preferences continue to shift to electronic format. As a result, published monographs increasingly are available electronically through library-acquired databases capable of delivering content directly to students' electronic devices. Stakeholders in online instruction should realize that e-book content is frequently indexed at the chapter or section level, allowing discrete units to be assigned and linked to within course management software or other manifestations of computer-mediated delivery.

In recent years, new e-book acquisition models have accompanied the developments in the quantity and accessibility of e-books. More publishers have developed hosting platforms for their e-books, including monographs, reference works, and edited collections. For example, Elsevier, a leading publisher of science and health information, hosts electronic journal articles, monographs, reference works, handbooks, book series, and other content on its ScienceDirect platform. Similarly, Oxford University Press now hosts several different platforms for different varieties of content, including Oxford Journals, Oxford Scholarship Online (monographs), Oxford Reference Online, Oxford Handbooks Online, Oxford Bibliographies Online, and others. Besides publisher-hosted sites, publishers and associations have begun making e-book content available through third-party aggregators like ebrary, NetLibrary, EBL, and MyiLibrary, allowing libraries to provide access to multiple publishers' titles through a single platform.

Many publishers and aggregators have formed partnerships with book acquisitions services for title-level distribution of e-book content. Through this mechanism for purchasing or leasing e-books at the individual title level, as opposed to packaging multiple titles as subject sets or subscription collections, libraries can respond more precisely to the dynamic needs of the online curriculum, acquiring titles based on specific instructor recommendations in support of online instruction. This model more closely resembles the print acquisitions model that evolved over the years to support traditional classroom instruction, and permits greater responsiveness by academic libraries to curriculum- and course-specific needs.

In addition to the access challenges and the varieties of content being addressed in their efforts to meet the demands of computer-mediated instruction, libraries have recently begun to confront the explosion of mobile computing devices gaining popularity among students and instructors. The most established of these devices, the laptop computer, presented relatively few challenges with respect to library resources, since laptops are essentially portable versions of desktop computers for which electronic resources were originally conceived. While library facilities and technology infrastructure had to adapt to laptops by introducing wireless networking and suitable workspaces, library resources themselves continued to be delivered in familiar formats, such as HTML, PDF, and Rich Text Format.

More recently introduced mobile computing technologies, however, pose additional challenges to the continued effective delivery of electronic resources. While the challenges presented by mobile devices have been discussed by librarians for several years (Roberts, 2006), e-book readers, smartphones, and newer tablet computers such as Apple's iPad are among the cutting-edge devices growing in popularity with students and some instructors who use library resources in online education. Each device introduces new or adapted formats required to view some desired content. For example, Amazon's Kindle, currently the most highly used e-book reader, relies to a great extent on a proprietary format and direct-to-consumer business model, making it difficult for libraries to incorporate the device into their support of online education. Smartphones and smartphone platforms like the Blackberry, iPhone, and Android platforms often require traditional content hosting sites to be optimized for the mobile platform in order to be viewed satisfactorily. As well, smartphones and tablet devices allow delivery of content through applications developed specifically for certain devices, and in order to support these devices in the context of online education, academic libraries need to explore possibilities, either locally or through collaborative ventures, for developing applications that facilitate access and use of electronic library resources. This development can be done most effectively by working with online instructors and curriculum designers to align library initiatives in this respect with the current and anticipated needs of computer-mediated instruction and the desires of the students it is meant to serve.

Libraries also must work with content providers, and the makers of the mobile devices themselves, to ensure that content delivery on all varieties of mobile computing devices includes consideration of library needs with respect to supporting online education. Proprietary formats, which currently predominate on many mobile devices, generally ignore the breadth of computer-mediated communication necessary to accommodate the diversity of students participating in online education. A greater measure of cross-platform compatibility needs to be introduced to these devices through inclusion of more open formats if they are to play a role in the future of online education and research. Libraries must continue to work with device makers and content providers to explain library roles in the delivery of online instruction and identify existing barriers to wider adoption of mobile computing technologies in the context of the library-publisher marketplace. Along similar lines, licensing terms need to be negotiated that allow the uses of electronic library resources to which students and online educators have become accustomed.

As they do with other innovative technologies like Web 2.0 (discussed below), academic libraries need to continue monitoring and influencing trends in mobile computing and content delivery in order to provide resources to online instructors and curriculum designers in ways most suited to the emerging technologies relevant to online education. Instructors and curriculum designers, for their part, can benefit greatly by communicating with their colleagues in the library not only about what technologies and delivery methods they are currently using in their online courses, but also about those they would like to use in the future. In this way, libraries can be prepared to meet future needs and instructors and designers can continue to take advantage of library resources in ways most amenable to the needs and goals of the curriculum.

Recent Trends in Scholarly Communication

Another important development for both libraries and online instruction has been a sea change in scholarly publishing leading to the proliferation of open access materials, as well as online repositories facilitating the distribution of that content. In recent years, the number and variety of open access journals has increased greatly, with more than 5,000 titles now included in the Directory of Open Access Journals. Additionally, government agencies are instituting mandates requiring publicly funded research to be published in online repositories. And as Harboe-Ree (2007) and others have documented, academic libraries have taken on a larger role in the emergence of open access content by becoming publishers themselves, situating them especially well to guide stakeholders in online instruction and curriculum development through the opportunities and challenges associated with emerging trends in scholarly communication.

The emergence of open access resources has vastly expanded the amount and varieties of information to which academic libraries are able to facilitate access. However, a recent study conducted by Jingfeng Xia (2010) shows that instructors' use of open access scholarship continues to be influenced by misperceptions of the quality. Academic librarians must continue to work with online instructors and course planners to build awareness of the availability and quality of open-access scholarship. Another recent study (Creaser, 2010) suggests this advocacy should take place through efforts by librarians to promote content held in local repositories, and to encourage deposit by online instructors for the purpose of making their own research more accessible to their students.

Indeed, while discussion thus far has focused on open-access resources within the universe of scholarly publishing, stakeholders in online curriculum and instruction can benefit by being aware of and, in some cases, involved with, libraries' roles in creating and publishing scholarly materials at their own institutions. Such roles involve digitization of print, microform, or audiovisual resources, and can include local and regional newspaper archives, government document collections, institutional records and archives, and unique collections of materials housed in libraries' special collections and archives departments. Academic library digitization and online access provision efforts may also involve collaborations with local businesses or campus and community organizations. At the University of Kentucky (UK) Libraries, for example, recent projects include a partnership with the Keeneland Association to reformat, preserve, and provide access to the *Daily Racing Form*, a unique publication documenting the history of the Thoroughbred racing industry, as well as collaboration between the UK Libraries' Louis B. Nunn Center for Oral History and the Kentuckiana Digital Library in developing a searchable interface to improve access to audio interviews with Bill Keightley, a significant figure in the history of the UK men's basketball program. Initiatives such as these may prove valuable to online instructors and student researchers who are able to involve them in computer-mediated instruction and research assignments.

In addition to traditional forms of scholarship and research openly available for online instruction, the evolution of scholarly communication has introduced early- and intermediate-stage research documentation and data into the universe of published scholarly materials. Such documentation can take the form of research notes and annotations, data sets published through websites or institutional repositories, and various informal communications among researchers. Academic libraries are formulating data management and sharing plans to preserve and provide access to data sets originating from a wide variety of sources, which holds considerable promise for enhanced educational use of such resources. While Barjak (2006) has discussed the value of informal scholarly communication to research productivity, librarians and online instructors may find that the published manifestations of these nontraditional research products can benefit the online classroom as well, serving not only as sources of research notes and data, but also teaching tools for use when discussing the research process in a computer-mediated setting.

LIBRARY SERVICES

Information Literacy

Teaching information literacy is a basic pillar of academic librarianship and, as such, part of the mission of every academic library. Information literacy is recognized as an essential skill-set not only for students obtaining education, but also for citizens who must make informed decisions in their post-graduation lives. The Association of College and Research Libraries (ACRL), recognizing the importance of information literacy, issued a set of standards which defines how those skills fit within the context of higher education. Those standards define an information literate person as someone who, after realizing an information need, can find, assess, and effectively use the chosen information (Association of College and Research Libraries, 2000). These skills have always been important, but now have reached a crucial level due to the overwhelming profusion of information available today. Academic libraries struggle to impart these skills to the students attending their institutions, regardless of whether they are taking in-person or online courses.

Teaching information literacy online is taken seriously by academic libraries, and is viewed as equal in importance to teaching it to in-person students. In fact, it can be seen as even more important, since online students cannot seek individual help as easily as face-to-face students. Also, many students taking distance education or online courses are undergraduates, with little knowledge of the library or college-level research (Thomsett-Scott & May, 2009). Yet, online information literacy efforts are still fairly new endeavors for libraries. Though some colleges and universities have been involved with teaching online information literacy since the mid-1990s, for most, it is in the early stages. As such, a consistent set of best practices has not been firmly established. Even the language used to describe facets of online learning differs from school to school.

Most colleges and universities have made the information needs of online and distance learners a priority and have tried to enhance the basic services available to them. Often, however, online students are unaware of many of the services available to them (Feeney, 2005). To remedy this, academic libraries began to create websites (as part of the main library site) that cater specifically to the needs of the online or distance learner – a centralized place explaining services and resources such as interlibrary loan, document delivery, or online reference services (Hines, 2008). These pages point out avenues of communication available to online and distance learners such as toll-free phone numbers, email addresses, and chat services. Some have a form allowing online students who live close enough to physically visit the library to set up appointments (Thomsett-Scott & May, 2009). Many academic libraries have a librarian whose primary duty is to serve the needs of online and distance learners (Hines, 2008). The distance learning or online librarian (as some are titled) provides help with reference, troubleshooting for technology or document delivery problems, and generally deals with any issues that arise.

Technological Tools

These informational measures, however, do not substitute for actual library instruction, in whatever form it takes. Studies show that online methods of teaching library instruction appear to be equally as effective as face-to-face methods (Zhang, Watson, & Banfield, 2007). Some online techniques mimic face-to-face classroom instruction. These approaches are known as web-based synchronous methods and include technological innovations such as virtual meeting rooms, webinars, whiteboards and virtual "environments" (such as Elluminate Live). Using these methods, students can follow along, in real time, with the librarian as he or she demonstrates methods online, watching on their computer screens as they are presented. In many cases, students can listen as the instructor talks and, if software allows, respond using a microphone, making for a fully interactive experience. In certain virtual environments, students can ask spontaneous questions after virtually raising their hands and participate in discussions or group work.

Yet, for all their potential, synchronous learning systems remain fairly uncommon as a method for delivering online information literacy in academic libraries. For most, the software is still too expensive and the learning curve too steep. Distance and online educators who have used synchronous systems, however, have found them to be more user-friendly and less cumbersome than videoconferencing technology, which was considered state-of-the-art only a short while ago. The new synchronous systems are more flexible, allowing students to use their own computers rather than gathering in a studio location. Libraries that make use of a web-based synchronous method usually combine it with other asynchronous methods.

Far more common at this point, for information literacy librarians, are asynchronous online methods. The most widely used tools in this area include online tutorials and research guides, or pathfinders. These methods have the advantage of potentially reaching large numbers of people with a relatively low drain on the limited personnel that most academic libraries can deploy for information literacy (Slebodnik & Riehle, 2009). In addition, because asynchronous instructional products are freely available on the web, they can serve the whole student population, both online as well as on-campus students. Free of time and place constraints, students can access them anytime and from any place, provided there is a computer (Anderson & Wilson, 2009).

One of the most popular asynchronous products is the online tutorial, which are self-guided mini-lessons on various topics ranging from a particular database or software to specific skills or concepts. These useful tools are relatively inexpensive to create, and, as a study by Silver and Nickel (2007) found, teach library concepts as effectively as in-class sessions. Not surprisingly, many students prefer them to in-class sessions, which require them to be in a particular place at a particular time.

Online tutorials have evolved greatly since beginning in the 1990s, moving from static HTML web pages, allowing for little interplay, to much more complex creations with many interactive parts and often having a game-like feel (Yang, 2009). Interactivity is an important feature; librarians have found that static, text-heavy pages are unappealing to an audience made up largely of Millennials, who have grown up with the web and have little tolerance for one-dimensional web sites. Teaching faculty can assign the tutorials to their students, with some even containing a built-in quiz at the end which can be turned-in.

Online research guides, sometimes called pathfinders, are also a very commonly used asynchronous method for delivering online information literacy information. Long a print staple of the academic library, research guides have stood the test of time, transitioning seamlessly into online environments. Research guides list resources that would be useful in researching a particular academic area. These can include whole disciplines or smaller subsets, such as resources for a particular class, assignment, or topic. Research guides can go beyond merely listing resources and explain how to use each item and describe its features and limitations. Some go beyond even that by giving tips in other useful areas, such as evaluating or citing the information found in the resources.

The online version of the research guide has the advantage of allowing students to work at their own pace, referring back when needed, and from wherever their computer is located. Research guides can also be quickly and easily updated or changed as newer material emerges or classes change. In the not-so-distant past, most academic libraries produced research guides using homegrown systems and hosted them on campus computers. The recent popularity of products such as Libguides from Springshare, however, shows that the trend seems to be moving towards hosted platforms, allowing use of multiple types of information. This approach has numerous advantages, including the presentation of a uniform look and feel for all research guides and easy-to-use and update templates, which can accommodate both static and interactive fields. Students are also very comfortable working with the Facebook-like format favored by these guides.

Facebook itself, along with many other Web 2.0 innovations, has been readily adopted by librarians for use in online information literacy initiatives. Librarians quickly recognized the potential in social networking sites to make information accessible, because they allow not only for back-and-forth communication between instructors and students, but also between the students. Blogs, wikis, and podcasts are other examples of applications that librarians use for disseminating information literacy information (Sodt & Summey, 2009). Online virtual worlds like Second Life represent a potentially rich environment for presenting information literacy content as well. The many interactive features and the multi-media nature of these tools make them ideal for engaging today's Millennial student population. The potential for student collaboration and participation in the teaching process encourages active learning, which keeps students engaged in the material and aids in the retention of information.

Online courses are often taught using a Course Management System (CMS), such as Blackboard or WebCT, which wholly facilitates communication between the course instructor and students and among the students. While librarians recognized the potential to use CMS in instruction, for the most part, they have been disappointed by the lack of information literacy-friendly features in the design of many CMS products (Gurney & Wilkes, 2008). Librarians have had to make do with existing systems and try to incorporate as much information literacy content as possible. The first step is often to simply add a link to the library's website in the CMS. Librarians have found that students are accustomed to having resources available at a mouse-click and will be less likely to search extensively for information not readily available. Often, librarians also add links to class and subject specific resources, such as databases, research guides, and tutorials (York & Vance, 2009), and in some cases, articles or e-books. Of course, providing students with a pathway to the materials does not make them information literate. One study found that even when links to databases were provided along with relevant search terms, the students still performed poorly (Gurney & Wilkes, 2008).

Collaboration

Clearly, the greater the librarian involvement in a class, the better the students will at mastering information literacy. The amount of librarian involvement in a particular class's CMS, is, however, a matter of negotiation between the teaching faculty and the librarian. Ideally, the librarian would be a full collaborator in the course. However, these partnerships are fairly rare for a variety of reasons, most common of which are time constraints of both librarian and teacher. There are other reasons as well, including teaching faculty members not wishing to lose autonomy or sacrifice lecture time and content, or not recognizing a need for information literacy instruction. Accreditation organizations have, however, stressed the importance of collaboration in recent years (Saunders, 2007). While the decision to partner with a librarian is the professor's choice at most schools, it is possible that university administrations will, in the future, mandate at least a minimum level of collaboration.

Librarians, as a rule, have always wanted to play an information literacy role in online classes, but have struggled to get their feet in the door. With upper administration and accreditation agencies recognizing information literacy's importance, librarians are finally getting what they always wanted. They are now more empowered to work with teaching faculty, which, in turn, should improve library relations with the rest of campus (George & Martin, 2004). When faced with these collaborative situations, teaching faculty recognize that librarians have something valuable to add to their classes that will help the students. For their part, librarians must live up to this responsibility and produce meaningful results.

When a librarian is integrated into the CMS at a level beyond merely adding links, the process is referred to as being an "embedded librarian" (York & Vance, 2009). (The term "embedded" can also mean other things such as a librarian being embedded in a department – i.e. having an office or office hours outside of the library and in a departmental building.) After the librarian finds a willing partner, the collaboration works best if the two can plan the class together, or at least allow for librarian input. The librarian can suggest research ideas that make the best use of the particular library's collection strengths. Librarians know from experience what types of assignments will be most successful because, often, they assist students with research assignments that fail to make the best use of the library resources. For example, an assignment might require students to research in an area in which the library does not have sufficient resources. With advanced planning, the librarian can learn the course objectives and topics and develop a viable project, perhaps even acquiring additional library resources to support the research. In some cases, the librarian will also help with the grading of the assignment.

Beyond collaborating on assignments, embedded librarians can create general information literacy content for a course, provide "point of need" assistance, create tutorials and class-specific research guides, and monitor the discussion board for the class, looking for issues to address with information literacy advice. When issues arise, the librarian can help (Lillard, Norwood, Wise, Brooks, & Kitts, 2009), as well as encourage students to ask questions. The librarian, of course, must be committed to checking the board regularly for this approach to work, and the professor's level of involvement is instrumental to the success of these efforts. He or she must make the students aware of the librarian's presence and encourage them to contact the librarian with research questions and to use the tutorials and research guides. Librarians have long known that encouragement from the professor is essential. What Hardesty noted back in 1995 is still true today in the embedded online environment: faculty must be involved for library instruction to succeed.

In terms of online information literacy, collaboration for the librarian does not end with teaching faculty. Often, teaching faculty are not experts in the technological aspects of the CMS. For help in this area, collaborators among the campus Information Technology department are important. Opportunities to collaborate with other campus community members also abound. Coordinators for General Education programs are excellent contacts for librarians wishing to be aware of over-arching policy about courses. Instructional support and instructional design teams can help with teaching strategies and pedagogical advice, as well as with creating online information literacy products, like tutorials. Writing centers and career centers are also good partners for librarians, because they have similar teaching goals.

Assessment

A byproduct of collaboration with faculty members is an enhanced ability to assess student learning in information literacy. Assessment of an information literacy program's effectiveness is considered essential, but is often neglected because it of its impracticality without faculty collaboration (Hines, 2008). And because most assessment methods are time-consuming and create extra work, teaching faculty have not often been particularly supportive of this endeavor, at least in face-to-face classes. As a result, many librarians have not pursued it. However, of late, it has become a very important issue, with many accreditation agencies adding assessment to their criteria needed for passing grades.

Many assessment methods become easier in an online setting, including pre-testing at the start of the course and then post-testing at the end to see what was learned, which can be accomplished with online surveys (Slebodnik & Riehle, 2009). Post-testing can take place immediately after an instruction session or at the end of the entire class to see if the knowledge was retained. Students can also be required to complete online course evaluations at the semester's end, rating their information literacy experience. In addition to surveys and evaluations, online products, such as tutorials and research guides can be assessed with usability studies (Slebodnik & Riehle, 2009). It is sometimes difficult to find participants for usability studies, but a very small sample (sometimes as few as five) can yield valuable results (Slebodnik & Riehle, 2009). Also, unobtrusive methods such as tracking use statistics on online tutorials and research guides can also be used (Hines, 2008).

Communication Issues

There are many challenges to successful collaboration between teaching faculty and librarians in online classes. The greatest is the lack of communication between the departmental teaching faculty and librarians. Because many colleges and universities coordinate online classes at a departmental level, not at a university or even college level, librarians often have trouble getting information about the online classes early enough to plan meaningfully. For a variety of reasons, departments often wait until nearly the last minute to assign instructors to classes, and as a result, librarians do not know which online classes are being taught until a catalog is available. Once aware, librarians can meet with professors and negotiate a level of collaboration, prepare research guides, tutorials, and specialized web pages for the class, or perhaps work together to formulate research projects for the students. Without advance information, librarians have to be reactive, creating products after classes begin and with little time allowed for thoughtful planning.

The entire blame for this situation cannot be placed on academic departments; communication issues go both ways. Teaching faculty members are often unaware of available information literacy services because libraries have done a poor job marketing them. This longstanding problem is a frequent topic in the professional literature. It is the librarian's responsibility to make services and products known and "sell" the faculty on what they can do (Ryckman, Long, & Yedinak, 2009). Studies have shown that, for the most part, teaching faculty do not exclude librarians because they are not satisfied with library services, but rather because the librarians have fallen off of their radar (Thomsett-Scott & May, 2009).

Part of the cause for the teaching faculty's lack of knowledge about library services is that personal contact is the most frequently used method for promoting information literacy efforts (Islam, 2005). This method is fairly inefficient, because it is difficult for one librarian to maintain contact with the many faculty members in a department. This problem is compounded by the fact that most subject librarians are not housed in the same building as the teaching faculty, so casual contact is fairly limited. Other avenues of communication include placing promotional materials on library websites, making presentations to departmental faculty, sending mass emails, posting messages to department email lists, and distributing printed materials to faculty mailboxes. Most libraries use at least one and often more than one of these marketing methods (Islam, 2005). More recently, librarians have begun using Web 2.0 applications such as blogs, RSS feeds, and social networking sites, to market information literacy services, which show a good deal of promise (Ryckman, Long, & Yedinak, 2009).

A side-effect of the lack of communication is the difficulty many librarians have getting professor buy-in. The term "professor buy-in" here refers to professors recognizing the importance of library learning and actively using information literacy services. Lack of buy-in has been a problem with libraries for a long time. A number of reasons contribute to professors sometimes resisting the incorporation of information literacy into their classes. As already discussed in terms of collaboration, some professors feel that involving a librarian in their class will impinge on valuable lecture time. Professors may also assume that because students can find a minimum level of information using internet search engines, that they are already information literate. Other professors also feel that adding an information literacy component to their classes would put an undue burden on themselves and their students.

Without professor buy-in, it is difficult to motivate online students to use library resources. While the librarian can provide useful, easily-accessible literacy materials, if the students are not required or strongly encouraged to, they will not, in most cases, use them (Thomsett-Scott & May, 2009). Unless instructors mention it, students may be completely unaware of the library website and the many resources available through it. Likewise, if a tutorial is not required as part of the coursework, students will have no incentive to complete it (Slebodnik & Riehle, 2009). However, with a professor's encouragement, the students can be made aware of the dearth of scholarly resources and the extreme variance in quality of materials available through internet search engines and begin to use the library-provided online resources (Gurney & Wilkes, 2008).

A number of the communication problems mentioned above can be alleviated by establishing a centralized coordinating office for distance and online education at the university or college level thats duties would include facilitating communication between the library and the academic departments. With such an office in place, librarians could be kept abreast of faculty changes, new courses, and other pertinent information (Ryckman, Long, & Yedinak, 2009). The distance learning librarian, in conjunction with an online library instruction coordinator, could work with the university-level office to make sure communication remains open between academic departments, library instruction personnel, and online students.

In addition to centralizing online course administration, another method for paving the way for information literacy collaborations between professors and librarians is to use the library's academic liaisons. The liaison's close relationship with departmental faculty and expert subject knowledge makes him or her an ideal partner for information literacy collaboration for both online and face-to-face instruction. In fact, the provision of information literacy is usually a central component of the liaison's job description. Many liaisons maintain close contact with departmental administrative personnel, participate on departmental email lists, and attend departmental faculty meetings. These methods are good conduits for communication about online classes. While liaisons are subject to the same lack of professor buy-in that affects all librarians, increased access to the academic department afforded the liaison can be useful in breaking down those barriers.

Pedagogical Implications

It is clear that there is a need for information literacy in online and distance education classes. Although many of today's students might disagree, it is clear that information literacy is not a skill set that can easily be picked up on one's own. In most cases, it must be taught (Shenton, 2008). Methods for teaching library instruction have evolved over the past 100 or so years since it first began. Many characteristics of good face-to-face instruction can be transferred to online instruction, including tying instruction to a specific assignment and then actively engaging the students in the learning process. Students are more motivated to learn when actively working on an assignment, as opposed to obtaining knowledge for a later, yet unspecified purpose. This characteristic can apply to an online setting with interactive tutorials, research guides, or synchronous teaching geared to specific class assignments. Many Web 2.0 products such as wikis and blogs include interactive features which allow the student to get involved in the learning process (Smith, 2006).

Students learn better when material is provided in more than one medium. This is why many colleges and universities provide a variety of methods to reach online learners. Visual learners benefit from tutorials, while those who learn best by reading can use research guides. Still others learn best when listening to a lecture, which can be accomplished using synchronous means. Clear objectives are always important; students will better be able to learn if they know what they are working towards. Objectives should be clearly stated in all online products. Students understand instruction best when being taught concepts that they can transfer to other areas. Collaborative learning is the most difficult tenet of good library instruction to achieve in an online setting. This is an area awaiting further technological development before it can become widespread, although some group work can be done within CMS systems and other synchronous software (Smith, 2006).

CONCLUSION

Betne and Castonguay (2008) recently published a study of library- and web-resource usage patterns among mathematics students at LaGuardia Community College. Among other findings, the authors observed a relationship between use of library and web resources and a general ability in students to engage productively with online communication and associated technologies. In response, the authors conceived a constructivist pedagogical model meant to exploit this relationship to improve both resource usage and student engagement with learning technologies and online communication. The survey on which the research was based was also intended to develop ideas for a library and web information portal to be used in conjunction with mathematics instruction at the college. While this research was limited to a single academic discipline, it serves as an example of the ways in which collaboration between librarians and instructors can lead to both theoretical and practical improvements to computer-mediated instruction.

The lessons of the preceding example also reinforce a more general imperative supported by the current chapter as a whole. As academic libraries conceptualize and execute their roles in online education, they must continue to anticipate, influence, and respond to changes in technology, scholarly publishing, and computer-mediated communication. As has been shown, none of these areas exist in isolation. Also, in the context of online education, none can be addressed without substantive dialogue and collaboration among the various stakeholders involved, including librarians, instructors, curriculum designers, and university administrators. If academic libraries and the institutions they serve are to successfully meet the challenges of online education, they must work together to understand the varieties of interdependence among these areas and the roles that institutional policy, pedagogy, and collaboration can play in facilitating the support and initiative necessary to employ each purposefully.

REFERENCES

Anderson, R. P., & Wilson, S. P. (2009). Quantifying the effectiveness of interactive tutorials in medical library instruction. *Medical Reference Services Quarterly*, *28*(1), 10–21. doi:10.1080/02763860802615815

Association of College and Research Libraries. (2000). *Information literacy competency standards for higher education*. American Library Association. Retrieved from http://www.ala.org/ala/mgrps/ divs/acrl/standards/ informationliteracy competency.cfm

Barjak, F. (2006). The role of the internet in informal scholarly communication. *Journal of the American Society for Information Science and Technology*, *57*(10), 1350–1367. doi:10.1002/asi.20454

Betne, P., & Castonguay, R. (2008). On the role of mathematics educators and librarians in constructivist pedagogy. *Education*, *129*(1), 56–79.

Brophy, J., & Bawden, D. (2005). Is Google enough? Comparison of an internet search engine with academic library resources. *Aslib Proceedings*, *57*(6), 498–512. doi:10.1108/00012530510634235

Creaser, C. (2010). Open access to research outputs--Institutional policies and researchers' views: Results from two complementary surveys. *New Review of Academic Librarianship*, *16*(1), 4–25. doi:10.1080/13614530903162854

Feeney, M. (2005). Centralizing information about library services and resources. *Internet Reference Services Quarterly*, *9*(1), 129–146. doi:10.1300/J136v09n01_10

George, J., & Martin, K. (2004). Forging the library courseware link: Providing library support to students in an online classroom environment. *College & Research Libraries News*, *65*(10), 594–597, 613.

Gurney, L. J., & Wikkes, J. (2008). Creating a library presence in online units. *Australian Academic & Research Libraries*, *39*(1), 26–37.

Harboe-Ree, C. (2007). Just advanced librarianship: The role of academic libraries as publishers. *Australian Academic and Research Libraries*, *38*(1), 15–25.

Hardesty, L. (1995). Faculty culture and bibliographic instruction: An exploratory analysis. *Library Trends*, *44*(2).

Hines, S. S. (2008). How it's done: Examining distance education library instruction and assessment. *Journal of Library Administration*, *48*(3), 467–478. doi:10.1080/01930820802289565

Islam, R. (2005). *Collaboration for distance learning information literacy instruction (SPEC kit 286)*. Washington, DC: Association of Research Libraries.

Lagace, N., & Chisman, J. (2007). How did we ever manage without the OpenURL? *The Serials Librarian*, *52*(1), 211–222. doi:10.1300/J123v52n01_17

Lillard, L. L., Norwood, S., Wise, K., Brooks, J., & Kitts, R. (2009). Embedded librarians: MLS students as apprentice librarians in online courses. *Journal of Library Administration*, *49*(1), 11–22. doi:10.1080/01930820802310544

Milliot, J. (2010). Consumers show growing satisfaction with e-books. *Publishers Weekly*, *257*(15), 5.

Roberts, G. (2006). So many mobile devices, so little time. *Computers in Libraries*, *26*(4), 29–31.

Ryckman, B., Long, V., & Yedinak, E. (2009). Establishing a baseline: History, evolution and evaluation of Grand Valley State University's off-campus library services. *Journal of Library Administration*, *49*(1), 75–88. doi:10.1080/01930820802310890

Saunders, L. (2007). Regional accreditation organizations' treatment of information literacy: Definitions, collaboration, and assessment. *Journal of Academic Librarianship*, *33*(3), 317–326. doi:10.1016/j.acalib.2007.01.009

Schonfeld, R. (2007). Getting from here to there, safely: Library strategic planning for the transition away from print journals. *The Serials Librarian*, *52*(1), 183–189. doi:10.1300/J123v52n01_15

Shenton, A. K. (2008). The information seeking problems of English high schoolers responding to academic information need. *Library Review*, *57*(4), 276–288. doi:10.1108/00242530810868715

Silver, S. L., & Nickel, L. T. (2007). Are online tutorials effective? A comparison of online and classroom library instruction methods. *Research Strategies*, *20*, 389–396. doi:10.1016/j.resstr.2006.12.012

Slebodnik, M., & Riehle, C. F. (2009). Creating online tutorials at your libraries: Software choices and practical implications. *Reference and User Services Quarterly*, *49*(1), 33–37, 51.

Smith, S. S. (2006). *Web-based instruction: A guide for libraries* (2nd ed.). Chicago, IL: American Library Association.

Snyder, C., Logue, S., & Preece, B. (1996). *Role of libraries in distance education (SPEC Flyer 216)*. Washington, DC: Association of Research Libraries.

Sodt, J. M., & Summey, T. P. (2009). Beyond the library's walls: Using library 2.0 tools to reach out to all users. *Journal of Library Administration*, *49*(1), 97–109. doi:10.1080/01930820802312854

Thomsett-Scott, B., & May, F. (2009). How may we help you? Online education faculty tell what they need from libraries and librarians. *Journal of Library Administration*, *49*, 111–135. doi:10.1080/01930820802312888

Xia, J. (2010). A longitudinal study of scholars' attitudes and behaviors toward open-access journal publishing. *Journal of the American Society for Information Science and Technology*, *61*(3), 615–624.

Yang, S. (2009). Information literacy online tutorials: An introduction to rationale and technological tools in tutorial creations. *The Electronic Library*, *27*(4), 684–693. doi:10.1108/02640470910979624

York, A. C., & Vance, J. M. (2009). Taking library instruction into the online classroom: Best practices for embedded librarians. *Journal of Library Administration*, *49*(1), 197–209. doi:10.1080/01930820802312995

Zhang, L., Watson, E. M., & Banfield, L. (2007). The efficacy of computer-assisted instruction versus face-to-face instruction in academic libraries: A systematic review. *Journal of Academic Librarianship*, *33*(4), 478–484. doi:10.1016/j.acalib.2007.03.006

KEY TERMS AND DEFINITIONS

Digital Rights Management (DRM): Technologies used to control or limit access to digital content and devices, often imposed by content owners on content users or licensees.

Directory of Open Access Journals (DOAJ): A directory and service facilitating access to free, open access, scholarly journal content.

Electronic Resources Management (ERM): The practices and software employed by libraries to manage electronic resources.

Embedded Librarian: In its broadest sense, this term refers to a librarian who provides a specialized service to a specific group and is located at least part of the time within the group rather than in a traditional library setting. However, in academia, this term refers to a librarian who collaborates to a significant extent with a teaching faculty member on a course, usually having access to the course management system and often helping with planning and grading. It can also refer to an academic librarian whose has an office or office hours outside of the library and in a departmental building.

Information Literacy: The skill set necessary to understand when information is required, and then to locate, evaluate, and effectively use the needed information.

Open Access (OA): Unrestricted access as applied to scholarly publishing.

Open-URL: A type of Uniform Resource Locator (URL) that serves to link library users to authorized content across multiple resources and databases.

Research Guides: In-depth guides usually prepared by librarians or teachers intended to help researchers locate information sources on a particular subject. Often a selective, rather than complete, listing designed to aid the researcher in getting started. Also called pathfinders.

Tutorials: A presentation providing individual instruction on a specific subject. Academic libraries often use the term to describe a short online instructional video, but tutorials can occur in other media as well.

Web 2.0: Easy-to-use second generation internet applications which promote interactive experiences for users. Hallmarks include information sharing, user-centered design and the ability to self-publish. Common applications include blogs, wikis, and social networking sites.

Chapter 4
The Usability of Online Quizzes:
Evaluating Student Perceptions

Tatyana Dumova
Point Park University, USA

ABSTRACT

This study focuses on assessment, an essential component of teaching and learning. It examines the usability of online quizzes and determines what features of Blackboard-based testing are most beneficial from a student perspective. A survey of 395 students provides new insights that can help teachers, administrators, course providers, trainers, and instructional technology specialists make informed choices in developing and implementing online assessments. The author concludes that instructors should move away from the "one-size-fits-all" approach towards flexible, learner-centered models.

INTRODUCTION

Advances in Internet-based educational technologies promise to create learning environments that could transform many aspects of the traditional relationships between instructors and their students, including the presentation of course content, learning activities, student-teacher and student-to-student interactions, as well as assessment (Oliver & Moore, 2008; West, Waddoups, & Graham, 2007). Currently, over fifty different web-based course management systems (CMS) are being used by higher education institutions around the world (Pieper, Edwards, Haist, & Nolan, 2010). Well-known online systems such as Blackboard, Desire2Learn, eCollege, Angel, Sakai, and Moodle have been widely adopted by educators and offer an array of embedded assessment tools (Adlakha & Aggarwal, 2009).

One study of instructors' use of course management systems revealed that 80% of respondents had used online testing in their face-to-face classes (Morgan, 2003). Another study of college students reported that 70% of them had taken

DOI: 10.4018/978-1-61350-077-4.ch004

online quizzes, 88.8% had used online practice exams, and 71.8% had obtained faculty feedback online (Kvavik, Caruso, & Morgan, 2004). Focusing on the educators' experiences, Liang and Creasy (2004) argued that technology provides numerous opportunities for online assessment to become more learner-centered, increase learner autonomy, and promote self-directed learning. It should be noted, however, that when educators integrate technology to assess student learning, they tend to overlook the significance of creating assessments that are both convenient and usable from the contemporary students' point of view.

In addition, the number of online learners in higher education continues to grow, according to the 2009 report by the Sloan Consortium. In the of Fall 2008, as the report states, 25% of all higher education students in United States, or over 4.6 million students, were taking at least one course online (Allen & Seaman, 2010). Finally, blended learning, which combines traditional face-to-face classroom instruction with online learning activities, is gaining increased recognition (Koohang et al., 2010). Given the prevalence of web-based technologies in educational practices, it is critical to adequately evaluate their effectiveness. However, despite the growing use of online assessments, there is insufficient evidence to indicate how suitable and useful these tools are from a student perspective. Regardless of the mode of instructional delivery, it remains unclear how much online testing and evaluation can add to the quality of students' overall learning experiences.

The purpose of this study is two-fold: (a) to examine the students' perceptions of the usability of web-based assessment tools offered by Blackboard, one of the most prevalent course management systems in higher education, and (b) to determine the formal features of online quizzes that are the most beneficial from a student perspective, as well as those that might need improvement. Thus, the chapter focuses on an essential component of computer-assisted teaching and learning—assessment. The author seeks to provide insight on the perceived usability of online quizzes, which can help teachers, administrators, course providers, trainers, and instructional technology specialists involved in the creation of e-learning applications, make informed choices in the development and implementation of online assessments.

The chapter is organized as follows: it first reviews the pertinent literature on e-assessment, and usability evaluation. The author then describes the study methodology, the results and procedures, and provides a discussion of the future trends. The concluding section contains recommendations for teachers, instructional technology specialists, and educational practitioners involved in the design and implementation of web-based testing.

REVIEW OF LITERATURE

Assessment of student competences is central to both teaching and learning. Regardless of the area of study and the mode of presentation, the evaluation of the effectiveness of classroom instruction is vital in today's system of higher education. A large volume of research has focused on student competencies, including measuring and reporting learning outcomes. The proliferation of Internet-based social interaction technologies (Dumova & Fiordo, 2010) such as social networks, blogs, wikis, and podcasts, and the growth of e-learning have revived interest in the assessment of student learning outcomes, but with an emphasis on the implementation of new and innovative ideas that may contradict the *status quo* and go against the traditional practices of evaluation of student learning.

In today's networked world, information sharing and the co-creation of knowledge have been increasingly viewed as the cornerstones of a new learning culture, one in which learners build connections, create a community, and engage in collaborative education. Steve Hargadon (2009),

Elluminate's Social Learning Consultant and the founder of the Classroom 2.0 social network (classroom20.com), introduced the term "educational networking" when referring to the pedagogical value and the use of social networking and other social interaction technologies in education. As detailed by Lee Rainie (2009), Director of the Pew Research Center's Internet & American Life Project, today's learners are:

- Skilled to capture inputs from a variety of information sources
- Self-directed and less dependent on top-down instructions
- Attuned to group outreach and group knowledge
- Reliant on sources of support, feedback and response.

The above characteristics of today's learners pose important challenges to the assessment of student competencies in the Web 2.0 age. Researchers need to identify ways of modifying educational assessment in order to address the gap between the everyday lives of contemporary students and traditional assessment practices (Elliott, 2007).

Given the growing penetration of information technology in educational practices, online testing has started to emerge as a distinctive area of research interest (Holms, 2002; Koohang, 2004b). Referring to the use of information technology for assessment purposes, e-assessment can be used in classroom-based, hybrid, and online courses. Advances in digital audio and video recording techniques, digital imaging, and the recent outburst of Web 2.0 technologies promise to continue to make e-assessment more dynamic and adjustable to individual student needs. From PBwiki to PennTags to TeacherTube—educational implementations that harness the power of a new generation of web-based tools flourish across the networked information ecology (Alexander, 2006), uncovering the emergence of a new wave of educational practices. The following qualities of assessment that meet the demands of the Web 2.0 age have been identified by Elliott (2007, p. 4-5) as:

- Problem oriented: original tasks requiring genuine problem solving skills
- Personalized: tailored to the knowledge, skills and interests of each student
- Socially constructed: using the student's social networks
- Authentic: involving real-world knowledge and skills
- Collaboratively produced: produced in partnership with fellow students.

This study focuses on the students' perceptions of the usability of online tests and quizzes, as well as the overall value that students attribute to online assessments. The most general definition of *usability* implies a delineation of the ways people are *able* to *use* products and systems. More specifically, usability has been defined as the capability of a particular system to satisfy the needs and specifications of the users (Holms, 2002; Koohang & du Plessis, 2004; Nielsen, 2000). Nielsen introduced a number of techniques for evaluating usability, ranging from "discount" usability testing (1989) such as focus group tests or "hallway" heuristic tests (1999) to professional laboratory testing (1993) equipped with cameras and microphones and including many advanced features, for example, eye tracking. At the same time, "cheap" usability testing administered for little or no cost and without special training can reveal up to two thirds of the problems identified by expensive laboratory evaluations simulating specific environments (Nielsen, 2007; Nielsen & Molich, 1990). Although methods of usability evaluation have been long established and well developed, there is lack of research on how online learners actually perceive usability.

METHODOLOGY

Study Design

The present study employed an online survey of 395 undergraduate students enrolled in introductory courses in physical and social sciences at a medium-size Midwestern university. In order to evaluate the usability of online assessments, the study focused on the level of students' perceived usefulness and satisfaction with online quizzes. An online quiz was defined as a mediated dialog between the students and instructor (Shackel, 1991; Shneiderman & Plaisant, 2005). More specifically, the study addressed the following research question:

RQ1: What are the student perceptions of the usability of online quizzes?

The first hypothesis predicted:

$H0_1$: There will be no difference in student perceptions of online tests and quizzes between students in physical sciences and social sciences.

The alternative hypothesis stated that there will be a difference in student perceptions of online tests and quizzes between students in physical sciences and in social sciences.

The second hypothesis posited:

$H0_2$: There will be no difference in student preferences in online testing between students in physical sciences and social sciences.

The alternative hypothesis stated that there will be a difference in student preferences in online testing between students in physical sciences and in social sciences. The study design and procedures were reviewed and approved by the institutional review board.

Study Participants

The study utilized convenience sampling, a non-probability sampling technique in which participants are selected because of their accessibility and proximity to the researcher. Participating in the study were 193 (out of 200) students who were enrolled in four introductory mathematics classes, and 202 (out of 205) students who were enrolled in a large lecture course in political science. Three quarters of all the participants were freshmen and sophomore students and one quarter of the students were juniors and seniors. Nearly 98% were between 18 and 25 years old, while less than 3% were between 26 and 40 years of age. About two thirds were male. Among the math students, females composed 33% of the participants; in the political science course, the women composed 31% of the total. All four of the math courses were taught by the same instructor, who utilized Blackboard-based take-home quizzes to help the students practice solving math problems and prepare for the tests. The political science instructor also used Blackboard for designing chapter quizzes, which were due before the each respective lecture and which students could take several times until they had the answers right. In both disciplines, most students (88%) were taking online quizzes on a weekly basis.

Survey Instrument

The author first examined the formal features of the Blackboard testing tools. Next, a measuring instrument aimed to rank the specific attributes of Blackboard tests was developed. Major usability attributes such as simplicity, navigability, user friendliness, comfort level, and control (Koohang, 2004a; Koohang & Harman, 2005) were integrated in the questionnaire. The questionnaire consisted of four demographic questions, four general questions on the use of online testing, twelve specific questions that asked students to rank their experiences with an online test or quiz that they took most recently on a 5-point Likert-type scale, six Yes-No-Don't Know questions regarding their preferences, and one open-ended question to accommodate any additional comments that they wished to make. Finally, respondents were asked which they would prefer if choosing between online or paper-based testing.

Procedure

The survey was administered by the course instructors asking the students to respond to the survey questions during a two-week period. The response rate to the survey was very high and reached 97.5%. Blackboard was chosen as the platform for hosting the survey, since it offered a familiar environment for the students and allowed the exporting of the collected data in a tab-delimited format directly into a spreadsheet. A *t* test for the significance of the difference between the means of two independent samples was applied to determine the probability level of incorrectly rejecting the null hypotheses stated above. A qualitative analysis of students' open-ended responses to the survey was performed, in order to obtain additional insights about the perceived strengths and limitations of online tests and quizzes and to identify common themes.

STUDY RESULTS

Student Perceptions of the Usability of Online Quizzes

RQ1 questioned the student perceptions of the usability of online quizzes. Results of the analysis indicate that the students considered online quizzes a useful and important contributor to their overall learning experiences (see Table 1). Measured on a 5-point Likert scale, the overall student satisfaction with online quizzes averaged 3.99 for the math students and 4.32 for the political science students. According to Table 1, there was a significant difference between the means among the math and political science students in all categories of perceptions of online quizzes. The independent samples *t* test yielded the probability values of $p < .05$, $p < .01$, and $p < .001$. Therefore, the first null hypothesis was rejected, and the alternative hypothesis – that there is a difference in student perceptions of online tests and quizzes between students in physical sciences and social sciences – was retained. The vast majority of students surveyed found online tests simple to use, user-friendly, and easy to navigate. Students also considered the use of text formatting features, as well as the integration of images, charts, and graphs as effective. Interestingly, students in political science consistently ranked most usability features higher than the math students. The only exception was the use of charts and graphs in quizzes, which the math students ranked higher. Finally, all students indicated a high degree of comfort they experienced while taking the online quiz with the mean of 3.94 for math students and 4.17 for political science students.

Student Preferences in Online Testing

The second hypothesis predicted that there would be no difference in student preferences in online testing between students in the physical sciences and the social science students. Table 2 provides descriptive statistics of student rankings of those formal features of online tests and quizzes that they found most important. As shown in Table 2, no significant difference was found across the four categories of preferences with the *p*-value for the *t* statistic exceeding.05 in all cases; therefore, the null hypothesis of no difference was retained. Students ranked the following benefits of online testing as most important: (1) keeping track of their progress through the use of an online gradebook, (2) taking quizzes at their own pace, without time pressure, and (3) receiving instant results and feedback on their tests and quizzes. The item that

Table 1. Student perceptions of the usability of online quizzes

	Math Students (n = 193)		Political Science Students (n = 202)		
	Mean	SD	Mean	SD	p
The online test interface was simple to use	4.17	0.76	4.43	0.82	.000***
The online test was user-friendly	4.02	0.79	4.21	0.81	.008**
I felt in control of using the online test	3.99	0.77	4.17	0.86	.019**
I was able to move around throughout the test with ease	3.93	0.92	4.29	0.83	.000***
Text boldfacing, italicizing, and underlining was used effectively	3.97	0.74	4.12	0.84	.034*
The use of images, charts, and graphs was effective	3.96	0.91	3.68	1.02	.000***
Indicate your overall degree of comfort with online tests #	3.94	0.73	4.17	0.67	.000***
Please rate your overall satisfaction with the online test ##	3.99	0.83	4.32	0.81	.000***

Note: The items above were measured on a five-point scale from 1=Strongly Disagree to 5=Strongly Agree.
This item was measured on a five-point scale from 1= Not At All Comfortable to 5=Very Comfortable.
##This item was measured on a five-point scale from 1= Not At All Satisfied to 5=Very Satisfied.
* p <.05; ** p <.01; *** p <.001

was ranked the lowest on the list—visualization of numeric information through graphs, charts, diagrams, etc.—was attributed less importance by all students, but particularly by students in political science. The lack of a significant difference between the preferences of the two groups of students indicates that these formal features of Blackboard testing tools are likely to be important for all students regardless of their major.

Students' Perceived Strengths and Limitations of Online Testing

Table 3 adds a qualitative dimension to the study and presents a summary of student perceptions of the strengths and limitations of online tests and quizzes obtained through open-ended responses to the survey.

Table 2. Student rankings of formal features of online testing

	Math Students (n = 193)		Political Science Students (n = 202)		
	Mean	SD	Mean	SD	p
Keeping track of my progress in the course using an online gradebook	4.59	0.65	4.62	0.56	.294
Working at my own pace	4.40	0.73	4.51	0.66	.070
Instant result/immediate feedback	4.42	0.79	4.48	0.72	.233
Graphs, charts, and diagrams to accompany statistical or other numeric information	3.88	0.74	3.83	0.84	.268

Note: The items above were measured on a five-point scale from 1= Not At All Important to 5=Very Important.

Table 3. Perceived strengths and limitations of online tests and quizzes

Strengths	**Math Students**	**POLS Students**
Gradebook allows keeping track of progress	x	x
Immediate feedback on tests/problems	x	x
Ability to work at your own pace and without the pressure of time	x	x
Ability to re-do a problem/assignment if the answer is wrong and learn from mistakes	x	x
Provide opportunity for extra practice before a test	x	x
Encourage to review the material and receive extra points	x	x
Convenient way to access from anywhere and from home	x	x
Efficiency and effectiveness	x	x
Limitations	**Math Students**	**POLS Students**
Feedback is available only after the whole test or quiz is submitted	x	
Lack of convenient printing options	x	x
Technical issues such as exiting or locking up	x	x
Platform/browser compatibility issues	x	x
Insufficient degree of control over font and color of text		x
Problems with displaying graphs/diagrams/math symbols	x	
Problems with integrating Java Script	x	
Problems with integrating multimedia	x	
Inability to review answers already submitted if needed	x	
Inability to view a test in its entirety if needed	x	
Too many confirming prompts	x	

Student satisfaction with online tests and quizzes was pervasive throughout the open-ended comments:

- "The ease of online tests is great."
- "Overall, the online testing has been an effective and easy way to learn."
- "Online tests are very useful."
- "I enjoy the online test format."
- "Blackboard makes good use of online quiz feature."
- "Online tests and quizzes are great!"
- "Overall, I have been satisfied with online quizzes and tests."
- "I like online quizzes and think that they are beneficial."
- "I think online assignments are easy to use and I enjoy doing them better than actual in class written assignments."
- "Online tests and quizzes are a great idea."
- "I am really happy with the online tests, and especially the online gradebook."

Besides, many commented positively about the ability to check their grades and self-monitor their progress by using an online gradebook offered by the course management system:

- "I enjoy having an online gradebook to check my grades as often as I wish."

- "I like that fact that you can see your grades online and see how you are doing after every assignment and test."
- "Online testing is a good way to keep grading simple."
- "I especially enjoy keeping track of grades on Blackboard. I would recommend that all instructors post grades on Blackboard."
- "It's a good system to use and more professors should use it for at least grades."

Another positive feature of Blackboard-based tests mentioned by students was the immediate feedback:

- "The instant feedback is the most important."
- "I like the quick feedback."
- "Overall, the online testing has been an effective and easy way to learn."
- "I liked using this for assignments very much, because the instant feedback helped spell out what I was doing wrong."
- "I enjoy using Blackboard. I think it has some features that allow for a better learning experience, especially the immediate feedback."
- "I learn from mistakes I made during the quiz, and the feedback helps to explain where I went wrong."

Among other strengths, students named the self-paced nature of online quizzes, the possibility to take multiple attempts and learn from mistakes, convenience, efficiency, and cost-effectiveness.

Some students indicated technical issues, ranging from problems with displaying graphs, diagrams, or math symbols such as the square root sign, to too many confirming prompts. Others pointed at difficulties with integrating JavaScript and multimedia. These issues would need to be addressed by software developers. Other concerns included the inability to review the answers already submitted, receiving feedback on individual questions, and viewing a test in its entirety if wanted. Instructors themselves could easily fix these problems, because they can control specific quiz settings within Blackboard.

Student Views of Online Testing for Future Development

The results in Table 4 break down student views of future developments for online testing. The vast majority of students stressed the importance of having an option to print the test, for reading it on paper or for studying. Most of the math students believed that visual representations of the material on the test were also critical. Slightly over half of all the students would prefer to have an option to control the font color or an ability to highlight the important parts of the text during the exam. Over one third of the students emphasized that they would benefit from being able to change the order of questions on the test or quiz based on the complexity of the questions.

Finally, when asked if they could choose between an online test and a paper-based test, which one they would prefer, 267 out of 395 students or nearly 68% said they would prefer online testing (Figure 1).

CONCLUSION AND DISCUSSION

The results of the study demonstrate a predominant preference for online testing among two groups of students enrolled in physical and social science courses. Overall, the analysis indicates that the majority of students perceive the simplicity of interface, quiz navigability, text formatting, use of digital images, and the level of overall comfort as the most important. A large number of students believe than the availability of audio presentation of information, such as prompts or directions, is also necessary. It can be recommended that in order to help students decrease test anxiety and gain greater control of their learning, educators

Table 4. Student recommendations

	Math Students (n = 193)			Political Science Students (n = 202)		
	Yes	**No**	**Don't Know**	**Yes**	**No**	**Don't Know**
Make an option to print out the test and read it off the printout available*	153 (79.3%)	15 (7.8%)	24 (12.4%)	173 (85.6%)	16 (7.9%)	13 (6.4%)
Make pictorial representations of the material to be examined or diagrams available	121 (62.7%)	28 (14.5%)	44 (22.8%)	98 (48.5%)	60 (29.7%)	44 (21.8%)
Change font color or highlight important parts of the text*	103 (53.4%)	58 (30.0%)	32 (16.6%)	108 (53.5%)	75 (37.1%)	18 (8.9%)
Change the order of questions depending on the level of complexity or your own preference	78 (40.4%)	41 (21.2%)	74 (38.4%)	69 (34.2%)	73 (36.1%)	60 (29.7%)
Adjust font face or size	63 (32.6%)	97 (50.3%)	33 (17.1%)	44 (21.8%)	140 (69.3%)	18 (8.9%)
Turn on spoken (audio) prompt or directions	43 (22.3%)	114 (59.0%)	36 (18.7%)	82 (40.6%)	101 (50%)	19 (9.4%)

* Indicates 1 missing value

need to focus on creating online tests and quizzes that are user-friendly, convenient, and simple to use. In addition, usable, carefully designed online assessments can help meet specific learning outcomes, facilitate problem solving, learning from mistakes and regular practice, as well as motivate and encourage students.

Furthermore, the results suggest that as instructors integrate online assessments in classroom-based courses, they should move away from the "one-size-fits-all" approach, and focus on learner-centered models that are tailored to fit individual students' needs. The ultimate goal is to match student learning styles (Brown & Liedholm, 2004) and personal differences with the flexibility and interactivity of the testing interface. Designers of educational assessments should provide test takers with flexible printing options, pictorial representations of the material, greater control over the font and color of text, the ability to view a quiz in its entirety, the ability to change the order of questions, and highlight important parts

Figure 1. Student preferences in online testing vs. paper-based testing

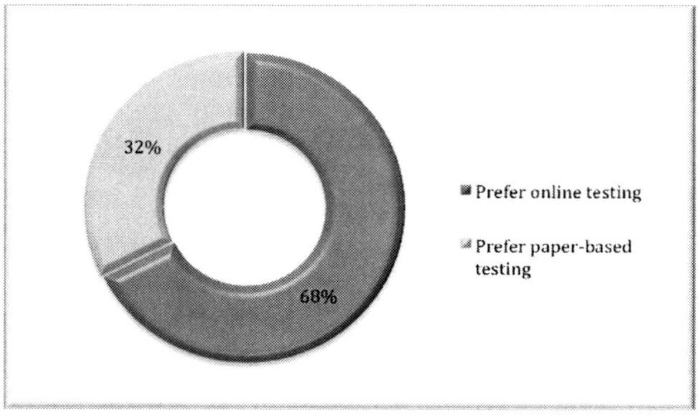

of the text. Finally, by implementing effective and reliable online assessments that are adjustable to the needs of individual students, educators can create far-reaching opportunities for students' self-evaluation, self-directed learning, and continuous growth and development.

In the future, focusing on student expectations of online tests, quizzes, or exams, and ways of integrating e-assessments will help to meet the demands of the interaction age. The next generation of assessment tools should be created through collaborative interaction with students. Students can be actively involved in the creation of assessment instruments through the collaborative review of course material, and by developing their own test questions. Collaborative test making can allow students to share their understanding of course concepts with classmates, increase self-esteem, integrate real-world knowledge and skills, and actively construct knowledge. Social interaction technologies provide multiple ways to support such initiatives through collaborative wiki-based platforms, virtual worlds, short messaging services, as well as more traditional synchronous collaborative applications such as group calendars, audio-video conferencing, and online chat (Araujo, 2009).

In addition, the proliferation of new communication devices and distribution platforms such as mobile phones, iPods, tablet computers, and e-book readers make it possible to extend the physical space of a classroom into personal learning spaces where students can continue their learning at their own pace, synchronously or asynchronously, receive "just-in-time" (Simkins & Maier, 2004; Wheeler, 2009) feedback and support, and hold "in-between" communications (Traxler, 2009) with their teachers and mentors when and where they need it. Further technological advances will require new assessment tools.

Online learning environments—classrooms without walls—present a different set of teaching strategies and consequently require a different set of pedagogical tools and assessment practices than face-to-face learning environments. What is the best way to communicate information to students online? How can instructors provide for students' effective interaction with the content, other students, and the teacher in an online course? How can we evaluate online learning outcomes and achievement of course goals online? These questions continue to remain at the center of attention of educational practitioners and academic researchers.

REFERENCES

Adlakha, V. G., & Aggarwal, A. K. (2009). Minimal functionalities of course management systems: A faculty perspective. *International Journal of Web-Based Learning and Teaching Technologies*, *4*(2), 26–42. doi:10.4018/jwltt.2009040102

Alexander, B. (2006). Web 2.0: A new wave of innovation for teaching and learning? *EDUCAUSE Review*, *41*(2), 33–44.

Allen, I. E., & Seaman, J. (2010). *Learning on demand: Online education in the United States, 2009*. Needham, MA: Sloan Consortium.

Araujo, A. L. (2009). Instrumental and social influences on adoption of collaborative technologies in global virtual teams. In Salmons, J., & Wilson, L. (Eds.), *Handbook of research on electronic collaboration and organizational synergy* (pp. 400–408). Hershey, PA: Information Science Reference.

Brown, B. W., & Liedholm, C. E. (2004). Student preferences in using online learning resources. *Social Science Computer Review*, *22*, 479–492. doi:10.1177/0894439304268529

Dumova, T., & Fiordo, R. (2010). Preface. In Dumova, T., & Fiordo, R. (Eds.), *Handbook of research on social interaction technologies and collaboration software: Concepts and trends* (*Vol. 1*, pp. xl–xlvi). Hershey, PA: Information Science Reference.

Elliott, B. (2007). *Assessment 2.0: Assessment in the age of Web 2.0*. Scottish Qualifications Authority. Retrieved June 23, 2010, from http://kn.open.ac.uk/public/ document.cfm?docid=10393

Hargadon, S. (2009, December 16). *Educational networking: The important role Web 2.0 will play in education* (White Paper). Retrieved April 24, 2010, from http://www.stevehargadon.com /2009/12/social-networking-in -education.html

Holms, M. (2002). *Web usability & navigation*. New York: McGraw-Hill.

Koohang, A. (2004a). Development and validation of an instrument for assessing users' views about the usability of digital libraries. *Issues in Informing Science & Information Technology*, *1*, 55–63.

Koohang, A. (2004b). A study of users' perceptions toward e-learning courseware usability. *International Journal on E-Learning*, *3*(2), 10–17.

Koohang, A., et al. (2010). Panel discussion: The current state of blended learning. In *Proceedings of Informing Science & IT Education Conference (InSITE) 2010*. Retrieved May 24, 2010, from http //www.proceedings. informingscience.org / InSITE2010/ InSITE10p705- 708Koohang774. pdf

Koohang, A., & du Plessis, J. (2004). Architecting usability properties in the e-learning instructional design process. *International Journal on E-Learning*, *3*(3), 38–44.

Koohang, A., & Harman, K. (2005). Usability of digital libraries and their reusable objects in e-learning settings. *Issues in Information Systems*, *6*(2), 272–278.

Kvavik, R., Caruso, J., & Morgan, G. (2004). *ECAR study of students and information technology, 2004: Convenience, connection, and control*. Boulder, CO: EDUCAUSE Center for Applied Research.

Liang, X., & Creasy, K. (2004). Classroom assessment in web-based instructional environment: Instructors' experience. *Practical Assessment, Research & Evaluation, 9*(7). Retrieved June 30, 2010, from http://PAREonline.net/ getvn.asp?v=9&n=7

Morgan, G. (2003). *Faculty use of course management systems*. Boulder, CO: EDUCAUSE Center for Applied Research.

Nielsen, J. (1989). Usability engineering at a discount. In Salvendy, G., & Smith, M. J. (Eds.), *Designing and using human-computer interfaces and knowledge based systems* (pp. 394–401). Amsterdam: Elsevier Science Publishers.

Nielsen, J. (1993). *Usability engineering*. Boston, MA: Academic Press.

Nielsen, J. (1999). *Users first: Cheap usability tests*. Retrieved March 11, 2000, from http:// www.zdnet.com/ devhead/stories/ articles/0,4413, 2224316,00.html

Nielsen, J. (2000). *Designing web usability*. Indianapolis, IN: New Riders.

Nielsen, J. (2007, January 2). *Fast, cheap, and good: Yes, you can have it all*. Jakob Nielsen's Alertbox. Retrieved June 24, 2010, from http:// www.useit.com/alertbox /fast-methods.html

Nielsen, J., & Molich, R. (1990). Heuristic evaluation of user interfaces. In J. Carrasco & J. Whiteside (Eds.), *Proceedings of the ACM CHI 90 Human Factors in Computing Systems Conference* (pp. 249-256). Seattle, WA: ACM Press.

Oliver, K., & Moore, J. (2008). Faculty recommendations for web tools: Implications for course management systems. *Journal of Computing in Higher Education, 19*(2), 3–24. doi:10.1007/ BF03033424

Pieper, S., Edwards, E., Haist, B., & Nolan, W. (2010). A survey of effective technologies to assess student learning. In Schreiner, C. (Ed.), *Handbook of research on assessment technologies, methods, and applications in higher education* (pp. 47–64). Hershey, PA: Information Science Reference.

Rainie, L. (2009). *Networked learners* (Keynote address at the 2009 Online Learning Symposium, Michigan Virtual University). Retrieved April 24, 2010, from http://www.pewinternet.org/ Presentations/2009/ 52-Networked- Learners.aspx

Shackel, B. (1991). Usability – Context, framework, design and evaluation. In Shackel, B., & Richardson, S. (Eds.), *Human factors for informatics usability* (pp. 21–38). Cambridge, UK: Cambridge University Press.

Shneiderman, B., & Plaisant, C. (2005). *Designing the user interface: Strategies for effective human-computer interaction*. Boston, MA: Pearson/ Addison Wesley.

Simkins, S., & Maier, M. (2004). Using just-in-time teaching techniques in the Principles of Economics course. *Social Science Computer Review*, *22*(4), 444–456. doi:10.1177/0894439304268643

Traxler, (2009). Mobile subcultures. In S. Wheeler (Ed.), *Connected minds, emerging cultures: Cybercultures in online learning* (pp. 17-28). Charlotte, NC: Information Age Publishing.

West, R., Waddoups, G., & Graham, C. (2007). Understanding the experiences of instructors as they adopt a course management system. *Educational Technology Research and Development*, *55*(1), 1–26. doi:10.1007/s11423-006-9018-1

Wheeler, S. (2009). *Connected minds, emerging cultures: Cybercultures in online learning*. Charlotte, NC: Information Age Publishing.

KEY TERMS AND DEFINITIONS

e-Assessment: Is the use of information and communication technologies for assessment purposes.

e-Learning: Refers to a specific mode of instruction based on electronic delivery of course material, learning activities, and assessments. Methods of delivery have evolved from videotape, CD-ROM, DVD, and interactive television to web-based platforms.

Online Gradebook: Is a formal feature of course management systems, such as Blackboard, which provides students with instant access to their current grades and allows them to keep track of their progress.

Online Quiz: Is viewed as a mediated dialog between students and instructor.

Social Interaction Technologies: Are Internet-based tools and techniques designed to initiate, maintain, share, and distribute interactive and collaborative activities and spaces online. The wide range of these tools includes online social networks, blogs, wikis, podcasts, RSS feeds, social bookmarks, folksonomies, mashups, and virtual worlds.

Synchronous Collaborative Technologies: Allow team members to interact in an instantaneous fashion without any time delay and enable real-time collaboration. Examples include instant messaging, audio-video conferencing, and online chat.

Usability: Is the capability of a particular system to satisfy the needs and specifications of the users.

Usability Testing: Refers to methods and techniques of evaluating usability of products and computer systems.

Chapter 5
ePortfolios and the Communicative Intellect in Online Education

Rich Rice
Texas Tech University, USA

ABSTRACT

According to Nielsen Internet ratings, YouTube totaled over 100 million unique users the month this chapter was written. What are those ratings today? Moreover, the number of times-per-day many individuals compose Facebook news feeds and profile content is equally staggering. Students inhabit these media-rich spaces, and if educators do not construct online teaching and learning environments, such as ePortfolios, in ways that effectively capitalize on students' literacies in computer-mediated communication, those educators are in danger of limiting students' communicative intellect. More to the point, educators are in danger of graduating functionally illiterate students according to what constitutes literacy in today's interactive age.

COMMUNICATION IN THE INTERACTIVE AGE

What our age needs is communicative intellect. For intellect to be communicative, it must be active, practical, engaged. In a culture of the simulacrum, the site of communicative engagement is electronic media. In the mediatrix, praxis precedes theory, which always arrives too late. The communicative intellect forgets the theory of communicative praxis in order to create a practice of communication. —Taylor & Saarinen, 1994, p.2

In her recent digital and media literacy recommendation report on behalf of the Knight Com-

DOI: 10.4018/978-1-61350-077-4.ch005

mission, Renee Hobbs (2010), suggests educators must create "respectful learning environment[s] where students' lived experience is valued and multiple points of view are encouraged" (p. 21). Hobbs' plan of action cites the National Association for Media Literacy Education (2007) and the Partnership for 21st Century Skills (2009). In so doing, she points out that "digital and media literacy education activates independent thinking, authentic dialogue, collaboration, reflection, creativity, and social responsibility as applied to the practices of responding to, creating and sharing messages" (p. 21).

According to Hobbs' report, "for all aspects of daily life, people today need a constellation of well-developed communication and problem-solving skills" that include essential competencies of digital and media literacy (p. 18). These competencies are synthesized from professional associations like the International Reading Association (IRA), the National Council of Teachers of English (NCTE), the National Center for Academic Transformation (NCAT), and the National Council for Accreditation of Teacher Education (NCATE). These competencies include the ability to act, access, analyze and evaluate, create, and reflect (p. 18). Hobbs quotes from the Common Core State Standards Initiative (2010) as well:

To be ready for college, workforce training, and life in a technology society, students need the ability to gather, comprehend, evaluate, synthesize, report on, and create a high volume and extensive range of print and nonprint texts in media forms old and new. The need to research and to consume and produce media [should be] embedded into every element of today's curriculum. (p. 18)

And Hobbs goes on to refer to policies of teacher education programs, such as the NCATE's *Professional Standards for the Accreditation of Teacher Preparation Institutions* (2007), which states that "teachers understand media's influence on culture and people's actions and communication; as a result, teachers [should] use a variety of approaches for teaching students how to construct meaning from media and nonprint texts and how to compose and respond to film, video, graphic, photographic, audio, and multimedia texts" (p. 57).

The work these agencies do to detail the role print and nonprint compositions play in education and beyond is significant when one thinks about designing online teaching environments and assessment tools. In essence, students must learn how to transfer knowledge from what is in their heads to real-world applications. In *The Design of Everyday Things* (2002), for example, Donald A. Norman makes connections between what he calls knowledge in the head and knowledge in the world. For Norman, the most effective communicative intellect is that which is "distributed—partly in the head, partly in the world, and partly in the constraints of the world" (pp. 54-55). According to Norman, people should determine best practices situationally, or *kairotically*. Norman goes on to point out that, to maximize knowledge-making in everyday situations, organizers must create environments that support the normal behavior of people who both combine internal knowledge with external information and also act within situational constraints (pp. 55-58). Within this context, hypermediating components of the teaching environment can be a helpful cue for learners to individually adjust learning styles in order to "transfer old knowledge to the new object" (p. 82). Sometimes, crucial information is in the head of the individual; sometimes it's available externally in other knowledge channels. Such information is also often determined by the limitations and ideals of the communicative situation and purpose or application of knowledge.

Just as the process movement of composition in the 1970s and 1980s advocated more reflection and recognition of students' own literacies and learning styles, today's students need varied online communication experiences because of the variety of communication tools used in any given workplace. Students constantly practice communication literacies in social media environments, and students have a right to their own modalities (differing modes of communication) and communicative environments.

EPORTFOLIOS AS MEDIA-RICH AUTHENTIC ASSESSMENT TOOLS

Simply put, technology is changing reading, writing, and thinking. As John Palfrey and Urs Gasser (2008) suggest with regard to students who inhabit online spaces, "just because Digital Natives learn differently from the way their parents did when they were growing up doesn't mean that Digital Natives are not learning" (p. 240). Hobbs (2010) goes further, defining digital media and literacy as an amalgamation of life skills needed in today's media-saturated world. These skills include the ability to

- Access information
- Analyze messages in a variety of formats
- Create content in a variety of forms
- Reflect over one's own conduct and communication behavior
- Take social action by working individually and collaboratively. (pp. vii-viii)

The key is to sustain media-rich teaching and learning environments that make use of multiple digital media and literacy types. To achieve this objective, educators must use media-rich assessment tools, such as ePortfolios. Moreover, educators must also pay attention to adults' learning needs while also investigating what makes media literacy so natural.

While pedagogy is the study of the process of teaching in general, andragogy is the study of teaching adult learners specifically. According to learning theorists Malcolm S. Knowles, Elwood F. Holton, and Richard A. Swanson (2005), adult learners have specific needs, such as the need to know, the need to a foundation of knowledge, the need to privilege a self-concept, the need for ready content, the need for immediately relevant orientation, and the need for personal motivation. That is, adults must know the reason *why* they are being asked to learn something. As a result, real-world experience should be a component of the learning environment in order to offer learning activities that provide real gains and failures.

As Knowles, Holton, and Swanson (2005) point out, within the context of such learning situations, adult learners need to play a direct role in planning and evaluating their own instruction. It is therefore important to note that subject matter must be problem-centered rather than content-oriented. Additionally, such subject matter must have immediate relevance to the lives of these adult learners. According to principles of andragogy, such learners also respond better to internal motivators rather than to extrinsic motivation. Educational environments that make use of principles of andragogy value more problem-centered environments and move away from teacher-directed toward more student-directed learning.

Arguably, portfolio assessment offers students these active roles and individually connecting problem-centered approaches. ePortfolios also afford increased design requirements, reflective metacognition skills, and rhetorically connective properties. (See, for example, Cynthia L. Baron's (2004) *Designing a Digital Portfolio*; Lorraine Stefani, Robin Mason, and Chris Pegler's (2007) *The Educational Potential of e-Portfolios: Supporting Personal Development and Reflective Learning*; Darren Cambridge, Barbara Cambridge, and Kathleen Yancey's (2009) *Electronic Portfolios 2.0: Emergent Research on Implementation and Impact*; and Darren Cambridge's (2010) *Eportfolios for Lifelong Learning and Assessment.*) Creating ePortfolio assessment systems that systematically embrace principles of new media and social media and that emphasize content more than "electronic schmoozery" can therefore be challenging. The media-rich communication needs educators should teach—and the tools they need to create and employ in order to adequately and accurately measure if what they are using to teach and how they are teaching is working—require significant course re-design.

COURSE RE-DESIGN MODELS AND NEW MEDIA

The six course redesign models outlined by the National Center for Academic Transformation (NCAT) demonstrate the complexity of designing media-rich learning environments. These six NCAT (2005a) course redesign models are as follows:

1. The *supplemental model* augments traditional delivery with technology-based, out-of class activities while emphasizing active learning in the classroom
2. The *replacement model* reduces the number of in-class meetings with out-of-class, online, interactive learning activities while making changes in the classroom
3. The *emporium model* altogether replaces class meetings with open or required attendance while using a collection or warehouse of online materials and on-demand, personalized support
4. The *fully online* model also eliminates in-class meetings; it differs from the emporium model in that all learning experiences are moved online with guided feedback
5. The *buffet model* for course redesign customizes the learning environment for each student while maintaining the same learning outcomes for each model
6. The *linked workshop* model provides remedial and developmental instruction by linking specific workshops for just-in-time supplemental support for students

Because an ePortfolio is an electronic, selected collection of work with clear goals and objectives and a unified yet flexible measurement heuristic, students building ePortfolios engage in individual and andragogically-motivated processes of selection or invention. The buffet model for course re-design, thus, affords learners a computer-mediated communication learning environment that is well suited for ePortfolio construction and assessment. This situation is the case because ePortfolios make use of a known-new andragogical contract approach, new media, and theories of "natural" media use.

In *The Language of New Media*, Lev Manovich (2002) describes principles underlying new media this way: all new media objects exist as data through numerical representation that is modular. Because new media objects are modular, they can be created and modified automatically. Automation allows new media objects to exist in multiple versions—what Manovich terms "variability." Through a postmodern process of connecting the computer, the cultural, and the ideological underpinnings and layers of content in different ways, the logic of the computer mediates, influences, and transcodes how individuals understand and represent themselves. This is what educators need to consider in redesigning courses to afford media-rich teaching and learning strategies and environments. NCAT (2005b), for instance, points out that the buffet model works to customize the learning environment "for each student based on background, learning preference, and academic/professional goals" (para. 7).

This model of instruction requires online assessment of learning styles and study skills, which values content as numerically-represented modular information. It offers students individualized strategies to achieve learning objectives. In so doing, it enables students to take content that is "automated" or standardized or available in a variety of formats online and make it, *invariably*, their own. Within this context, the buffet style in the form of ePortfolio provides an individualized learning contract or makes clear, module by module, how each learning component can be completed. The ePortfolio buffet style can, moreover, offer an array of teaching and learning modalities through artifact representation. These could include lecture, individual and group discovery labs and review and projects, small-group study sessions, videos, various training module types, or oral and written presentations. The content is modularized, eliminating duplication of effort for teachers, and enabling the institution to focus more on the measurement student success.

COLLABORATIVE OWNERSHIP AND "NATURAL" CMC

The scholarship on portfolio pedagogy has largely focused on construction, presentation, and assessment measurements (Rice, 2002). Authentic assessment, in particular, refers to developing tools to measure students' work that are both generative and re-usable in scope. That is, that the heuristics used to determine the quality of work can create new knowledge and might be the type of tool or approach used in the workplace.

Computer-mediated communication technologies have opened new worlds of authentic assessment practices due to thinking through the transcoding or multi-layering of computer and culture. For instance, some teachers value offering final exams through Twitter instead of essay format. The reason is because, ultimately, the workplace culture of the job talk or the elevator discussion or dialogue with colleagues or other students might value distilled content of core concepts more so than sustained arguments for generalists. Being able to define "computer-mediated communication" in 140 characters or less just-in-time may be a more authentic assessment measurement than composing a 5000-word essay on the topic if the intent of knowing the value of CMC is rooted is contextualized in specific workplaces. Regardless, if students are using Twitter in their own discourse, it is arguable that educators should use it to construct, present, and/or assess in their classes.

Similarly, the text-based essay is not a primary genre used in most workplaces. Collaboratively produced compositions include a variety of selected genre and media types to best convey or persuade specific audiences. Philippe Aigrain (2005) talks about such information exchanges in a collection of essays called *CODE: Collaboration Ownership and the Digital Economy.* Aigrain points out that "information, creative works in all media, software, and other intellectual or knowledge entities are easier to put together individually and even more collectively. Such synergy can be done, to the point that commons-based peer production "have been explored by scholars across the disciplines" (p. 289). Aigrain goes on to suggest that media offers opportunities for meaningful concentration and focus, vertical integration, and access control that can be "counterbalanced by explicit efforts to ensure diversity, freedom of choice, empowerment, and openness" (pp. 289-291). The result is that the cooperative usage of composing and presenting technologies in the workplace gives rise to increased reusing of artifacts and a greater reliance on interconnected and interlocking intellectual property.

Today, students are composing, presenting, remediating, remixing, and mashing content in these ways outside of the classroom. And the results of such mixing and experimenting have major implications for issues like ownership, distribution of content, and intellectual property. Additionally, evidence suggests students will be required to produce variable and modular content in these ways in the workplace. Such natural media look a lot more like the genre of ePortfolio than the traditional text-based essay. But the prospects for and the complexities of collaboration go far beyond this. In fact, collaboration between instructional design principles of academic faculty and student affairs professionals create another important dialectic created by ePortfolios (Johnson & Rayman, 2007).

Media naturalness theory is an important concept, too, that informs this argument. Ned Kock (2005) developed media naturalness theory to investigate how the ways in which humans make knowledge influence their behavior toward the use of electronic communication tools. Koch begins his investigation by deconstructing media richness theory, which is a concept explained in Richard L. Daft and Robert H. Lengel's 1986 article in *Management Science*. According to Daft and Lengel, media richness is a way to describe a medium by its ability to reproduce information. One example is that nonverbal communication cues from a face-to-face conversation cannot be as readily conveyed over phone as well as over iPhone's FaceTime or some other video chat application. Because nonverbal cues are not present, the message could be easily misunderstood. More media rich tools or environments are thus better when the communication situation might involve a greater degree of complexity or ambiguity.

Kock (2005) goes further, however, to point out a communicative medium (such as an ePortfolio) that is more *natural* or similar to a face-to-face or authentic medium is likely to generate a more accurate response to the conveyed meaning. According to Kock:

The media naturalness hypothesis argues that, other things being equal, a decrease in the degree of naturalness of a communication medium (or its degree of similarity to the face-to-face medium) leads to the following effects in connection with a communication interaction: (1) increased cognitive effort, (2) increased communication ambiguity, and (3) decreased physiological arousal. (p. 117)

If the amount of cognitive effort it takes for a student or any communicator to understand a communicative act is too great, the student is more likely to misunderstand a message, or disengage. Kock calls this process of understanding *effort*. According to this concept, when a media platform is used to deliver a message, its level of naturalness can ultimately muddle the transmission and compel more effort, requiring more on-task time. Time on task, or what Kock calls *fluency*, can make communicating more difficult or can make it inefficient and confusing.

Communication ambiguity can be cultural, or it can involve a lack of understanding of the appropriateness of a specific medium. In either case, such ambiguity can cause multiple interpretations of the same message. Such gaps can thus result in different conclusions about the meaning of a message. To adapt, the brain looks for other signals to process the message. Kock (2005) explains, "several of the stimuli we automatically search for are those present in actual face-to-face communication, such as contextual cues (available in co-located communication), immediate feedback (available in synchronous communication) in the form of facial expressions and body language, and vocal intonation" (p. 122).

Interestingly, educators often ask their students to read their own writing aloud, so both the student and those around them hear what they can't see. This approach represents a form of media naturalness that helps remove ambiguity in writing. In essence, educators do such exercises to help students realize a medium can impact the potency of a message in several ways. Thus, in teaching students to write more effectively (i.e., become "literate"), educators are also teaching students that literacy inherently involves the selection of an appropriate medium.

Within such pedagogical contexts, assessment tools like ePortfolios work well because every student can be required to include the same number of artifacts (i.e., entries in the portfolio). Every student can also be required to provide the same metareflective content. At the same time, the media types used to make the portfolio enable students to draw upon the best or most natural means of communicating according to the student's own understanding of what literacy is in today's wired world. In terms of media naturalness theory, a medium must be relatively transparent. This factor is essential, for it allows the related message is easily understood and usable, or in cases where this is not possible, it allows for including discussions about the medium as part of the message.

EPORTFOLIO MEDIA TYPES

More recent scholarship on ePortfolios has focused on how ePortfolios development can be used for individual and social transformation, authentic assessment in digital environments, lifelong learning, deep reflection in social mediated spaces, identity in online environments, course design and inclusive learning, software solutions, and globalization. Four recent books on ePortfolios are relevant here, including Cynthia L. Baron's (2004) *Designing a Digital Portfolio*; Lorraine Stefani, Robin Mason, and Chris Pegler's (2007) *The Educational Potential of e-Portfolios: Supporting Personal Development and Reflective Learning*; Darren Cambridge, Barbara Cambridge, and Kathleen Yancey's (2009) *Electronic Portfolios 2.0: Emergent Research on Implementation and Impact*; and Darren Cambridge's (2010)*Eportfolios for Lifelong Learning and Assessment.* What is clear with computer-mediated communication is that assessment of student learning in digital learning environments offers many challenges.

For instance, there are many types of portfolios. There are showcase or best practice portfolios, development or process-based portfolios, reflective portfolios, and assessment portfolios. Like every type of communication, the rhetorical situation is of primary importance. A portfolio in a discipline-specific class has a much different purpose than one for a capstone course, which again has a much different purpose than a portfolio designed to accompany a job application. Also, when discussing portfolios, it is important to note that digital portfolios have many benefits over print-based ones. Such advantages include the ability for students to

- More readily connect between and reflect over artifacts
- Scale and bring in multiple authors more readily, an easier time of storage for long-term assessment purposes
- Connect to formative and summative assessments through course management tools, production and marketing, and through repurposing and optimizing.

In discussing other benefits of ePortfolios, Stefani, Mason, and Pegler (2007) cite Helen Barrett and others in focusing on the idea that the primary purpose of an ePortfolio is to foster learning and document growth over time. According to this perspective, the ePortfolio that affords greater immediate reflection and peer and teacher response than a print-based portfolio can more readily serve as a constructivist model of learning, becoming a story of learning owned by the learner (pp. 74-86).

In addition to many types of ePortfolios, which can be used in a variety of ways for a variety of purposes, the new media skills that individual students have expertise in can be used both systematically and selectively. For instance, as in the buffet model of course design, teachers can create online courses that require specific goals and objectives and measurement plans, but the tools students use can be flexible. Using Lev Manovich's concept of variability and Malcolm Knowles' andragogical values of decision-making and choice, students can connect tools that are more natural to the rhetorical situation or more familiar to their own technological literacies through an ePortfolio design structure. Such tools could include a course management system, a weblog design, an ongoing wiki, a Google Docs space, a series of podcasts with an RSS feed, or a website with specific navigational scheme requirements. In addition to the reader, the writer, and the text, time and location play a role in any rhetorical situation.

What defines an ePortfolio includes the type of portfolio and the type of software and output tools used to display and distribute the content. These items influence both the nature of the portfolio's artifacts themselves and the rhetorical purpose of each entry and the overall portfolio that contains them. As a result, ePortfolios are repositories of information about learners, the learning process, the content being learned, the application of the learning, documenting learning successes, and the scalability or transferability of learning to other contexts. These factors mean ePortfolios could be used for course, program, or institutional purposes. They are principally owned by the learner, but they are also often co-owned in sophisticated ways through database management and complex connective properties. Through variability processes, the artifacts and presentation focus of each artifact can be determined in part by who the audience is.

Moreover, because of the ease of scalability, a student might have a portfolio of portfolios for lifelong learning or large summative assessment purposes, or a teacher may have a portfolio of student portfolios for assessment purposes. As Trent Batson (2005) and others have pointed out, ePortfolios can track student work over time with other student or teacher feedback. ePortfolios can also be dynamically updated for formative assessment purposes over time, and they can aggregate many students' work. Finally, they can be used to assess different courses, majors, or entire programs of study.

Darren Cambridge (2010) offers a thorough discussion of strategies to provide "genuine eportfolio assessment" that capitalizes on the capabilities of the genre (p. 87). In *Eportfolios for Lifelong Learning and Assessment*, Cambridge points out that "eportfolios produce a more comprehensive and flexible collection of data than standardized assessments do" (p. 91). Over time, such data can provide very meaningful trends in patterns of performance in programs. ePortfolios, ultimately, can serve as a bridge between quantitative measurement demanded by administrators in institutions of higher education and the more situated and dynamic and creative qualitative needs of teachers and employers in the workplace.

For years, the term "authenticity" is one that has flummoxed face-to-face, hybrid, and online teachers who use computer-mediated communication tools. What is *engfish* (writing composed more to receive what students think teachers want rather than conveying authentic voice) in academic discourse can manifest itself in PowerPoint presentations, in asynchronous discussion board posts, or in comments on blogs or wikis. It can also appear in synchronous exchanges where students are assimilated into the conversation of mankind as well as in podcasts where every jingle and soundbite is present but true voice and conviction is missing. Even in graphical representations where the rhetoric of schmooze reins king, ePortfolios are no different. The idea is to create an assessment tool that teaches. But designing a form that is both standardized and personalized is difficult, especially when so many media options are available. Projects that are too personalized do not demonstrate that the goals and objectives of the course have been achieved. Conversely, projects that are too standardized limit self-discovery, and "by expressing who we are, we are defining ourselves, calling ourselves into being. Thus, creativity becomes essential" (Cambridge, 2010, p. 13). The goals of lifelong learning might be larger than completing a portfolio to pass a course or a degree program. A Facebook page or a blog, for example, might be a lifelong ePortfolio of sorts. ePortfolio praxis should include the idea that people write for life. Individuals write to making meaning of life, and the type of writing people do today includes multiple media types and rhetorically reflective positions.

SITUATING YOUR EPORTFOLIO MODEL

The capstone MA in Technical Communication (MATC) ePortfolio at my institution, Texas Tech University (TTU), originally required a homepage that included a student's own navigational scheme. In the process of building the scheme, as many portfolio scholars have recommended, students would often make connections between their own goals and those of the program. Or, they would create a metaphor that represents their learning as a whole. Some metaphors are better than others. Some are more difficult to assess with a nod toward validity and reliability and consistency across assessors. TTU's MATC ePortfolio model has moved toward a jump-off page that is an essay, and at least one of the artifacts in the portfolio should demonstrate facility with technology. Yet, because the members of the MATC faculty value authenticity in design as well as in writing, and because they value students' right to their own multimodal literacies, the ePortfolio model used in the MATC program is tied directly to program objectives, but it also affords opportunities to make personally meaningful and directed connections that are helpful in job placement.

The TTU MATC ePortfolio (see http://www.english.ttu.edu/tcr/MATC/matc_portfolio.asp for an overview of the portfolio) begins the moment students are accepted into the program. Students are admitted with the end in mind; specifically, students must learn to

- Create effective technical documents justified with relevant theory.
- Analyze and respond appropriately to rhetorical situations and key issues in the field, including the differing goals and agendas of audiences, organizations, and societies.
- Employ and justify user-centered approaches to technical communication.
- Demonstrate sensitivity to the ethical, professional, and cultural issues that face technical communications.
- Use communication technologies and media appropriate to users and situations.
- Write clear, correct, and stylistically effective prose.

The faculty in the program distinguish between job search portfolios and what faculty members are looking for in this final ePortfolio project. In so doing, the MATC faculty emphasize differences among rhetorical situations, reflection, and delivery and privacy issues.

What students produce in the MATC program serves them well as a draft for a job search portfolio. Moreover, nearly every student reworks his or her ePortfolio into an artifact linked through a professional home page. The faculty evaluators do ask students to include a reflective essay in the ePortfolio, and this essay, in effect, is a thesis statement. The thesis is that the student has met the objectives of the program, and the student has included three to six learning artifacts as evidence. The artifacts, in turn, are like body paragraphs in an essay, and each artifact requires contextualizing or metareflection to situate how each was a process and helped meet one or more objectives.

Of the entries in these ePortfolios, at least one artifact must be an academic essay or research paper, and at least one artifact must be a practical technical communication project. The objective for requiring such entries is to balance theory and practice in the program. The faculty in the program value revision in artifacts as well as connections between artifacts by way of program objectives. For this reason, they offer graduate students a mid-program review of ePortfolio development. Very few students fail this review process, in part because the faculty in the program offer ongoing workshops, and they also cultivate an ethos of portfolio keeping throughout every course. To do so, faculty members identify assignments that would work to demonstrate meeting the requirements of program objectives, and they teach students to focus rhetorically rather than chronologically when selecting and discussing portfolio entries. Additionally, the program's model ePortfolio requirement is designed to allow creativity within a framework that can be assessed objectively. (See http://www.english.ttu.edu/tcr/MATC/matc_portfolio.asp for examples.)

CONCLUSION

It is clear that students are growing quite skilled at reading, writing, and thinking in digital environments. It is also clear that the processes involved in the types of compositions required in the workplace resemble socially-mediated spaces more so than they do print essays. The rhetorical purpose, the narrative style, and the mediation are all quite different. As Renee Hobbs' (2010) digital and media literacy action plan suggests, formal education and lifelong learning must converge as related types of literacy, including media literacy, health media literacy, and digital literacy. As Hobbs points out, students need to know how to use various types of texts, tools, and technologies. They also need critical thinking and analysis skills, and they need to know how to compose messages effectively and creatively. Finally, students need to engage in reflection and ethics as well as know how to actively participate individually and collaboratively.

In face-to-face, in hybrid, and especially in online teaching, educators should consider the ePortfolio as a genre that supports the communicative intellect of students in these ways in terms of media naturalness, authentic genre and assessment. Educators should also view it as a mechanism that embodies the principles of new media and andragogy—principles that students will find motivating and transferable to workplace situations. Cynthia L. Baron (2004) concludes the introduction to *Designing a Digital Portfolio* this way:

A well-designed portfolio is a way for an illustrator or photographer to boost his or her work out of the excruciatingly competitive world of clip art and stock images. It's a way for an animator or other moving image artist to present his work more accessibly and with finesse. It raises the visibility and the stock of fine and performing arts. Most importantly, it allows you to show the world that you value and respect your work--the first step in making sure that others do too. (p. xxiii)

In this way, ePortfolios in online courses that value computer-mediated communication, as well as ePortfolios in programs which do the same, can function as purposeful bridges between the academy and workplace in theoretical and practical ways that prepare students and employees to be active, to be practical, and to be engaged in order to communicate effectively today.

REFERENCES

Aigrain, P. (2005). Positive intellectual rights and information exchanges. In Ghosh, R. A. (Ed.), *CODE: Collaborative ownership and the digital economy* (pp. 287–316). Cambridge, MA: MIT Press.

Baron, C. L. (2004). *Designing a digital portfolio*. Indianapolis, IN: New Riders.

Batson, T. (2005). The electronic portfolio boom: What's it all about? *Campus Technology*. Retrieved February 15, 2011, from http://www.campustechnology.com/article.asp?id=6984

Cambridge, D. (2010). *Eportfolios for lifelong learning and assessment*. New York, NY: Jossey-Bass.

Cambridge, D., Cambridge, B., & Yancey, K. (2009). *Electronic portfolios 2.0: Emergent research on implementation and impact*. Sterling, VA: Stylus Publishing.

Common Core State Standards Initiative. (2010). *English language arts standards*. Retrieved February 14, 2011, from http://www.corestandards.org/the-standards/english-language-arts-standards/introduction/key-design-considerations

Daft, R. L., & Lengel, R. H. (1986). Organizational information requirements, media richness and structural design. *Management Science*, *32*(5), 554–571. doi:10.1287/mnsc.32.5.554

Hobbs, R. (2010). *Digital and media literacy: A plan of action*. Whitepaper. New York, NY: The Aspen Institute. Retrieved March 30, 2011, from http://www.knightcomm.org/digital-and-media-literacy-a-plan-of-action

Johnson, G., & Rayman, J. R. (2007). E-portfolios: A collaboration between student affairs and faculty. *New Directions for Student Services*, *119*, 17–30. doi:10.1002/ss.246

Knowles, M., Holton, E. F. III, & Swanson, R. A. (2005). *The adult learner: The definitive classic in adult education and human resource development* (6th ed.). Burlington, MA: Elsevier.

Kock, N. (2005). Media richness or media naturalness? The evolution of our biological communication apparatus and its influence on our behavior toward e-communication tools. *IEEE Transactions on Professional Communication*, *48*(2), 117–130. doi:10.1109/TPC.2005.849649

Manovich, L. (2002). *The language of new media*. Cambridge, MA: MIT Press.

National Association for Media Literacy Education. (2007). *Core principles of media literacy education in the United States*. Retrieved February 2, 2011, from http://namle.net/wp-content/uploads/2009/09/NAMLE-CPMLE-w-questions2.pdf

National Center for Academic Transformation. (2005a). *Institution D: Developmental mathematics*. Retrieved March 10, 2011, from http://www.thencat.org/PlanRes/ProposalExD_DevMath.htm

National Center for Academic Transformation. (2005b). *Welcome to the National Center for Academic Transformation*. Retrieved March 30, 2011, from http://www.thencat.org/R2R/R2R%20PDFs/Six%20Models%20for%20Course%20Redesign.pdf

National Council for the Accreditation of Teacher Education. (2007). *Professional standards for the accreditation of teacher preparation institutions*. Retrieved February 15, 2011, from http://www.ncate.org/public/standards.asp?ch=4

Norman, D. A. (2002). *The design of everyday things*. New York, NY: Doubleday.

Palfrey, J., & Gasser, U. (2008). *Born digital: Understanding the first generation of digital natives*. New York, NY: Basic Books.

Partnership for 21st Century Skills. (2009). *Professional development: A 21st century skills implementation guide*. Retrieved March 1, 2011, from http://p21.org/documents/p21-stateimp_professional_development.pdf

Portfolio, M. A. T. C. (2010). *Texas Tech University Department of English*. Retrieved March 1, 2011, from http://www.english.ttu.edu/tcr/MATC/matc_portfolio.asp

Rice, R. (2002). *Teaching and learning first-year composition with digital portfolios*. Dissertation. Retrieved February 1, 2011, from http://richrice.com/dissertation.pdf

Stefani, L., Mason, R., & Pegler, C. (2007). *The educational potential of e-Portfolios: Supporting personal development and reflective learning*. New York, NY: Routledge.

Taylor, M. C., & Saarinen, E. (1994). *Imagologies: Media philosophy*. New York, NY: Routledge.

KEY TERMS AND DEFINITIONS

Andragogy: The study of teaching adult learners. Adult learners have specific needs, such as the need to know, the need to a foundation of knowledge, the need to privilege a self-concept, the need for ready content, the need for immediately relevant orientation, and the need for personal motivation.

Engfish: While teachers want students to create sophisticated writing in appropriate grammatical and stylistic form, students often relate writing which lacks personality, emotion, and conviction and write what they think the teacher wants. The result is forced "academic-sounding" writing that blocks the possibility of sharing developed ideas.

ePortfolio: An electronic, selected collection of work with clear goals and objectives that incorporates hyperlinks to and between artifacts and uses an authentic assessment guide.

Hypermediation: Drawing attention to or reinforcing the content through an over-emphasis or hyper-identification of the form or medium.

Kairos: Stating just the right thing in just the right amount at just the right time.

Rhetoric: The art of persuasion, including attention to the reader, the writer, the text, the modality, and the location of the communicative moment and situation.

Transcoding: The interpretation of the "cultural layer" of anything in terms of computer logic, and the technology in terms of human cultural terms. Each layer influences the other and is not separable from the other.

Variability: Media and therefore meaning that can exist in infinite versions, as in hypertext or interactive media with multisequential choices.

Chapter 6
Using Web-Conferencing to Increase Learner Engagement:
The Perspectives of a Librarian and Educational Technologist

Kristin Whitehair
University of Kansas Medical Center, USA

Kathy Tally
University of Kansas Medical Center, USA

ABSTRACT

This chapter examines themes in scholarly literature regarding web-conferencing as applied at the University of Kansas Medical Center (KUMC). The chapter profiles how applications of web-conferencing can dramatically change the "classroom" environment for both users on campus and at a distance. We conclude that successful implementations of web-conferencing offer new opportunities in the classroom, increasing learner engagement, but also require adequate resources to support faculty and students.

INTRODUCTION

Web-conferencing is an instructional technology that can be used in education to connect participants across any distance synchronously, to create an interactive exchange of information. Educators can employ a variety of technology tools via web-conferencing to enhance learner experience, which in turn creates an environment that facilitates and increases learner engagement. Web-conferencing, used in an educational setting, can also be used for webcasting. Web-conferencing communication occurs synchronously, when faculty and students are present at one time in the conference, whereas webcasting occurs asynchronously, with faculty or students accessing a recorded session at a time convenient to them. In his article, "Cyber-Guest Lecturers: Using Webcasts As a Teaching Tool,"

DOI: 10.4018/978-1-61350-077-4.ch006

Steven Bell defines web-conferencing as "the transmission of live or pre-recorded audio and/or video to computers that are connected to the Internet" (2003, p. 4). This type of learning environment can mirror interaction that previously could only be accomplished in a face-to-face (f2f) classroom. Today, distance education faculty and students benefit from web-conferencing technology. Synchronous technologies connect faculty and students at the "same time, anytime and anywhere" (Patcha & Scales, 2005, p. 278), allowing them to learn from each other despite geographical separation. Web-conferencing can link learners and instructors synchronously, creating a distance learning platform that is not text-based. The web-conferencing environment conveys the tone of voice, verbal cues and immediate feedback, and enriches learning by maximizing engagement and increasing community between students and faculty (Baecker, et al., 2007).

Prior to text-based distance education classes, lecture was the most ubiquitous method used by instructors to transmit content and information to students (Baecker, Moore, & Zijdemans, 2003). The article "Reinventing the Lecture: Webcasting Made Interactive" states that "webcasting gives us the opportunity to reinvent the lecture" (Baecker, et al., 2003). With the emergence of Web 2.0 technologies, "defined as the philosophy of mutually maximizing collective intelligence and added value for each participant by formalized and dynamic information sharing and creation" (Hoegg, Meckel, Stanoevska-Slabeva, & Martignoni, 2006, p. 13), and increased access to the Internet, students can participate in synchronous learning with a high-speed Internet connection and a personal computer in their home or any other location where these resources are available. This "dynamic information sharing and creation" can be accomplished using web-conferencing in a distance/online educational setting (Hoegg, et al., 2006). With the continued growth of the resources and technologies accessible through the world-wide-web, "schools and universities all over the world are continuously exploring ways to use the web technology in improving teaching effectiveness" (Yunus, et al., 2006, p. 916). The article "Use of webcasting in teaching higher education" reported findings from a study comparing a control group that attended f2f lectures with students who used video on demand (VOD). The modes were live streaming (web-casting) or pre-recorded VOD lectures archived and made available to students anytime and anywhere, and "the analysis showed that the students who went through the VOD group showed the most gain in the tests" (Yunus, et al., 2006, p. 916).

INSTITUTIONAL HISTORY AND ADOPTION OF WEB-CONFERENCING PLATFORM PROGRAMS

The University of Kansas Medical Center offers traditional on site f2f classes, online classes, and hybrid classes that combine elements of both f2f and online classes. For the academic year 2009-2010, the school reported 80-85% fully online courses in the graduate school. The degree Master of Science in Nursing has five advanced practice tracks with Psychology Mental Health Nurse Practitioner completely online. The other four tracks, Family Nurse Practitioner, Adult Gerontology Nurse Practitioner, Adult Gerontology CNS, and Nurse/Midwife are a blend of hybrid courses, which encompass an f2f component and an online component, and fully online courses. The total number of students enrolled in these five tracks was 223 in Spring 2010. There are four leadership tracks: Clinical Research Management, Organizational Leadership, Public Health, and Healthcare Informatics. These degree plans have seventy-nine students and are fully online.

The two doctoral programs, Doctor of Nursing Practice, and Doctor of Philosophy in Nursing, are fully online with seventy-seven students for the 2009-2010 academic year. There are four

Post Master Certificate programs that are fully online with 21 students enrolled for the Spring 2010 semester. With over 300 students enrolled in hybrid and fully online courses in the master and doctoral programs, and an initiative to increase learner engagement, the use of web-conferencing provides the needed tools to facilitate synchronous live sessions as well as recordings for students to view who cannot join the real-time sessions.

SOFTWARE

The Kansas University Medical Center (KUMC) School of Nursing's (SON) distance education programs adopted the Elluminate web-conferencing tool in an effort to increase learner-engagement. Implementation began in 2006, when three schools, the School of Nursing, School of Allied Health, and the School of Medicine, combined resources to acquire a total of 436 seats for remote students with a maximum of 18 concurrent sessions. In the second year, 2007, KUMC doubled its seat license to 1,344. By the fourth year, 2009, KUMC purchased an unlimited seat license.

Prior to the fall 2009 installation of the bridge, which gave faculty access to create their own sessions within the learning management system (LMS), ANGEL, a web request form had to be completed by the faculty for each session. The sessions and student logins were created manually by an Elluminate Administrator in the Department of Teaching and Learning Technologies. In the spring of 2010, the bridge became available to all faculty who used the LMS. This automation and integration was a catalyst for increasing innovations and broader use of the web-conferencing technology. As predicted by the Teaching and Learning Technologies staff, this increased faculty autonomy and decreased the level of administrative duties.

Like the West Virginia University School of Nursing (WVUSON), KUMC's SON has enjoyed a long history of using distance education technologies to deliver graduate nursing education (Ostrow & DiMaria-Ghalili, 2005, p. 6), such as interactive-television (I-TV). Although I-TV is still used for some courses at KUMC, Elluminate is interfaced in the multi-point connections for student access. Students can access their courses via live streaming, for a synchronous connection to lectures, and students unable to join the live class session have access to the recorded lectures, provided via asynchronous video-on-demand (VOD).

As a team of educational technologists who work with faculty using web-conferencing, the authors have observed first-hand the use of the tools and features that facilitate synchronous live interaction, and observed increased learner engagement. Some of the web-conferencing features that faculty are encouraged to use in order to increase learner engagement are delineated using the following examples.

Using Elluminate, by opening two to three microphones, multiple simultaneous speakers can communicate easily. Multipoint video allows up to six session participants to send a live video stream directly from web-cameras installed on their computers, to other participating students. Students who do not have microphone capabilities can enter questions throughout a session, which can be read by a facilitator and by all participants, creating synchronous/real-time feedback and input. Faculty can use a polling feature to solicit immediate feedback on content, questions or issues being discussed during a live session. The interface allows for communication that expresses a number of cues and features. For example, emoticons are provided, and are used by participants to allow non-verbal responses, such as a smile or even applause, to the presenters anytime during the session. Additionally, faculty and students can be granted moderator privileges, allowing faculty and students to load PowerPoint presentations to replace the out-of-the-box generic whiteboard, and the interface allows all of the participants to write, highlight, or draw emphasis marks on the PowerPoint – feedback that is visible on the screen. Alternately, a simple whiteboard can become an interactive drawing space for participants to work through formulas or brainstorm.

Any participant can initiate a web tour, which opens a web browser where an Internet address can be entered. Once s website is displayed, participants are able to follow along as someone within the session drives the web tour to display specific points of interest. Another CMC tool that is commonly used by KU's School of Nursing faculty and students is application sharing. This option is used for a variety of educational purposes. For example, faculty members can share a software application that is available to them on their computer, in order to demonstrate how to use it, without the students needing the software downloaded on their computers. Participants can also be placed in breakout rooms at anytime during a live session, to discuss topics. After such discussions, the facilitator/moderator of the session can return all participants to the main room, where discussion can continue, with each group reporting what resulted from their respective discussion. These features of Elluminate greatly enhance traditional text-based online education, adding communication features the engage students in new ways.

Because many of KU's School of Nursing courses are offered via online distance education, it is essential that students who cannot attend the live/synchronous sessions have access to the recorded session available for viewing at a later time. These teaching and collaborative sessions are recorded at the discretion of the facilitator/faculty and made available for asynchronous/not-live post viewing as needed.

In his white paper, *Unified Learning and Collaboration: Meeting the Needs of Education and Training in the 21st Century*, Gary Dietz, Senior Product Manager for Elluminate, Inc., defines this unique approach, Unified Learning and Collaboration (ULC), as "the delivery of a cohesive set of technology solutions to bring content, instruction, and community expertise to teachers, trainers, and learners, regardless of who or where they are, what or when they need to learn, or their individual learning styles" (Dietz, 2010, p.1). He also puts forth that "ULC makes communication instant, collaboration continuous, and learning accessible, personal, and meaningful for today's 21st century learners" (Dietz, 2010, p. 3). See Figure 1 to view the user interface for the Elluminate web-conferencing tool.

Several alternative web-conferencing systems are available in addition to Elluminate. For instance, Adobe Connect is a competitor to Elluminate, offering a robust suite of web-conferencing tools. An open source web-conferencing system called ePresence was developed to "facilitate scalable communications and knowledge sharing at a distance" (Baecker, Wolf, & Rankin, 2004) with a similar mission; to create an engaging interactive software useful for knowledge building and sharing.

TRAINING AND SUPPORT

In order to maximize learner engagement through web-conferencing, with unlimited sessions and seats, the faculty and students must have adequate support and training. The software offers the potential to create a university with unlimited classrooms, unlimited seats in each classroom, using point-to-point video, simultaneous speakers, a variety of bandwidth connections, audio input and output devices (internal microphones, external speakers, and headsets), multiple operating systems and more. With an ongoing mission and commitment to increase learner engagement, KUMC has multiple departments and staff invested in the success of web-conferencing.

In addition to the need for hardware and software support, the faculty must be trained and supported during their scheduled sessions, especially as they become comfortable teaching with the web-conferencing software. Effective teaching and learning with distance education requires the teacher to become conversant with the new technology, developing new instructional

Figure 1. Student applied demonstration example from an Elluminate session Spring 2010 by Paul Bennetts

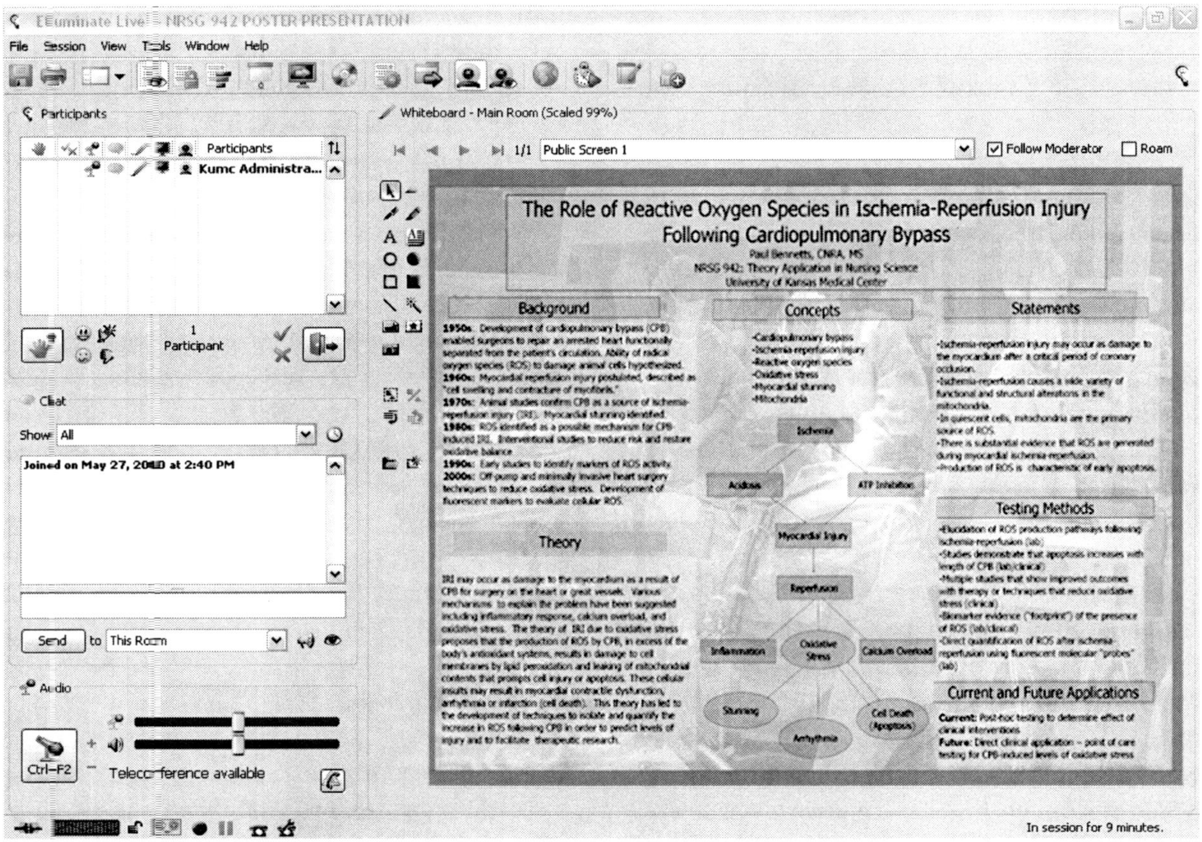

styles (Strain, 1987). In the article "Effective Teaching in Education," Dan Mielke describes distance education as a "method of education in which the learner is physically separated from the teacher and the institution sponsoring the instruction" (Mielke, 1999, p. 2). Web-conferencing does not overcome the physical separation, but the enhanced communication options and features provide it with the ability to connect the instructor and students in ways that surpasses text-based interactions prevalent in early distance education modes of delivery.

At KUMC, the Department of Teaching and Learning Technologies (TLT) provides the human resource capital to ensure that the scheduled sessions are successful to the greatest degree possible, taking into consideration that technology can fail for unforeseen reasons that cannot be avoided (e.g. severe weather, loss of Internet connection by one or more members in the session, etc.). Each school has two-team members, and educational technologist and an elearning support specialist, that train the faculty how to use the tool to maximize the student learning experience and support students entering the session, so that the students achieve the technical skills and successfully connect using their input and output devices (audio, video, and sound). The educational technologist focuses on how the various tools can be used, and collaborates and explores the new ideas with faculty, in order to implement the new ideas at the start of each semester. This exploration is usually piloted by one faculty member, and then shared with all faculty members if it is successful in engaging students, enhancing their learning experience, and is accomplished without excessive technical problems. The Elluminate system administrator is also a member of the TLT team, and works with each team as needed to support sessions, troubleshoot, and communicate with the vendor if there are problems with the web-conferencing software. A priority of the TLT team is to work with the faculty so that they are comfortable with the technology and the tools that enhance their teaching experience as well as the learners'.

LEARNER ENGAGEMENT

One of the features in distance learning conducted online is the asynchronous nature of the interactions between faculty and students, and students' interactions with each other. Asynchronous interactions are those that occur at different times, via a discussion forum, blog, or other media, where each thread of a discussion has a date and time stamp indicating when each post was submitted. These asynchronous discussions often replace the live discussions that are commonplace in a face-to-face classroom, where the faculty and students are participants in synchronous dialogic behavior. A recent study conducted by Marcia D. Dixson suggested "that multiple communication channels may be related to higher engagement and student-student and instructor-student communication are clearly strongly correlated with higher student engagement with the course, in general" (Dixson, 2010, p. 1). Additionally, the research advises online instructors to incorporate meaningful ways to interact with students, and for students to interact with each other. Another study concluded that "online instruction required strong methodology and opportunities for students to interact with each other and the instructor" (Maki and Maki, 2007, p. 532).

In the article "Patterns of Engagement in Authentic Online Learning Environments," the authors provide 10 characteristics of authentic activities based on a wide literature review of recent research theory (Herrington, Oliver, & Reeves, 2003). Among the 10 characteristics, web-conferencing addresses the opportunity for students to collaborate, and this collaboration is considered an integral task. The opportunity for students to view a recorded web-conferencing session more than once addresses a second characteristic, the opportunity to reflect, "enabling learners to make choices and reflect on their learning both individually and socially" (Herrington, Oliver, & Reeves, 2003, p. 62-63).

In the article "Engagement, Excitement, Anxiety, and Fear: Learners' Experiences of Starting an Online Course," Dianne L. Conrad poses the question, "How, though, in the absence of traditional classroom physicality, is the sense of engagement and togetherness that bonds groups of strangers in a learning community established?" (2002, p.205). Live synchronous web-conferencing opportunities are proposed as a way to engage students in learning where the instructor and students' physicality is projected across the distance via microphones and web cameras into the homes, offices, libraries or coffee shops where distance learners reside.

ALIGNMENT WITH NATIONAL BEST PRACTICE STANDARDS TO INCREASE LEANER ENGAGEMENT

Web-conferencing is one tool available to faculty and students of distance courses. While this tool has been warmly embraced by the KU Medical Center teaching faculty, the University of Kansas Medical Center's School of Nursing has also recognized the value of quality assurance tools, used to monitor the quality of its distance education offerings. The School of Nursing became subscribers of Quality Matters (QM) in 2009. Earlier in 2003, with funding from a Fund for the Improvement of Postsecondary Education (FIPSE) grant, MarylandOnline, Inc. began developing a "replicable pathway for inter-institutional quality assurance and course improvements in online learning" (http://www.qualitymatters.org/FIPSE.htm). The QM project was developed to provide a tool set and process to promote quality in online and hybrid courses. The tool set an extensive rubric consisting of eight categories and 40 standards. Of the eight categories, Category 5 in the QM Rubric addresses Learner Engagement and crosses two additional categories, Resources and Materials (Category 4) and Course Technology (Category 6). The standards addressed include Essential Standard 5.2, stating that "Learning activities foster instructor-student, content-student, and if appropriate to the course, student-student interaction," and Essential Standard 6.2, which states, "The tools and media support student engagement and guide the student to become an active learner."

Web-conferencing technology addresses the standards that align with engaging students with the instructor, each other, and the course content. In using QM, the School of Nursing provides measures by which to evaluate the use of technologies to increase student engagement. Adoption of the QM framework provides a guideline for the faculty and administrators to evaluate and guide the use of web-conferencing.

Illustrating the breadth of uses for web-conferencing, the School of Nursing faculty uses the web-conferencing tool for a variety of purposes, from holding office hours to alternative assessment. The technology provides opportunities for faculty to become resourceful and creative when measuring objectives or assessing outcomes. More specific applications of the tool include:

1. Course tours for first time and returning students at the start of a semester provide an opportunity for an introduction to the faculty, student peers, and an introduction to the learning managements system. Quality Matters Specific Review Standard 1.1 is met at this point: "Instructions make clear how to get started and where to find various course components" (http://qminstitute.org/home/Public%20Library/About%20QM/RubricStandards2008-2010.pdf).
2. Students can attend online lectures and interact with faculty, other students, and experts in their field using the chat, audio and video features. Lectures are recorded and available for students who are unable to attend and for post-lecture viewing and reviewing.
3. Web-conferencing has expanded the options for alternative assessments. Applied demonstrations are used for students to present case studies, special projects, and research throughout a semester or as a culminating final project presented to the faculty.
4. Standardized "patient" interviews are conducted using two-way video, where a student and a prepared actor following a script to evaluate the patient-actor and make recommendations for treatment. These interviews are also recorded, so that the student can review the interview and submit a post analysis and self-critique.
5. Group work sessions are created by the faculty for the students to use throughout the semester for collaboration, accommodating a variety of schedules.
6. Test preparation and review sessions are commonplace, as faculty members make themselves available to students with equal opportunities for attendance no matter the distance. See Figure 2 for an example.
7. Office hours are scheduled using the web-conferencing system at regular intervals for student to faculty interaction.

Expanding on the brief descriptions of the CMC web-conferencing applications above, it is imperative to explore the value that these applications offer students, faculty, and staff. First, Elluminate provides a means for delivering guest lectures. Web-conferencing greatly expands opportunities for guest lecturers. This is true for both on-site classes that invite remove guest lecturers

Figure 2. Radiologist uses Elluminate to prepare students for an upcoming exam by using one of the tools to draw the image of a bat over a slide in the presentation to show a "bat-wing" deformity

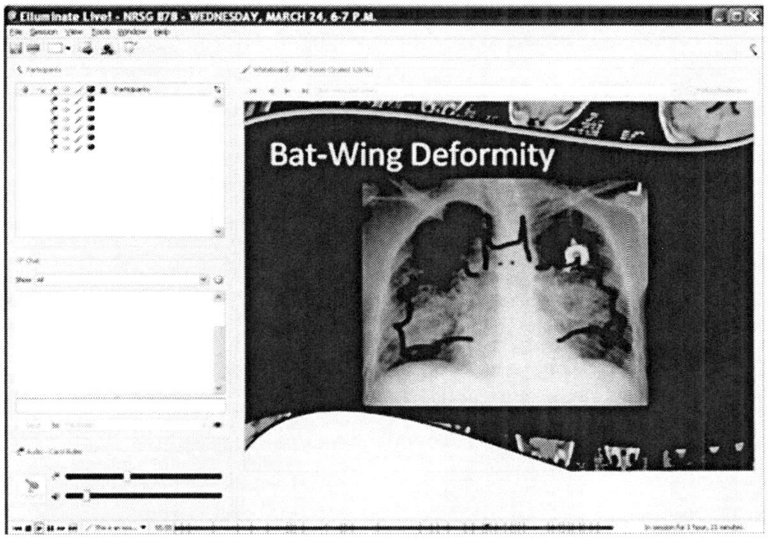

to virtually visit their on campus classes using Elluminate, as well as for distance classes that do not physically meet. Topical experts, such as medical professionals specializing in the specific area related to the class' content, can visit the class as guest lecturers. Often, web-conferencing is significantly easier to coordinate, since it does not require physical travel to campus. Guest lecturers may simply access the Internet and proceed through the relatively simple steps required to present using Elluminate. Building on the ease of one guest lecturer visiting class using web-conferencing, bringing groups of experts together for class presentations is also significantly easier.

The Department of Teaching and Learning Technologies staff actively works with guest lecturers to provide instruction and guide them in using Elluminate. This level of support is necessary for successful guest lectures, especially when the invited presenters have not had past experiences with web-conferencing. As well, guest lecturer presentations can easily be recorded, including the interactions with students throughout the lecture. These recordings can be useful for students unable to attend the synchronous lecture, as well as for review prior to an exam.

Moreover, web-conferencing has proven to be a valuable CMC tool that enriches faculty training. For example, outside of the KUMC School of Nursing, the School of Medicine uses web-conferencing to enrich regularly-scheduled staff development sessions called Grand Rounds. In this setting, a specific concept is presented by an expert. Traditionally, Grand Rounds have taken place with a group of residents or doctors in one room, discussing a specific patient or condition. KUMC's School of Medicine includes the main campus in Kansas City and a second smaller campus in Wichita. Using web-conferencing, faculty members that are not on the same campus can easily attend staff development sessions like Grand Rounds. This application of web-conferencing connects experts to faculty.

Similar to the advantages that web-conferencing offers in facilitating guest lecturers, web-conferencing also allows for dynamic and interactive presentations by other guests. For example, librarians are able to use it to assist students in conducting their literature searches. Traditional library instruction for on campus classes often involves a librarian visiting the classroom, or the class visiting the library. In traditional library instruction, the librarian provides instruction regarding literature searching, usually tailored for the specific course or assignment. When classes moved to an electronic format, these traditional sessions also change delivery mechanisms. Providing support for students at a distance to assist them with developing advanced searching skills can be challenging. Often, this task requires both an introduction of theoretical concepts related to search strategies and navigating several online databases. Web-conferencing provides a solution to this challenging situation. By allowing both real-time audio communication and visual demonstrations, the librarian and student can interact similarly to in class demonstrations.

Web-conferencing using Elluminate has proven to be a valuable CMC tool for efforts centered on information literacy instruction for students at KUMC. Defined by the Association of College and Research Libraries as "a set of abilities requiring individuals to recognize when information is needed and have the ability to locate, evaluate, and use effectively the needed information" (2000), information literacy has traditionally been an effort that has been led by the Kansas University biomedical librarians. Navigating numerous information seeking interfaces such as library catalogs and databases can be challenging, and beyond simply locating information, information literate students must understand the cycle of information use that begins with identifying when information is needed, continues through the finding and use of information, and ends with understanding legal and social standards for using information.

Library instruction sessions focus on transferrable information literacy skills, rather than navigating through one system. By tailoring the information literacy instruction content to the specific needs of the discipline of study, students are able to develop the skills finely tuned to the needs of their profession (Grafstein, 2002). However, these transferable information literacy skills can also be difficult to convey at a distance. The interactive nature of web-conferencing, allowing users to share their desktop with others, among many other technologically functionalities, is an effective method to demonstrate concepts and assist students with their journeys navigating literature searching interfaces and in becoming information literate.

Research indicates that the instructional design of information literacy sessions is crucial to their success, or failure. Active learning and assessment are critical elements in providing effective information literacy sessions to distance students (Dewald, Schotz-Crane, Booth, & Levine, 2000). Unlike an exclusively text-based environment, web-conferencing offers tools that allow presenters to incorporate both of these elements. For this purpose, web-conferencing provides many of the benefits of traditional face-to-face interactions. In addition to these benefits, web-conferencing sessions can be recorded to share with those who are unable to attend them synchronously, or used for review by students. This recording can also be used by the librarians to review and evaluate their instruction skills.

Illustrating the need for librarian instruction for distance students at KUMC, many School of Nursing faculty members highly recommend using citation management software to students. For many students, combining the intricacies of literature searching with using citation management software is a challenging prospect. Based on the experience of the authors, text-based instructions and phone conversations often lack the type of interaction needed to provide clear answers to student questions about these topics. Citation management software, for example, requires users to navigate between several electronic resources. When possible, the librarian liaison responsible for providing instruction in this area arranges for face-to-face meetings with students, to provide individual instruction. However, face-to-face meetings are often not a realistic option, because School of Nursing students are located both in the immediate area and nationwide. Additionally, students are often employed full-time and work varied schedules, which complicates finding a meeting time even for local students. Thus, web-conferencing is a useful solution for teaching content like this, which combines theoretical concepts and technical procedures. Since web-conferencing can connect any two locations with Internet access, it provides flexibility for both the student and librarian liaison. Often, the librarian liaison holds individual web-conferences with students in the evenings and during the weekend. Moreover, the advantage of being able to record sessions is a benefit that face-to-face meetings lack. Students can review content and processes after the initial session.

Beyond interaction with individuals, web-conferencing has also proven to be a valuable tool for working with classes whose instructors are requesting instruction for using citation management software. In the spring of 2010, the KUMC librarian who delivers on site classes for beginners and intermediate users of citation management software delivered the beginning citation management software management material to a distance class, by using web-conferencing. On-site classes covering citation management software are regularly scheduled for ninety minutes to two hours. To prepare for the web-conferencing class, the librarian revised her content, building more structured opportunities for feedback and clarification. The class was also revised to account for the reduced presentation area, her shared computer desktop, for delivering content. This demonstrates the importance of revising existing class outlines and developing content and presentation techniques that capitalize on web-conference features and account for the differences in delivery methods. The development and delivery of the web-conferencing version of the citation management class required significant planning, a revision of the class outlines, and the allocation of adequate time during the web-conferencing sessions to allow for clarifications and questions. In addition to highly positive feedback from students and the teaching faculty member for the course, recordings of this session were requested by other School of Nursing faculty to serve as a resource their students could consult. Overall, the delivery of this traditionally face-to-face class via web-conferencing expanded access to this level of highly specified training that previously was not available off campus.

To successfully deliver library instruction sessions through web-conferencing, preparation is essential. To ensure that the students are prepared, it is helpful to ask them to login a few minutes early, in order to work through any technical problems that may arise. It is also helpful provide alternate contact information, such as a telephone number, to provide a means of communication that aids in working through technological problems. At KUMC, the Department of Teaching and Learning Technologies staff members provide the technical assistance. Customer support services from Elluminate are also used to supplement support provided by KUMC.

For the librarians delivering content through web-conferencing, the same technical support concerns apply as discussed above. Especially for librarians who are less familiar with web-conference or the specific web-conferencing system, institutional support and support from the vendor are critical. A familiarity with the tool and features are needed for the presenter to effectively interact with the students. In regard to content, it is important to communicate with the students to identify their goals for the session and to learn about specific concerns they may have. This preparation is similar to how one would prepare for traditional face-to-face library instruction classes. However, for web-

Figure 3. Measures of effectiveness bar graph shows the increased use of the web-conferencing system across three academic years

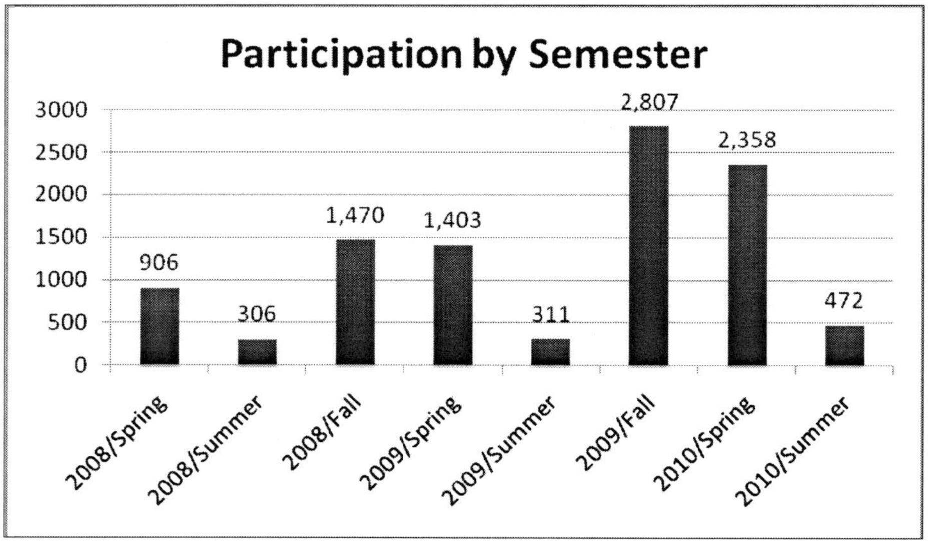

conferencing, the delivery of content is limited to the web-conferencing application. Presenters who rely heavily on non-verbal communication from students they will need to incorporate alternative feedback measures to gain the information they are used to picking up on through non-verbal cues. For example, asking students for responses or the use of activity indicators or emoticons can provide this type of information to the instructor. Overall, the effective delivery library instruction via web-conferencing is more involved than simply delivering the same classes that are presented face-to-face. Instead, classes need to be revised to address limitations and functionalities available in the web-conferencing session. To prepare for this transition, the authors have observed that it is useful for presenters who are preparing to use the technology to attend web-conferencing sessions as a student, in order to gain experience as a student in this online environment.

The use of web-conferencing is only one aspect of a larger shift in the delivery of library instruction. Online students have reduced barriers to interacting with the library and library staff (Markgraf, 2005). Although geographically, distance students may be less likely to visit the library building, web-conferencing is an example of how librarians can capitalize on technological advancements to improve their services.

With unlimited access to a web-conferencing service, and technological bridges in place allowing faculty to initiate their own web-conferencing sessions, the use of web-conferencing at KUMC has dramatically increased (see Figure 3 for details). While at first, web-conferencing was largely used to bring traditional class meetings to distance students, with easy access and no limitations on use of the software, faculty, students, and staff are using web-conferencing in new and evolving ways.

One example of a new way in which the technology is being applied is the use of it for meetings between only two individuals. With unlimited access, an instructor can schedule a meeting with a student without affecting other users. Compared with previous technology for distance meetings and communication, these meetings take on more similarities to the face-to-face meetings often scheduled by instructors. Prior to web-conferencing, these types of meetings were often difficult for distance students. Meetings via telephone or instant messaging (IM), for example, lacked the ability to use visual aids to demonstrate concepts, and email messages lacked the immediate interaction of synchronous communication. As observed through the authors' experiences, web-conferencing allows for geographically separated individuals to meet, with both attendees interacting with each other and the ability for visual aids or examples to be demonstrated for both attendees to see.

Similar to scheduled student meetings with instructors, faculty began using Elluminate to hold office hours. Prior to the use of web-conferencing, faculty teaching distance education courses often had virtual office hours where they would be available by phone or IM. However, the same limitations discussed above of neither of these options providing both immediate interaction and an ability to use visual aids. Faculty members can create office hour sessions in Elluminate and sign in to be available at the stated hours. Students can then drop in to communicate with their instructor.

In addition to meeting with the course instructor, web-conferencing facilitates meetings between students and other experts. For example, as mentioned above, students can schedule meetings with librarians, in which a librarian can advise research strategies and demonstrate searching techniques. This same model could be used for meeting with other experts on campus, such as representatives of a writing center, or academic advisors. The opportunities available for distance students to meet with topic experts enhance distance students' educational experience.

The use of web-conferencing on campus is not exclusive to meetings with students. With many faculty members and professional educational support staff involved in projects, professional organizations, and other activities that include professionals off campus, web-conferencing is a tool that can assist staff and faculty with their professional activities outside of the classroom, and KUMC has made use of this potential. Recently at KUMC, for example, Elluminate was used to hold a meeting of an area librarian's professional organization. With the time and financial constraints required for in-person meetings, web-conferencing allows for more efficient and cost-effective meetings. The librarians used web-conferencing technology to facilitate the planning for an upcoming conference. While it can be difficult to identify meeting times when individuals from several educational organizations must allow for significant travel time, web-conferencing eliminates travel expenses and substantially reduces the time commitment required. By supporting professional meetings, like the example discussed, the communication benefits web-conferencing technology extend beyond students in the classroom. The features of the web-conferencing extend beyond the conference calls and other technology that has previously facilitated meetings of geographically dispersed groups. For example, a valuable benefit that web-conferencing offers is the ability to record the meetings for those unable to attend at the time of the meeting. While notes or minutes are traditionally recorded of meeting activities, web-conferencing allows for recording the entire meeting, so discussions on specific topics, which are often omitted from the minutes, can be viewed. In this setting, web-conferencing offers attendees multiple ways to interact through text chat, audio communication, and visual aids (Figure 4).

Beyond the ability that web-conference provides students to meet with their instructors and with experts, it allows the students to meet with each other. For example, group projects and reviewing for exams with study groups can be challenging for distance students. Although mainstream technologies like email, IM, and traditional video chat can be helpful, the full suite of tools available with a web-conferencing tool like Elluminate better aide students in their collaborations. By using the Elluminate bridge, faculty can

Figure 4. Web-conferencing is used for professional development and recording faculty responses to polls and questions incorporated into the training

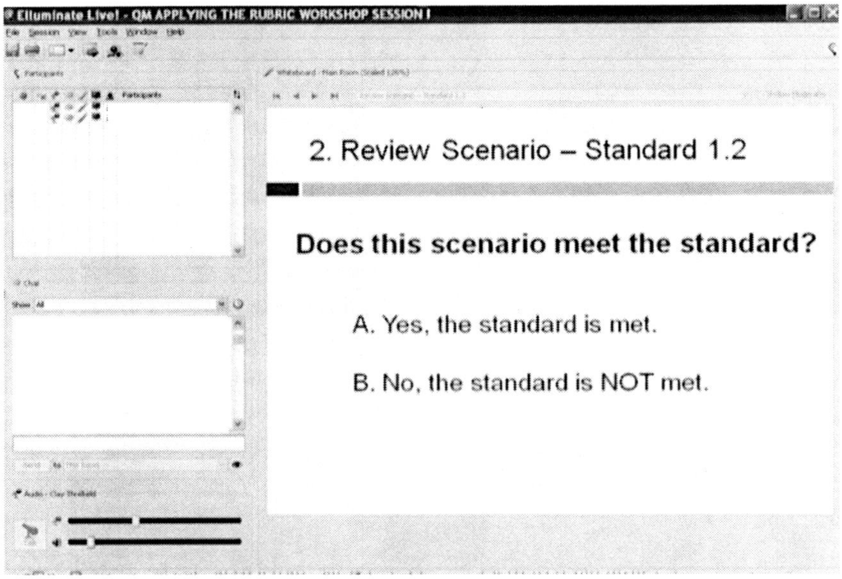

create sessions for specific groups or teams from their class to meet. Then students can communicate with each other using the features in Elluminate. In addition to voice and text chat, students can share documents or full desktops to work on projects together. Meetings for groups like this can be scheduled to be ongoing over a period of time, so student groups can sign in as it fits their schedule without requiring coordination by the instructor to set up a specific date and time.

According to faculty members, these student-to-student interactions are a particularly important aspect of education for graduate students in distance education programs. Networking that occurs in traditional face-to-face classes can be more difficult in an online environment, but with web-conferencing, students are able interact more fully with each other. This aids building professional network contacts for students.

CONCLUSION

Distance education is looking toward new and emerging technologies to increase student engagement. Web-conferencing tools have evolved as technological advancements have widened possibilities for online interaction. In the future, it can logically be expected that web-conferencing tools to enhance engagement of students and faculty will continue to expand. To fully take advantage of technological opportunities, adequate technical support from educational technologists is necessary. Additionally, staff serving students on campus can also use web-conferencing to reach students off campus. Outside of the classroom, these technologies are often employed in unexpected ways. The experience of the University of Kansas Medical Center School of Nursing illustrates the value of close collaboration between instructional faculty, teaching and learning technology staff, and the librarian liaison. Moreover, the implementation of a quality assurance framework like Quality Matters is an effective means to guide implementation of new technologies.

REFERENCES

Association of College and Research Libraries (ACRL). (2000). *Information literacy competency standards for higher education.* Retrieved from http://www.ala.org/ ala/ mgrps/ divs/ acrl/ standards/ informationliteracy competency.cfm#

Baecker, R., Birnholtz, J., Causey, R., Laughton, S., Rankin, K., Mak, C., … Wolf, P. (2007). Webcasting made interactive: Integrating real-time videoconferencing in distributed learning spaces. *Lecture Notes in Computer Science, 4558*, 269-278.

Baecker, R., Moore, G., & Zijdemans, A. (2003). Reinventing the lecture: Webcasting made interactive. *Proceedings of HCI International 2003.*

Baecker, R., Wolf, P., & Rankin, K. (2004). The ePresence interactive webcasting and archiving system: Technology overview and current research issues. In J. Nall & R Robson (Eds.), *Proceedings of World Conference on E-Learning in Corporate, Government, Healthcare, and Higher Education 2004* (pp. 2532-2537). Chesapeake, VA: AACE.

Bell, S. (2003). Using webcasts as a teaching tool. *TechTrends*, *47*(4), 10–14. doi:10.1007/ BF02763506

Conrad, D. L. (2002). Engagement, excitement, anxiety, and fear: Learners' experiences of starting an online course. *American Journal of Distance Education*, *16*(4), 205–226. doi:10.1207/ S15389286AJDE1604_2

Dewald, N., Scholz-Crane, A., Booth, A., & Levine, C. (2000). Information literacy at a distance: Instruction design issues. *Journal of Academic Librarianship*, *26*(1), 33–44. doi:10.1016/ S0099-1333(99)00121-4

Dietz, G. (2010). Unified learning and collaboration: Meeting the needs of education and training in the 21st century [White paper]. Retrieved from http://www.elluminate.com/ Resources/ White_Papers/?id=95

Dixson, M. D. (2010). Creating effective student engagement in online courses: What do students find engaging? *Journal of the Scholarship of Teaching and Learning*, *10*(2), 1–13.

Grafstein, A. (2002). A discipline-based approach to information literacy. *Journal of Academic Librarianship*, *28*(4), 197–204. doi:10.1016/ S0099-1333(02)00283-5

Herrington, J., Oliver, R., & Reeves, T. (2003). Patterns of engagement in authentic online learning environments. *Australian Journal of Educational Technology*, *19*(1), 59–71.

Hoegg, R., Meckel, M., Stanoevska-Slabeva, K., & Martignoni, R. (2006). Overview of business models for Web 2.0 communities. *Proceedings of GeNeMe*, *2006*, 23–37.

Maki, R., & Maki, W. (2007). Online courses. *Handbook of Applied Cognition, 2,* 527-552.

Markgraf, J. S. (2005). Librarian participation in the online classroom. *Internet Reference Services Quarterly*, *9*(1-2), 5–19. doi:10.1300/ J136v09n01_02

Mielke, D. (1999). *Effective teaching in distance education*. ERIC Digest.

Ostrow, L., & DiMaria-Ghalili, R. (2005). Distance education for graduate nursing: One state school's experience. *The Journal of Nursing Education*, *44*(1), 5–10.

Patcha, A., & Scales, G. (2005). Development of an internet based distance learning program at Virginia Tech. *Proceedings of the 6th Conference on Information Technology Education.*

Quality Matters. (2010). *Quality Matters program.* Retrieved from http://www.qmprogram.org/ new-website- welcome-page

Strain, J. (1987). The role of the faculty member in distance education. *American Journal of Distance Education*, *1*(2), 61–65. doi:10.1080/08923648709526585

Yunus, A., Kasa, Z., Asmuni, A., Samah, B., Napis, S., & Yusoff, M. (2006). Use of webcasting technology in teaching higher education. *International Education Journal*, *7*(7), 916–923.

KEY TERMS AND DEFINITIONS

Computer Conferencing: Computer conferencing is the term used by WorldCat's thesaurus to describe online conferencing.

Distance Education: Students taking classes that do not physically meet on campus.

Educational Technology: Broader category that describes web-conferencing.

Information Literacy: The skills of knowing how to locate, evaluate, and use information.

Interactive Multimedia: Multimedia that allows interaction.

Libraries and Distance Education: Thesaurus term from EBSCO addressing library services for distance students.

Library Instruction: Sessions taught by librarians about using library research resources and tools, also addresses searching skills.

Multipoint Video Conferencing: uses a series of video cameras to synchronously transmit video and audio from each location participating.

Webcasting: Occurs asynchronously, when faculty or students access the recorded session after-the-fact at a later date.

Web-Conferencing: Occurs synchronously, when faculty and students are present at one time in the conference.

Chapter 7
Visual Arts Online Educational Trends

Laura M. Rusnak
Ohio University, USA

ABSTRACT

The intent of this chapter is to understand the implications of online education for the visual arts and how the objectives of a traditional art education can be adapted to computer-mediated learning. The focus is on three trends affecting the arts: visual culture, cultural production, and originality in art and practice.

INTRODUCTION

According to the Sloan Consortium, the number of postsecondary students taking advantage of online courses rose 17% to 4.6 million from 2008 to 2009 with 25% of postsecondary students having taken at least one online course (Sloan Consortium, 2010). As more students migrate to the online classroom to fulfill their educational goals, there are a variety of students entering the realm of visual arts that are not generally part of the target demographic for on-ground schools of art. It has been noted that the current competition for recruiting students occurs within a particular educational sector, i.e., the traditional institutions competing amongst themselves, separately from the online institutions, concluding that recruiters for online institutions do not target the usual student groups, such as high school seniors who will go directly to college post-graduation, and instead focus on working adults (Maeroff, 2003; Sull, 2008). While this may change in the future due to various pedagogical and economic strategies, such as a rise in the number of blended or hybrid course offerings, students that do not fit the traditional visual art education demographic include, but are not limited to: artists and designers who are updating their skill sets, e.g. a print designer moving into web media; artists and designers who are exploring alternative media

DOI: 10.4018/978-1-61350-077-4.ch007

and fields, e.g. a painter interested in learning about digitally printing on textiles; the entrepreneur starting a new business based on creative services; the individual who wanted to pursue art in their youth, but did not have the chance due to life circumstances (Sull, 2008); and, the hobbyist who would like to know more about a particular subject matter, such as photography.

In many of these cases, students are not committed to the four-year degree structure of an on-ground institution, because they already have a degree, want an accelerated program, or they are looking for a program that can be completed based on their schedule (Schroeder-Moreno, 2010), and from any Internet-connected locale (Buckley & Smith, 2007). Knowles and Kerkman (2007) have outlined a general discussion on the motivational factors of students who are taking online classes, listing grades and funding—external forces—and "desire to learn"—internal motivation—as some of the incentives that encourage student performance and commitment to coursework (p. 71). While external and internal motivations play a significant role in student attendance and achievement regardless of field of study, internal motivation is predominant in the list of non-traditional visual art students seeking online courses previously mentioned. Thus, the intent of this chapter is to understand the implications of online education for the visual arts and how the objectives of a traditional art education can be re-evaluated and re-configured to adapt to a computer-mediated model of learning, while attempting to meet the needs and desires of a variable and non-traditional audience.

BACKGROUND

Some visual arts educators acknowledge the benefits of Internet-based art courses as effective means for distributing practice-based techniques and theories to students that had been excluded in the past (Maeroff, 2003; Stewart, n.d). Yet, there are others who hesitate to migrate arts courses to the Internet, due to the impracticality of teaching hands-on techniques and concepts—ones customarily taught in a studio setting with artist-teacher demonstrations—via a distance format, synchronous or asynchronous. The use of CMC for teaching many arts techniques is challenging, because it lacks the type of art experience in which students engage directly with artistic materials, the actual artist-instructor, or an original work of art, such as a class excursion to a museum or gallery (Bastos, 2010; Knowles & Kerkman, 2007). Therefore, visual arts educators in the online environment need to assess the roles that traditional visual art pedagogies play, as they develop online curricula, and to look potential technologies to deliver visual art content in an efficient, but meaningful way.

Paralleling the onset of the World Wide Web, Dunn (1996) has recognized that technology is and will continue to have an effect on education, with its full incorporation inevitable, while linking its successful integration to learning models based on *active student participation* (p. 7-10). The latter indirectly addresses a reoccurring theme within digital divide forums that questions whether simply having access to information is enough to create an "informed society" (O'Hara & Stevens, 2006). Hence, the need to package information communication technologies and media literacy skills with accessibility, in order to make full use of available information. While this is a starting point, it is not enough to promote the active learning Dunn speaks of. Active participation can be facilitated by technology that connects the student through interactive modes with the course materials (Dunn, 1996) and encourages their desire to learn (Knowles & Kerkman, 2007; Maeroff, 2003). However, content is not always presented in both a coherent and interactive fashion that stimulates and constructs knowledge.

Content organization and interactivity should depend on the field of study in question. The ways in which technology is incorporated, as well as the ways in which course materials are presented to students are important considerations in the development of online pedagogies. Lai and Bell (2004) have stressed that online education should be acknowledged as a distinct form unto itself, but not without considering its relationship and similarities with face-to-face learning environments (p. 22). From this vantage point, visual art and online educators have emphasized the *organization* of information, typical of all curriculum development—traditional or online—over *interaction* with information (unique qualities of technological tools). Here, the priority for the online learning environment is based on the assembly and interconnections of course content and less on the aesthetic experience of interacting with course content (Mayo, 2007). Likewise, Roland (2010) has written that technological tools "become worthwhile educational tools only when driven by challenging curriculum goals" (p. 17), while others have argued that an online learning experience based on "traditional components" is more significant than the technological tools that make it feasible (Drummond, 2008).

The literature recognizes the need to incorporate technology into educational pursuits, but to variable degrees, and in ways that lean toward established methods. Educators who embrace technological tools and welcome their interactive capabilities to facilitate knowledge creation and well as content organization, embrace the notion that the tools, such as interfaces, must not function as secondary features or add-ons (Roland, 2010), but as integrated systems that reformulate existing educational frameworks (Lai & Bell, 2004) and perspectives (Semchuk & Tien, 2004). The equitable union of the interaction of information in the form of technological tools, with the organization of information, assists in the creation of online learning environments that are rich with the potential for active student participation. However, some online learning environments remain static page-turners, text-based and lacking the engagement necessary to create a dynamic learning experience. The text-based format that works well for some fields is less compatible with the tactile quality and hands-on nature of visual arts training, especially the studio arts like painting and sculpture (Maeroff, 2003). This is also a good reason for the hesitation by some art administrators to migrate to online environments, as it potentially reduces artworks to text descriptions and poor reproductions due to image compression and bandwidth limitations. In regards to using rich media in an online visual arts classroom, Drummond (2008) has argued that the institutional process generally uses available technologies as bells and whistles, with little consideration for the content, in order to wow digitally savvy users (p. 44). As collaborative models promoted through Web 2.0 concepts have become the norm outside the online classroom, institutions cannot rely on technologies as context-lacking enhancements (Buckley & Smith, 2007), rather they need to focus on integrated modules for constructing knowledge "based upon the new affordances digital interfaces offer" to the online learning environment (DePew & Lettner-Rust, 2009, p. 175).

The challenge for online course designers is not to dazzle the digital natives, but to recognize that the hypermediated spaces students are exposed to in their daily lives inform the way they approach the online classroom (Roland, 2010). Recognizing the manner in which students use the Internet and other technological tools, online visual arts courses should support an interdisciplinary perspective where various media, arts, and cultures intersect in a multi-media/-modal space, and not act simply as a place to post digitized images of art assignments for grading purposes. Educators, content experts, and course designers must make technology choices that fit the intended course goals. Web 2.0 concepts provide visual arts educators with the ability to create new forms of content, while taking advantage of its social aspects to build inclusive communities (Buffington, 2008). While art educators are divided between a view of technology and interactivity as secondary elements or as a means to reconstitute online art curricula altogether, the collaborative environment fostered by current social media sites and Web 2.0 concepts is beneficial to the success of the online classroom. The social aspect of the format gives students a space to "reflect" and "validate" their artistic decisions (Overby, 2009, p. 19), sharing their opinions and working together to advance the art-making process. According to Mayo (2007), various stakeholders and the "cultural objects" with which they interact are significant factors for development and "their contributions should be cultivated in art education models that integrate arts-technology experimentation" (p. 48). The online environment has the ability to engage the representations of cultural objects that students use in their work as a place of convergence because of the variety of media the technological tools can facilitate.

For the arts, this particular environment can also act as a living record where artistic development is continually recorded through the activities in the classroom. In art education, documenting the art-making process and the conceptual development of an art project is just as important as the final piece presented for critique. There are technological tools that exist as third-party applications today that can be integrated into the arts online classroom and curricula of the future, for example graphic art software[1] that allows for the recording of image manipulation and playback as a process evaluation instrument for a project produced by a single student; revision control systems[2] implemented on servers to maintain a historical record of various file types, which can be used in a collaborative environment where multiple students work as a unit on a single project; collaborative brainstorming sessions can be held using whiteboard services[3] that allow input through a synchronous feed; and screencasting software[4] with which users can create narrated videos of real-time actions on their desktops. These screencasts can be utilized for technical tutorials and artist presentations as well as collaborative demonstrations. The aforementioned technologies do not have to be relegated only as teaching tools for the instructor in a one-way communication context, but can be viewed as multidirectional channels that afford reciprocity between all class members.

The online space in which visual arts is taught should be as dynamic as the field itself—a field that responds to a multitude of social, economic, political, and ideological influences that create the context in which student artists find themselves (Boughton, 2000). Students attending online courses hail from diverse backgrounds and create artistic works that reflect their distinct viewpoints, affected by their cultural surroundings. Thus, art educators must be able to assess a wide range of new media, including video games, blogs, podcasting, and more, as they help students "realize their power as media manipulators" through a range of artistic outputs (Mayo, 2007, p. 47). As art students formulate their own interpretations of media, art, and technology, the objective of a visual arts curriculum is for them to acquire the skills to reflect on the cultural contexts that relate to their work as artists and to the communities they interact with as members of society (Gude, 2004). To promote student inquiry and reflection concerning cultural representations within the art-making process, Gude (2004) has outlined three criteria for visual arts education based on practical teaching experiences: 1) "art as investigation" encourages students to understand the larger context in which their own artwork is viewed, along with a greater appreciation for work created by other artists; 2) "curriculum based on generative themes" creates relevant classroom experiences that acknowledge local practices in students' lives; and, 3) "studio art projects based on diverse practices" that incorporate traditional and contemporary art production techniques and concepts (p. 8). Using Gude's principles as a guide, the following sections examine the synthesis of visuals arts education and the online learning environment based on concepts of visual culture (arts investigation), cultural production (generative themes), and originality in art and practice (assorted art practices).

VISUAL CULTURE

The idea of art as an investigative strategy resonates with the impact that visual culture studies (VCS) is having on art education in general. Since the mid-1990s, VCS has steadily gained momentum as a field of study (Duncum, 2003; Zimmerman, 2009), while paralleling the rise of the Web and growth of online learning. The field focuses on visual objects and the way in which we perceive and interact with them in our environment (Duncum, 2003). It examines and interprets cultural references that are embedded within visual materials, but the field is not limited to visuality, as images are often contextualized by text, sound, and motion graphics presented through multi-mediated channels (Duncum, 2003). As Knight, Keifer-Boyd, and Amburgy (2005) have commented, "most contemporary theories of visual culture [...] are not grounded in a separation of the intellectual and the ethical from visual aspects of life" (p. 263). This comment by Knight et al. indirectly addresses differences in visual arts pedagogies that center on popular art versus fine arts (Boughton, 2000), and the question of whether popular art is an acceptable means of practice within the elite art world. This line of thinking extends to art-making in conversations of traditional versus digital formats, and into the educational process of on-ground versus online classrooms. The common denominator in these discussions is often the roles that media and technology play, and what effects it has on art when it becomes part of the production and delivery process.

Visual art educators recognize the reciprocal nature of popular media borrowing visually and conceptually from the art world and vice versa (Duncum, 2003), and further, that the accessibility of visual content via media and technology for use as reference materials and as actual imagery embedded in artwork grows exponentially with the use of the Web. Students no longer have to rely on slides and photographs stored away in a fine arts resource library, when they can simply search the Internet for an image of a desired item. Yet, attached to any appropriated representation are the cultural circumstances in which it was created, and why art as investigation leads to an appreciation of visual culture studies. The representations that art students utilize in their work are made up of objects that contain "symbolic forms" that signify a particular reality (Carey, 2009). By studying the vast array of visual materials, scholars come to understand the environments in which these symbolic forms are produced along with their reception amongst audiences (Duncum, 2003). To acknowledge the multiple contexts in which historical art images may be found, on the Internet and in other resources, instructors have an opportunity to open a dialogue with students about how these images' meanings are altered when the venue in which they are seen changes. One way to do so is to perform an online image search for an artistic work, such as Munch's painting *The Scream*, and then note the variable instances in which the painting exists. Comparing these findings with the original intent of the artist, or with available art criticism for the chosen piece, informs students' understanding of how the reality of cultural objects can shift based upon their context.

Visual arts students are both producers (senders) of and interpreters (receivers) of symbolic forms. As content producers and consumers, understanding the context surrounding their art in relation to other artists, and visual culture in general, is part of the investigation that art students face during the learning process. The online learning environment has the capability of acting as hub for the investigative process through the presentation of visual and aural materials that would be tedious to orchestrate in a physical location. The generated discussions would be less about whether a student demonstrated a particular design principle accurately in an assignment that was created in a vacuum, and instead concentrate on how that design principle has been used within a variety of artworks and media, showing its integration with other concepts as well as its infinite opportunities to facilitate meaning (Bryant, 2010). The online classroom offers a form of synthesis that acknowledges the multi-faceted or hypermediated nature of the world in which our students navigate their daily lives; a forum for the development of their artistic identity. The artist's identity is understood by viewers through the artist's artistic practice and visual style; for students to find their place amongst their peers, they must understand the visual landscape in which they disseminate their works.

CULTURAL PRODUCTION

With the Web to exposing them to countless images and ideas, students gain access to a variety of symbols from within and outside their surroundings, which can only be synthesized through a critical eye. Media technologies, like the Web, act as a feedback mechanism for culture, in which visual imagery is recycled and reconstituted, influencing or contesting established customs (Carey, 2009). By incorporating their perceptions of the world around them, artists act as social agents as they comment and reflect upon various societal issues through their work, often repurposing the symbolic forms that permeate their settings. Duncum (2003) has commented, "Since everyday life is now a forest of media signs, it is to be expected that artists are recycling media images" (p. 23). When working with student-artists in the development of their work, the concern is that the students make informed decisions within art processes that affect the production and cultural relevance of newly constructed images, which have the potential to penetrate back into the environment. To address this concern, art pedagogies should acknowledge students not only as "cultured subjects," but also recognize their "cultural performances, interpretations, assumptions, and backgrounds" that are enacted through communication technologies – aspects that may differ greatly when the student faces their peers in a physical location versus a mediated, virtual arena (Lai & Bell, 2004, p. 21-3).

While the development of students' artistic identities and skills are the goals for both on-ground and online visual arts education, unlike a face-to-face classroom in which students must gather in a central locale, and in many cases leave their communities to reside near a particular institution, the online classroom provides both a central, albeit virtual, location for art students to meet and experience intercultural perspectives as they remain in their communities. The latter changes the dynamic of time and space in which one receives arts training and performs everyday rituals. Within the context of community ties, students make connections in their artwork between the micro patterns of everyday life and the macro themes of society-at-large (Giovannella & Graf, 2010) that can be brought to the online classroom for evaluation by a multitude of peers, as opposed to receiving comments in studio by a select few. This hub of information affords dialogue from assorted perspectives that are based on varying circumstances ranging from social settings to gender differences to generational viewpoints, as we consider the non-traditional working adults that are part of an online cohort. The process of peer evaluation can be taken one step further, by opening group forums where, not only a variety of students from a single classroom can comment on a particular project, but multiple sections of the same course, or various levels of students, can provide comments, increasing feedback on a project as well as interconnections amongst the student body.

It is useful for online educators to be aware that technologies are not neutral, as they too, are embedded with symbolic forms (Ess, 2003). Educators must consider how technologies play a role in students' cultural performances online. However, the lack of attention to individuals' physical characteristics through mediated channels in the online classroom affords students a place where they can speak openly and feel like they have equal opportunities to develop their artistic voices (Akins, Check, & Riley, 2004; Maeroff 2003), because visual cues are more readily controlled through the text and images that students provide about themselves (Lai & Bell, 2004). The multi-vocal feedback students receive helps them understand the ways in which their use of symbolic forms communicates particular messages to others. At the same time, it teaches students how to write coherently about their artwork in response to constructive criticism (Lai, 2002). Oral and written communication skills will continue to be an asset in the post-Information Age, an era that is said to focus on conceptual models requiring persons who can understand, use, and produce symbolic thinking (Bryant, 2010, p. 45). Although some arts educators believe skill-based courses will translate better to the online learning environment than abstract ones (Lozar Glenn, 2007), it is this same environment that can provide for analytical and metaphorical thinking through a synthesis of visual arts discourse and interactive tools that engage with cultural objects and the communities in which they are created.

ORIGINALITY IN ART AND PRACTICE

In the visual arts, traditional/hand-made forms of art-making have been normalized and previously assumed to be the only pure methods for creating art (Knight et al., 2005), but as the boundaries between fine and popular arts diminish (Boughton, 2000), working digitally has become an accepted practice, as demonstrated by the multitude of arts and technology departments that have come into being in the last fifteen years. Hence, it appears to be an obvious step to migrate digital art courses in a traditional setting to an online classroom, where the tools used in the creation of digitally enhanced artworks would be embedded into the interfaces for instructional purposes. A less comprehensible step is moving traditional courses with traditional techniques, such as sculpture, to the digital classroom. But non-digital programs of art study have been incorporating computer technologies, such as digital looms in the textile arts and rapid prototype printers in sculpture that construct plastic models from 3D images. With the growing use of digital technologies in various art media, the distinctions between art techniques blend together into interdisciplinary genres, and the endless reproducibility of digital files (Ligon, 2010) continues to question the authenticity of art. Online assignments can take advantage of this hybridity in artistic practice by encouraging students to merge elements that are constructed by traditional methods with those that are digitally formulated. For instance, a project may start with a hand-drawn object that is scanned into the computer and overlaid with a wash of color using digital tools. Such an image could then be printed on art paper[5] and reworked, using more traditional materials like oil pastels, before it is scanned again for more digital augmentation, including existing materials or found objects. At each step, a student could post the progress of the piece, and when completed, have a record of the developmental process. But is the final piece a one-of-a-kind, traditional work of art or a reproducible, digital construction?

Walter Benjamin first inquired about art's authenticity in his essay *The Work of Art in the Age of Mechanical Reproduction*, when the mechanical process of photography was shown to capture the reality of everyday life more accurately than any painting made by hand. However, instead of competing with photography's realism, painting explored other expressive outlets based on abstract themes and styles that utilized Modernist design principles, such as the dynamic motion of Futurism and the flattened multiple perspectives of Cubism. What was introduced with the new mechanical process was the ability to create multiple reproductions of the same image without degradation. This capability to create multiples of visual representations is emblematic of mass-produced popular artworks that have garnered less value over time than one-of-a-kind works of art, because they lack a true original touched by the hand of a master artisan. In the contemporary art world, there is still a large market for one-of-a-kind art objects, but as digital technologies become established in art practices, the market has transformed to accommodate these additional types of artworks.

Teaching visual arts online will only encourage the continued transformation, as more artists leave art school with training in digital technologies being one of many skills available to them for producing artwork. Art acquisition paradigms will need to shift attention from acquiring works of a specific medium to works of a particular conceptual and cultural significance, regardless of the tools used to create them, because the prevalence of the computer in the visual arts will be inconsequential (Mayo, 2007). Thus, the online learning technologies that converge with visual arts pedagogies will be rightly positioned to educate artists who produce "novel ideas in cultural contexts" (Zimmerman, 2009, p. 386), versus those based on pure aesthetics of a particular medium. From an educational perspective, digital technologies for the online environment will afford the storage, documentation, and retrieval of vast quantities of visual images and artistic works that can be re-situated to discover new, expressive forms (Husarik, 2007).

DISCUSSION

In the past, Lai (2002) has found that "computer-enhanced and online learning are concepts that focus on the technological tools themselves and not on the specific ways such tools can be used to affect 'classrooms' and 'schedules'" (p. 35). Tool *uses* versus tool *affect* signals several ways in which the online environment has informed visual arts education and ways in which it has yet to inform visual arts education. The Internet's ability to grant instant access to catalogs of images and information blurs the line between popular and fine arts, because considerable numbers of images residing online are equally available for consumption via search engines. Thus, interactive tools are needed to bring a variety of imagery (as well as aural and kinetic materials) together in the virtual classroom in a malleable fashion that can be re-organized on a whim—creating a mashup of sorts—to facilitate class discussions about relevant themes and cultural implications embedded in visual objects. The narratives derived from such experiences could enable students to cultivate cultural sensitivities as a part of their artistic voices, while delivering artwork expressing local perspectives to larger audiences using skills adapted from the communicative channels of the online classroom, such as blogging. From this synthesis of learning and creating in a computer-enhanced environment, visual arts education online will eventually be as natural as an on-ground course with hands-on demonstrations, transcending the argument of what an authentic art practice and/or artwork could and should be.

REFERENCES

Akins, F., Check, E., & Riley, R. (2004). Technological lifelines: Virtual intimacies and distance learning. *Studies in Art Education, 46*(1), 34–47.

Bastos, F. (2010). A 21st-century art room: The remix of creativity and technology. *Art Education, 63*(1), 4–5.

Boughton, D. (2000). *The shaping of visual arts in education* (pp. 956–969). Routledge International Companion to Education.

Bryant, C. (2010). A 21st-century art room: The remix of creativity and technology. *Art Education, 63*(2), 43–48.

Buckley, W., & Smith, A. (2007). Application of multimedia technologies to enhance distance learning. *RE:view, 39*(2), 57–65. doi:10.3200/REVU.39.2.57-65

Buffington, M. L. (2008). New media art education. *Art Education, 61*(3), 36–41.

Carey, J. (2009). *Communication as Culture* (rev. ed.). New York, NY: Routledge.

DePew, K. M., & Lettner-Rust, H. (2009). Medaitising power: Distance learning interfaces, classroom epistemology, and the gaze. *Computers and Composition, 26*(3), 174–189. doi:10.1016/j.compcom.2009.05.002

Drummond, G. (2008). Success in online education. *Distance Learning, 5*(4), 43–48.

Duncum, P. (2003). The theories and practices of visual culture in art education. *Arts Education Policy Review, 105*(2), 19–25. doi:10.1080/10632910309603458

Dunn, P. (1996). More intergrated inter and art. *Art Education, 49*(6), 6–11. doi:10.2307/3193617

Ess, C. (2003). Liberal arts and distance education. *Arts and Humanities in Higher Education, 2*(2), 117–137. doi:10.1177/1474022203002002002

Giovannella, C., & Graf, S. (2010). Challenging technologies, rethinking pedagogy, being design-inspired: The grand challenge of this century. *eLearn Magazine.* Retrieved from http://www.elearnmag.org/ subpage.cfm? section= articles&article= 114-1

Gude, O. (2004). Postmodern principles: In search of a 21st century art education. *Art Education, 57*(1), 6–14.

Husarik, S. (2007). The impact of digitalization upon the arts and humanities. *International Journal of the Humanities, 5*(7), 119–126.

Knight, W. B., Keifer-Boyd, K., & Amburgy, P. M. (2005). Visual culture explorations: Un/becoming art educators. *Journal of Social Theory in Art Education, 25*, 255–277.

Knowles, W., & Kerkman, D. (2007). An investigation of students' attitude and motivation toward online learning. *InSight: A Journal of Scholarly Teaching, 52*, 70-80.

Lai, A. (2002). From classrooms to chatrooms: Virtualizing art education. *Art Education, 55*(4), 33–39. doi:10.2307/3193966

Lai, A., & Ball, E. L. (2004). Students online as cultured subjects: Prolegmena to researching multicultural arts courses on the Web. *Studies in Art Education, 46*(1), 20–33.

Ligon, S. (2010). *Digital art revolution.* New York, NY: Watson-Guptill Publications.

Lozar Glenn, J. M. (2007). Mastering the science and the art of teaching online. *Business Education Forum, 61*(3), 10–15.

Maeroff, G. I. (2003). *A classroom of one: How online learning is changing our schools and colleges*. New York, NY: Palgrave.

Mayo, S. (2007). Implications for art education in the third millennium: Art technology integration. *Art Education, 60*(3), 45–51.

O'Hara, K., & Stevens, D. (2006). *Inequality.com: Power, poverty and the digital divide*. Oxford, UK: Oneworld Publications.

Overby, A. (2009). The new conversation: Using weblogs for reflective practice in the studio art classroom. *Art Education, 62*(4), 18–24.

Roland, C. (2010). Preparing art teachers to teach in a new media landscape. *Art Education, 63*(1), 17–24.

Schroeder-Moreno, M. S. (2010). Enhancing active and interactive learning online: Lessons learned from an online introductory agroecology course. *NACTA Journal*, *54*(1), 21–30.

Semchuk, S., & Tien, L. (2004). Telling story! Voice in photography: An online visual art studies program evaluation. *International Review of Research in Open and Distance Learning*, *5*(3). Retrieved from http://www.irrodl.org/ index.php/ irrodl/ article/ view/ 206.

Sloan Consortium. (2010). *Learning on demand: Online education in the United States, 2009* (pp. 1–29). Babson Survey Research Group.

Stewart, M. (n.d.). Teaching art at a distance. *eLearn Magazine*. Retrieved from http://www.elearnmag.org/ subpage.cfm? section=case_studies& article=39-1

Sull, E. C. (2008). *Teaching online with Errol* (pp. 33–38). Online Classroom.

Zimmerman, E. (2009). Reconceptualizing the role of creativity in art education theory and practice. *Studies in Art Education*, *50*(4), 382–399.

KEY TERMS AND DEFINITIONS

Digital (Mechanical) Reproduction: Technological processes that produce multiple versions of a file, image, etc. without degradation leading to the question (in art) as to which is the authentic original.

Graphics Software: applications used in art and communication processes, these tools are designed to function in particular ways and thus influence knowledge creation.

Hypermediation: The excessive focus on communication channels.

Mashup: A collage or derivative work created from multiple sources; a concept that could be used for flexible classroom interfaces.

Remediation: Media used within other media, e.g. a photograph of a painting that is viewed through the lens of the Internet.

Revision Control System: RCS is a collaboration mechanism that would operate as part of the content management system handling the materials in the online classroom.

Screencasting: Allows users to capture computer desktop actions in a static screenshot or video along with audio narration.

ENDNOTES

1 Painter™ by Corel® is an example of graphics software with recording and playback features.

2 Github (http://github.com/) is an online open-source revision control system that includes hosting, while Subversion (http://subversion.apache.org/) is an open-source revision control system application that can be installed on one's server.

3 Online whiteboard services include: Dabbleboard (http://www.dabbleboard.com/) and Skbrl (http://www.skrbl.com). Class Live Pro is a similar whiteboard service built into online classrooms managed by eCollege. Google's 3d Warehouse also allows for asynchronous collaboration using its Sketchup tool for three-dimensional applications, which could be used for Interior Design and Sculpture.

4 TechSmith's Jing® and Camtasia® software are screencasting applications with various levels of recording and editing capabilities.

5 Artists can print on the variety of surfaces with the help of Inkaid (http://www.inkaid1.com/), which is a coating painted on to surfaces, such as watercolor paper, ceramic tiles, etc. that allows ink jet images to adhere properly.

Section 2
Social Implications

Chapter 8
Promoting Global Citizenship through Intercultural Exchange Using Technology:
The Travel Buddies Project

Laurie A. Henry
University of Kentucky, USA

Clarisse O. Lima
Educational Technology Consultant, Brazil

ABSTRACT

This chapter presents a critical instance case study that describes the implementation of an international, telecollaborative project between elementary level students in rural Kentucky and those located in Rio de Janeiro. Learning activities focused on the development of cultural knowledge and understanding, with the main goal of increasing the students' global citizenship characteristics by comparing cultural backgrounds with an emphasis on cultural similarities and differences between the two groups. This was accomplished through the Travel Buddy Project, a new pedagogical approach to learning that combines blog exchanges with photographic documentation couched in culturally oriented lessons.

INTRODUCTION

The purpose of this chapter is to present a critical instance case study (Davey, 1991) in which the researchers were interested in examining the implementation of an innovative and engaging way to establish international, cultural exchanges through the use of Internet-based communication tools. To this end, we studied two groups of students engaged in a Travel Buddies Project (see http://www.globalschoolnet.org/programs/travelbuddies/). This online, international exchange had two main objectives: 1) to provide an

DOI: 10.4018/978-1-61350-077-4.ch008

opportunity for young children (ages 6-8) to begin to understand and appreciate diverse cultures while developing characteristics of global citizens, and 2) to introduce young children to Internet-based, information and communication technologies (ICTs) that would help them develop new literacies for success in the 21st century.

This chapter highlights a telecollaborative exchange between early elementary level students in a small, rural Kentucky school district (population of approximately 1,100) and those in a school located in Rio de Janeiro, a large metropolitan city in Brazil (population of approximately 6.3 million). The exchange focused on the development of cultural knowledge and understanding through a variety of curricula-based activities with the main goal of increasing the students' global citizenship characteristics by developing an awareness of cultural similarities and differences between the groups. This was accomplished through a Travel Buddy Project blog exchange, which included photographic documentation, culturally oriented lessons, and introduced students to basic practices of computer-mediated communication.

We begin by providing a statement of the problem followed by a description of the two theoretical frameworks, new literacies and global citizenship, which supported and guided this cultural exchange. Next, an explanation of the format for the telecollaborative exchange, including descriptions of the participating classrooms is provided. Then we illustrate the curricula-based artifacts that were collected throughout the four-month collaboration. Various instructional lessons are highlighted that allowed the students to learn about similarities and differences between themselves and their international partners. Lastly, we discuss implications for the classroom related to teaching and learning at the early elementary level.

STATEMENT OF THE PROBLEM

The Internet is rapidly becoming the defining technology for today's youth (New Literacies Research Team, 2007). The most recent Pew Internet and American Life Project report shows that among teens (aged 12-17) in the United States, 93 percent go online and 73 percent use social networking sites (Lenhart, Purcell, Smith, & Zickuhr, 2010). However, this trend is not isolated to the United States. Internet access and use worldwide is exploding with a 444 percent increase in Internet use worldwide over the past decade (Internet World Stats, 2011). Looking specifically at the two countries of interest for this international exchange, the United States shows Internet growth from 2000-2011 at 156.9 percent with a population penetration rate of 78.3 percent; Internet usage in Brazil has grown by an astounding 1,419.6 percent during this same time period with a population penetration of 37.4 percent. The Nielsen report (2009) shows that teenagers (ages 13-19) in the United States are ranked third worldwide with monthly Internet use at 24 hours and 54 minutes, and Brazilian teens are the world's top Internet users logging 43 hours and 50 minutes monthly. Clearly, Internet use worldwide is on the rise, but what about Internet usage among younger populations?

When we focus specifically on children, including younger aged children, in relation to Internet use statistics, a similar trend is found. In a recent Kaiser Family Foundation Study that surveyed children in the Unites States aged 8 to 18, in which 9 percent of the respondents were third graders, 70 percent indicated that they used a computer on a daily basis (Rideout, Foehr, & Roberts, 2010). This report also shows that 8-18 year olds use the Internet an average of 1 hour and 29 minutes per day. Similarly, the Norton Online Living Report (2009) that included children aged 8-17 in 12 countries indicated that children use the Internet an average of 39 hours per month.

Children in the United States are slightly higher than this average at 42 hours per month, and children in Brazil spend the most time online of the 12 countries studied at 70 hours per month. Children in Brazil were viewed as the "most open" online with their parents with 70 percent including a parent on their buddy list, 80 percent in their email contacts, and 60 percent who connect with parents through social networking sites. Brazilian children also socialize online more than children in other countries at a rate of 13 hours per week. Children in the United States ranked highest for the number of online friends at 83, and they reportedly use texting at a much higher rate (10 hours per week) than the global average of 4 hours per week. Finally, a Nielsen Online study reported, "children aged 2-11 comprised nearly 16 million, or 9.5 percent, of the active online universe" (The Nielsen company, 2009, para. 1). This report shows that in five years, the number of children in this age category who are online has increased by 18 percent. Further, among this age group, the amount of time spent online has increased from an average of 7 hours per week in 2004 to 11 hours per week in 2009. Thus, this is an important age group of Internet users to observe.

Even with these growing rates of Internet usage, there is still an issue related to Internet access in rural areas and for those who are socio-economically disadvantaged, commonly referred to as the digital divide (Anderson, Bikson, Law & Mitchell, 1995; DiMaggio, Hargittai, Celest, & Shafter, 2004; Norris, 2001). As illustrated in the above statistics, there are differential patterns of Internet access worldwide as shown by the penetration rates of the Internet in the United States compared to Brazil. However, these differences are much smaller when we look at access to the Internet in schools. In the United States, by the year 2005 it was reported that 94 percent of instructional classrooms in public schools in the country had Internet access. That number has grown to 98 percent today. Brazil shows a similar trend in which the Ministry of Education began a program in 2008 to "provide broadband Internet connectivity to all 56,000 federal schools by 2010" (Glaser, 2009, p. 18). Thus, more and more children have access to the Internet from school.

Access to the Internet provides individuals with entry into a seemingly unlimited amount of information flow about the most exotic topics with a click of the mouse or the point of one's finger. But how this information is processed and used is an important aspect for educators to teach. A secondary level digital divide presented by DiMaggio and Hargittai (2001) focuses on how the Internet is being used by different populations of people. Henry (2008) built on this earlier work to present a tertiary level digital divide that focuses on three levels, including issues related to Internet access, Internet use, and Internet reading skill. It is at this third level that Henry looked specifically at how schools may or may not prepare children for success in online spaces using specific lessons focused on Internet-based reading. The importance of education related to the Internet becomes fundamental for providing the skills people must have to retrieve information and the ability to use it in order to participate, take advantage of, and be creative in new technological environments supported by increasing levels of Internet access. With Internet access more prevalent in schools than ever before, it becomes important to consider how teachers are using the Internet in their classrooms and how technology can be used to support student learning.

THEORETICAL FRAMEWORK

New Literacies Perspective

With the increasing access and use of the Internet previously described, an important implication for classroom instruction is to determine how classroom teachers can help students develop the skills necessary for reading, writing, and communicating with digital media and Web 2.0 technologies. As more teachers turn to Internet-based technologies to enhance teaching and learning, it becomes apparent that additional support for these technologies is required as many teachers are not always comfortable with or adept at using these digital tools (Zhao & Bryant, 2006).

Several instructional models have been introduced over the past two decades that help scaffold teachers (and students) in the development of these important new literacies, or 21st century skills. They include WebQuests, Internet Scavenger Hunts, and Internet Projects to name a few. These instructional models often combine a variety of web-based tools such as search engines for inquiry-based learning, wikis, blogs, and information-based web pages, all of which help with the development of online reading, writing, and communicating. However, many of these models do not address one critically important component of new literacies, which the Partnership for 21st Century Skills (P21, 2004) identifies as "global awareness" or the development of cultural awareness and globalized perspectives. A globalized perspective for language and literacy (Castek, Leu, Coiro, Gort, Henry, & Lima, 2007) considers how to support students' development of 21st century literacy skills "by drawing on their experiences in both language and technology as a stepping-stone to a successful future in a global society" (p. 132).

There are many research reports that highlight how the integration of technology can have a positive impact on teaching and learning, however they do not indicate how the use of electronic medium might help teachers use technology to broaden their global perspectives. With the increased diversity of school populations and increased emphases to improve cultural understanding across nations, teachers must develop expertise to help their students become more culturally aware and globally minded (Cushner & Brennan, 2007; Friedman, 2005; Gay, 2002; Hollins & Guzman, 2005). According to Merryfield (2002), "global educators help students explore other cultures through literature, history, news, and websites from other parts of the world [that include an ability to] write about and understand a new, more comprehensive history that includes diverse perspectives and knowledge bases" (p. 19). Accessing new technologies and knowing how to use them appropriately is an important focus of education today in order to prepare global citizens. Citizenship in a "flat world" requires different skill sets, increased knowledge of the world, and diverse perspectives that were not required by earlier generations preparing to enter an industrialized workplace (Friedman, 2005; Gates, 2007; Heyl & McCarthy, 2003; P21, 2004). Thus, the skills and strategies related to the new literacies of the Internet, which include searching for and locating information, critically evaluating information, synthesizing information across multimodal text formats, and effectively communicating with various Internet based tools (Leu, Kinzer, Coiro, & Cammack 2004), become an increasingly important component of educating children for their role in an information based, global economy.

Global Citizenship

The emergence of ICTs is seen as one of the leading forces of globalization, linking people through large networked infrastructures that have flattened the world (Hajela, 2005). However, it is argued that the consequences brought by globalization to the world can be viewed as both positive and negative. At the same time that the world is getting smaller and becoming widely connected by the Internet, some nations and groups of people are being excluded from this dominant system (see Internet World Stats, 2011). Concurrently, in an interdependent world, the concepts of education and citizenship must be redefined in order to lessen the exclusive divides. We need to shift our focus to not only equip individuals with the necessary new knowledge and skills needed to be active participants in the globalized society, but also to teach individuals how to be culturally responsive and tolerant to others' differences; in essence, how to become global citizens.

The new globalized era leads to new concepts and new practices, and education is at the center of a fair and inclusive globalization movement as a core element for development at the individual, community, and society levels (Bakhtiari & Shajar, 2005). The implementation of new educational practices and strategies that are aligned with the requirements of the ever changing world we live in are fundamental for advancing the progress of a country, as well as offering opportunities for its people to participate actively in the society as conscious, powerful, and responsible citizens. Additionally, two of the main resources in the world today that provide a competitive edge are knowledge and information (Boisot, 2002); therefore, countries need to build strong technological infrastructures to process and use information to assist in their development. For example, a range of networks link the most remote areas in Africa to big centers such as Tokyo and New York, thus providing individuals in these areas that have been geographically isolated an opportunity to participate in the global economy for the first time in history (see the Fair Trade Federation at http://www.fairtradefederation.org and Baskets of Africa at http://www.basketsfromafrica.com).

Lapayese (2003) acknowledges that a timely topic within the context of globalization is the understanding of how nations and schools around the globe are preparing young students to assume the role of global citizen. Banks (2004) argues that citizenship education needs to be transformed due to the increasing racial, ethnic, cultural, linguistic, and religious diversity in nation-states throughout the world. This author reiterates that citizenship education should assist students to "develop global identifications and a deep understanding of the need to take action as citizens of the global community to help solve the world's difficult global problems, such as conflict and war, the AIDS/HIV epidemic, global warming, and world poverty" (Banks, 2003, para. 11). Similarly, the Partnership for 21st Century Skills (P21) highlights interdisciplinary themes related to civic literacy, health literacy, environmental literacy, and economic literacy in which all of these themes include an emphasis on global awareness (P21, 2004).

According to some authors, in order to adjust to the globalized times in which we live, the concept of citizenship needs to overcome its traditional and political notions and should aggregate more affective and loyal definitions as well as multiple identities presented by the multiculturalism and diversity seen in the world today (Munck, 2005; McIntosh, 2005; Noddings, 2005). Diversity encompasses a desirable mix of people representing racial, ethnic, cultural, and religious differences and, according to Noddings (2005), "human life is certainly enriched by the existence of different cultural practices, and cultural practices are influenced by race" (p. 13). In this scenario, a new definition of global citizenship is demanded, one that takes into account the new societies that are being reconfigured because of globalization, one in which citizens' rights prevail over every form of social exclusion (Torres, 1998). Global citizens need to be aware and concerned with what is happening not only in their nation and geographic region but throughout the world as well. Global citizens also need to develop a voice to promote social and economic justice for themselves and their fellow countrymen by demonstrating care and respect for other's welfare.

According to Lima (2006), a global citizen should be defined as a person who possesses the following characteristics:

- Awareness and concern about globalization and its impacts on individuals and in the world;
- Awareness about rights and responsibilities of a citizen;
- Meaningful and socially valued use of ICTs;

- Basic skills needed to fully operate and participate in a globalized society (including the new literacies needed for accessing Internet-based information).

As was mentioned previously, ICTs constitute a fundamental tool that, when mastered, can allow individuals to live and participate in their societies as global citizens. With the continued flattening of the world, acquiring the skills to operate and manipulate ICTs effectively and efficiently is essential for the development of a nation and its individuals (Boisot, 2002). Thus, an additional shift for educators is the need to teach the new literacies of online reading, writing, and communicating, which focus on the fundamental skills and strategies that are necessary for individuals to take full advantage of the information flow available at their fingertips.

INTERCULTURAL EXCHANGE

Our intercultural exchange was based on the original Travel Buddies Project, as described below. The buddies were part of a cultural exchange between students in the United States and Brazil in which toy mascots, or representatives, were sent to a host location. As guests, the mascots participated in activities with the children both inside and outside of school. Students kept in touch throughout the exchange by recording events and activities using blog posts, photographs, email exchanges, and diary/journal entries. At the end of the exchange, the buddies were returned home to share their adventures and what they learned about their host location with the original group of children.

The Travel Buddies Project History

The original Travel Buddies Project began with an exchange between groups of elementary level students in Clayton, Missouri and Bonita, California in the United States and Monash, Canberra in Australia. This classic exchange included the documentation of the travels of two small, stuffed animal mascots (i.e. Woodsy Woodchuck and Al Rogers the koala), with the exchange including photographs and electronic journal entries exchanged via email communications that were posted and shared publicly on the Global SchoolNet Foundation website (http://www.globalschoolnet.org/GSH/pr). This successful project quickly led to other collaborative projects crisscrossing the globe. Currently, there are 2,573 archived projects and approximately 30 "in progress" exchanges hosted in the Project Registry of the Global SchoolNet Foundation website[1]. Other communities of teachers such as oz-Teachernet (http://www.oz-teachernet.edu.au) adopted the same idea and implemented the Travel Buddies Project. This community has been working with and for teachers since 1995 and is managed and maintained by academics at the Queensland University of Technology in Australia. Their online projects offer essential national and international networks to teachers and students. The project that inspired the authors to run the same type of project was the one from Keith, the Aussie koala that completed an exciting trip from Australia to Japan.

Classroom Descriptions for U.S.-Brazil Exchange

The school in Kentucky was located in a very small, rural, socioeconomically depressed community. Approximately 78 percent of the families live below the poverty line. This pre-kindergarten through 8th grade school with approximately 100 students is located in an independent school district in which graduates attend a county high school in a neighboring school district. This rural community can be described as monolingual and non-diverse in which approximately 98 percent of the inhabitants are Caucasian, native English speakers. There were a total of 20 students involved in the exchange from Kentucky, which included 7 first grade students and 13 second grade students.

The school in Rio de Janeiro was a private catholic K-12 school in the heart of the city within a wealthy neighborhood that borders the biggest *favela* (i.e. slum or shanty town) in the country called Rocinha. The school has about 1,200 students. The families that compose the school's community are considered part of the Brazilian elite coming from privileged backgrounds. All students begin learning English in Kindergarten. There were 112 students participating in the project, which included 42 first grade and 70 second grade students.

U.S.-Brazil Travel Buddies Exchange

Each class of students chose a stuffed mascot that was a representative of the local area or connected to the school curricula in some way. First grade students in Kentucky selected a horse, named *Chocolate Cocoa Derby*, which represented the rich equestrian history and world-renowned Kentucky Derby horse race. Second graders in Kentucky chose a University of Kentucky (UK) "cheer bear" basketball mascot (named *Scratch*) that played an audio recording of the UK fight song when squeezed, to show their loyalty as UK basketball fans. In exchange, first graders in Rio de Janeiro selected *Little Wolf*, which was a hand puppet used in the classroom to study the fairy tale Little Red Riding Hood. The second grade students in Rio de Janeiro selected a stuffed toucan, *Tuc-Tuc*, since it is a native bird of Brazil.

There were four classroom blogs set up on Blogspot.com, one for each participating classroom. This provided a virtual space fore documenting the exchange, by posting photographs and journal-style entries that were crafted as a shared writing experience, with the teacher and students composing the text together as a group (Button, Johnson, & Furgerson, 1996). Entries from the American classrooms were posted in English, whereas those from the Brazilian classrooms were posted in Portuguese with an English translation. Comments were made in response to posts between the partner classes. Due to the small number of students in Kentucky, daily pencil and paper entries were made by individual students in a journal notebook to document activities both inside and outside the classroom as the children took turns bringing the mascots home with them. Alternatively, the large number of Brazilian students required project activities in Rio de Janeiro to be restricted to the school and classroom, including field trip excursions away from the school setting. On occasion, the mascots visiting Brazil had an opportunity to experience life outside the classroom. For example, *Chocolate* went with a first grade class to a student's birthday party to experience and document cultural traditions outside the school setting.

The duration of the exchange was four months beginning in April 2009 and ending in July 2009, when the mascots were returned to their countries of origin. It should be communicated that the exchanges in this project were made in person by the two project facilitators in each country. Due to these individuals' involvement in national and international research conferences, they were able to travel with the mascots to make the two exchanges, one at the beginning of the project and one at the end of the project. The first exchange took place in the United States (see Figure 1 for the Brazilian mascots' trip to Kentucky), and the second exchange was completed at an international conference site in Porto, Portugal, which provided additional opportunities for lessons related to world geography.

However, this in-person exchange is not the norm for most Travel Buddies Projects. Most telecollaborative exchanges utilize international postal services (e.g. United Parcel Service or Federal Express Worldwide Services) to perform the exchange of the mascots.

Figure 1. Tuc-tuc and Little Wolf traveling to Kentucky

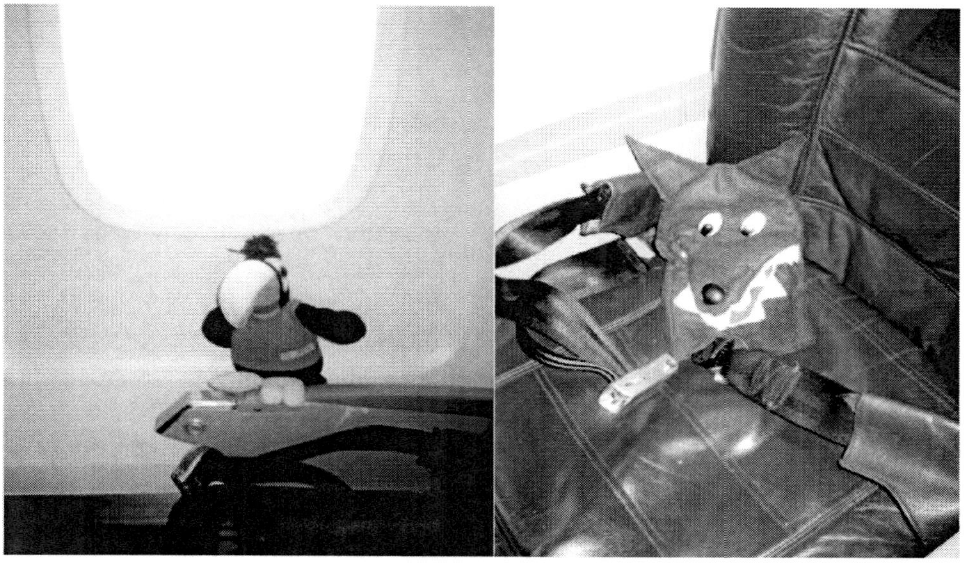

ARTIFACTS

Blog Posts and Photographic Documentation

Blog posts were used as the central mode of communication between the classes and often initiated impromptu lessons based on the content that was posted by the partner classes. All the activities done by the classes and the mascots that were documented by the photographs and selected to be posted on the blog were done so according to safety and ethical rules established by each school in order to protect the students' identities. For example, pictures depicting only one student facing the camera were avoided, and group shots were preferred. In Kentucky, parental permission was obtained for the students to have their photos posted on the Internet, following the school policy. The pictures that were taken depicted the moments of collaboration and exploration among the students and their visiting travel buddy. The students were eager to share all that they were learning with their mascot friend as a way to inform their foreign colleagues what they were experiencing in school.

An initial post was created that included a welcome message and a photograph of the exterior of the school building in each location. Then, in a subsequent post, the students in Brazil introduced their buddy during a technology class. They had a lot of fun pretending they were teaching their mascot how to work with computers, and they were happy to present it to their new friends from the northern hemisphere. When the Brazilian students saw the first picture of their peers from Kentucky at school, they were very curious to know why they did not wear a uniform, since in Brazil most of the schools, including theirs, have a particular uniform, usually with the school name and emblem. These became the first of many question-and-answer exchanges in which the students learned about many differences between their schools. One big difference the children noted was related to the school day. The American 1st and 2nd graders spend both the morning and afternoon at school, while in Brazil the same class goes to school during one time period only, either morning or afternoon, but not both.

Classroom Lessons

This intercultural, telecollaborative exchange began with an introductory lesson that provided an explanation of the Travel Buddies Project and a basic geography lesson that oriented students from both countries to the location of their new "friends" who were located on a different continent. In Brazil, the technology teacher, who is fluent in English, coordinated the project, taking advantage of the computer classes to introduce and develop most of the activities related to the travel buddies. Although the technology teacher took the lead, the classroom teachers were also involved in the project; taking the mascot to their classes and having it become part of their daily activities, including English and Math classes, recess, play time, and field trips. The activities described below demonstrate that the collaboration between the computer teacher and the classroom teachers resulted in the use of technology as an integral part of many of the lessons.

In Kentucky, the classroom teachers facilitated the instructional activities within their own classrooms with little involvement from other teachers in the school building, although they did share some of the activities from the project with the other teachers during planning times and staff meetings. One of the first activities that the American students completed were self-portraits, drawn and colored by the students, each including an "I like..." statement (e.g. I like chocolate ice cream; I like to play soccer; etc.). These self-portraits were delivered to the children in Brazil as part of the initial mascot exchange approximately two weeks after the project began. The Brazilian second graders also completed the "I like ..." activity, but unlike the U.S. students, theirs was done directly on the computer, as was the case with most of the activities within the project. Because there were such large numbers of students in Rio, their self-portraits were shared electronically with their friends in Kentucky. The topics of this activity were shared between the students so they could learn more about each other's personal interests and identify similarities between the groups of students.

Social Studies and Geography Connections

Students in Kentucky engaged in several lessons focused on geography. They mapped out the travel routes that the mascots took, which included visits to California, Ohio, and Portugal before arriving back to their homeland. Questions erupted about how far away Brazil was, whether you could drive to Rio de Janeiro from Kentucky, and the size of Brazil compared to the United States. Students were surprised to learn that these two countries were relatively the same size geographically.

There was also an emphasis on lessons that taught the children about historic landmarks, local points of interest, and cultural celebrations (e.g. Easter, Carnival, and the Thunder Over Louisville Festival). One of these activities included student rendered computer drawings of the Brazilian mascot presenting places the students liked to go in Rio de Janeiro using the software program Microsoft Paint© (see Figure 2 for a blog post with the drawings).

Another computer-based graphic activity had the Brazilian students take the American mascots on virtual field trips to visit some of the most famous tourist attractions in their city (see Figures 3a and 3b). The students used Microsoft PowerPoint© to superimpose photos of the mascots onto images of the city's soccer stadium and famous statue of Jesus Christ that overlooks the city. Presentation of these tourist attractions to the students in Kentucky quickly identified similarities they shared with their Brazilian peers. Many of the students became excited when they realized they had a common sport that they all enjoyed, and several students commented on the statue of Jesus. Upon seeing this image, one first grade student in Kentucky jumped from her chair and emphatically declared "I love Jesus!"

Figure 2. Electronic drawings of Little Wolf visiting local attractions in Rio de Janeiro

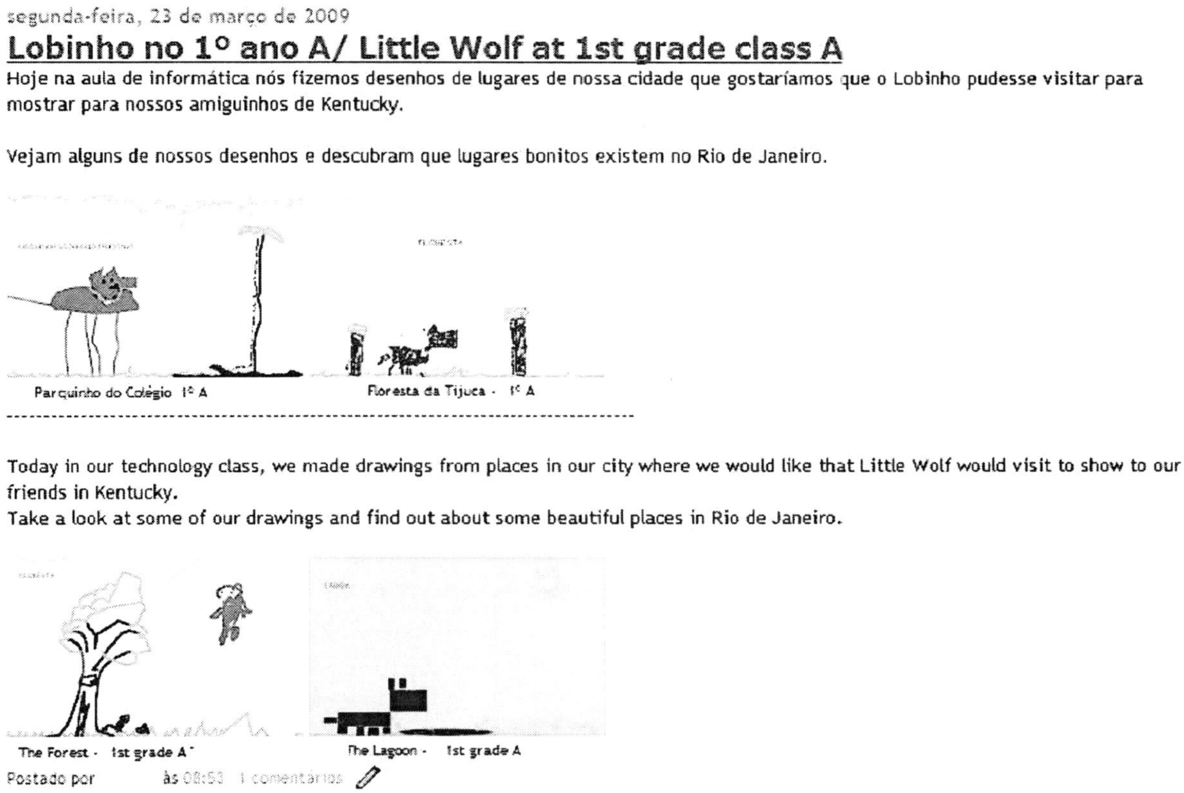
segunda-feira, 23 de março de 2009

Lobinho no 1º ano A/ Little Wolf at 1st grade class A

Hoje na aula de informática nós fizemos desenhos de lugares de nossa cidade que gostaríamos que o Lobinho pudesse visitar para mostrar para nossos amiguinhos de Kentucky.

Vejam alguns de nossos desenhos e descubram que lugares bonitos existem no Rio de Janeiro.

Parquinho do Colégio 1º A   Floresta da Tijuca - 1º A

Today in our technology class, we made drawings from places in our city where we would like that Little Wolf would visit to show to our friends in Kentucky.
Take a look at some of our drawings and find out about some beautiful places in Rio de Janeiro.

The Forest - 1st grade A   The Lagoon - 1st grade A

Postado por às 08:53 1 comentários

A small internationalization grant funded by the University of Kentucky provided an opportunity for the technology teacher from Brazil to visit the school in Kentucky. During the visit she showed some materials from Brazil such as post cards and pictures of the Brazilian students interacting with the travel buddies. The students were curious to know about the trip, how many hours the flight lasted, etc. The visiting teacher used a world map as a support to show the route she had taken to get from Rio de Janeiro to Kentucky (see Figure 4).

The teachers in Kentucky provided their Brazilian visitor with brochures about many of the attractions in their state (e.g. Dinosaur's Cave, The Kentucky Derby and opening day celebration known as Thunder Over Louisville, the amusement park Six Flags, and other local attractions) that they collected from a visitor's center. These brochures were taken back to Brazil and shown to the students, who identified many connections to points of interest in their own country.

Throughout the exchange, the students continued learning about cultural similarities and differences. At Easter time, Brazilian students discovered that in the U.S. Easter is not celebrated as such a big holiday as it is in Brazil, although both countries do share customs such as the Easter bunny and egg hunt as representative traditions of Easter. Students in Kentucky were surprised to learn that Carnival in Brazil is a national holiday, and that schools are closed during this festive time of year. Other connections that were made included topics related to religion (i.e. recognition of the famous Jesus Christ statue in Rio de Janeiro), popular culture (e.g. Disney *Cars*), favorite books (e.g. *The Grouchy Ladybug* by Eric Carle), and common sports activities (e.g. soccer).

Figure 3. Graphic representations of mascots visiting local attractions a. Chocolate at Maracanã (soccer field), b. Scratch at the Christ statue

Science Learning

Weather was a very interesting topic discussed by the new friends. The American class reported one day that they woke up with snow on the ground and that in the afternoon it was gone. Most of the students from Rio had never seen snow and explained that in their city it never snows, since the temperature is usually around 70°F, even during winter and when it rains. The Brazilian children were intrigued that although they were having very hot days at the time, their friends in the U.S. were facing a lot of cold temperatures and snow. This resulted in an impromptu lesson by the teacher in which it was explained that due to their geographic location, one country is in the northern hemisphere and the other is in the southern hemisphere, the seasons were different: when it was winter in the North, it was summer in the South; spring in the North meant it was fall in the South; and vice versa. This topic came up again when students in Kentucky were preparing for their summer break, yet students in Rio were in the early stages of their school year cycle.

Additional science lessons focused on the Amazon Rainforest and the exotic animals that are found there. The American students quickly realized that there was a mural painted on the wall in their cafeteria of a rain forest that showed many of these animals, including a toucan, spider monkeys, and iguanas. It was fun to see the excitement on their faces when they made this connection between their rural school in Kentucky to their faraway friends. The Brazilian mascots visited a nature center in Kentucky to learn about the animals that are native to Kentucky, like the black bear seen in Figure 5 below. This event was documented through blog posts with photographs and descriptions of the animals.

Subsequent posts emphasized additional curricular activities that highlighted similarities between the groups of students. For example, Figure 6 depicts students in Rio de Janeiro learning about gardening, germination, and the cycle of plants. Students in Kentucky connected this content to the strong agricultural base in their local community, where farming is one of the largest industries.

Developing Characteristics of a Global Citizen

As previously stated, one of the characteristics of a global citizen is developing an awareness and concern about globalization and its impacts on individuals and in the world (Lima, 2006). We found that participation in this project did bring a level of awareness and concern for individuals located in a different part of the world for this

Figure 4. Brazilian technology teacher visits students in Kentucky

very young group of children. For example, at one point, there was a blog post that reported *Little Wolf* had suffered an accident while at home with one of the students in Kentucky. The Brazilian children were a bit nervous and concerned about their little friend and did not rest easy until they received news that their mascot was doing well and recovering. The mascot was damaged when he lost an eye, and the teacher had to find him a new set of eyes in order to repair him. The children were relieved when they finally saw new photos of *Little Wolf* visiting the nature center in good repair.

Similarly, the end of the school year for the American students, in June, brought feelings of sadness and disappointment to the children as they realized their exchanges would come to an end. The end of the year for the students in Kentucky occurred almost simultaneously to the Brazilians' end of the first semester, in July, when both groups of students would have extended time away from school. Thus, this was a natural breaking point to end the Travel Buddies Project. In one final exchange, the students went to the blog to say goodbye, to ask their new friends about what they were doing during their vacation time, and to share what they usually do for fun when on vacation themselves. Students in both locations also said goodbye to their foreign visitors as they prepared for the long journey back to their native countries.

Teacher Responses

The teachers, both in the U.S. and in Brazil, were very involved in the project at the beginning. The Travel Buddies exchange started with two classroom teachers in Kentucky, but the second grade teacher struggled with many other school responsibilities and was not able to find time to

Figure 5. Little Wolf and Tuc-Tuc visiting nature center

keep up with the blog posts. However, the first grade teacher stepped up and took on a leadership role for the project, helping to coordinate the various activities for both the first and second grade classrooms. At first, she struggled with the technology, because she had never used a blog before, and did not have access to a digital camera on a regular basis, as there was only one camera at the school for all of the teachers to share. However, with some guidance, she soon realized that she could locate many photographic images of landmarks and points of interest in Kentucky on the Internet. She learned how to save images and post them to the blog space along with comments by the students. Once she became accustomed to using the blog, she was able to make it a more integral part of her classroom lessons. She stated that The Travel Buddy Project had a positive impact on her teaching. This teacher reported that she would participate in a Flat Stanley Project exchange (see http://www.flatstanleyproject.com) during the following school year because of the positive impact the exchange had on her students. Flat Stanley is a similar type of exchange that documents the travels of a Flat Stanley mascot that is mailed to a variety of locations, collecting artifacts along the way.

In Brazil, the technology teacher counted on the collaboration of the classroom teachers since she had only one class with the students per week. The teachers embraced the project's ideas and contributed with materials such as pictures of children with the mascots in their classroom activities, during field trips, and at recess time. They also had informal conversations with the technology teacher sharing their opinions about the project and reporting how the children were responding to it.

Challenges

The implementation of this project was met with several challenges aside from the use of new technology tools previously mentioned. First and foremost, we quickly realized that differences in the school calendars for each location introduced some hurdles. As one school year was winding down for the summer break, the other school was still in full swing. This resulted in a reduction in the amount of time that the exchange took place, because in the U.S. there would be new students following the summer break as those taking part in the exchange would move on to the next grade level. Additionally, the teacher in Kentucky was required to administer the standardized state assessment for two weeks during the telecollaborative exchange. This resulted in two weeks during which there was little to no activity posted on the blog. The students in Rio were left eager to learn more about their American friends for a

Figure 6. Blog post from second graders in Rio de Janeiro

Aula na Horta do Colégio / School Garden Class

Tuc-Tuc acompanhou a turma do 2° ano B na horta do colégio. Nessa aula, nós aprendemos sobre o ciclo das plantas. Nós começamos preparando a terra, e nessa hora vemos muitas minhocas! Depois plantamos as sementes e a cada nova aula vamos regando e observando a germinação. Quando a plantinha está grande nós colhemos e usamos como tempero na comida.
É uma experiência muito interessante e o Tuc-Tuc se esbaldou brincando na terra.

Tuc-Tuc joined the 2nd graders from class B at the school's garden. In this class, we learn about the cicle of the plants. We begin preparing the soil, and at this time we find lots of worms! Then we plant the seeds and in each class we water them and observe its germination. When the plant grows we use it as seasoning in our food. This is a very interesting experience and Tuc-Tuc had a blast!

period of time in which no new information was exchanged. This resulted in disappointment as well as a need to rekindle interest once the assessment period was over.

We also realized that language differences could be a barrier in this type of international exchange. Although the students in Brazil begin learning English at an early age, it is critically important to have at least one member of the exchange who is fluent in both languages to assist with the translation and explanation of the content that is presented. For the exchange described here, the computer teacher in the school in Rio de Janeiro was able to facilitate the translation from Portuguese to English and vice versa. Without a fluent English speaker to assist with this translation, the project would not have been as successful because the depth of knowledge about various topics would have been lost.

The final challenge that we would like to present is related to expectations for the exchange. It is important to plan the activities based on common topics from the curricula of the grade levels involved as well as to establish a basic timeline for the frequency that blog posts will be done for each classroom. Without these guidelines, there may be times when the students are eager to view the blog and see what new information was presented only to find out that a new post or comment had not been made, which turns eagerness into disappointment as previously described. In addition, having a list of topics for lessons ahead of time would make their integration more thorough and ultimately more meaningful for the students, thus having a greater positive impact on student learning.

IMPLICATIONS FOR LEARNING

Development of Cultural Knowledge and Diverse Perspectives

This type of shared experience helps students understand that culture is a system of beliefs, knowledge, institutions, customs/traditions, languages, and skills shared by a group of people, and that there are a variety of factors that promote cultural diversity in a community (Ogbu, 1992). Therefore, students begin to perceive that an understanding and appreciation of the diverse complexity of cultures is essential to interact effectively and work cooperatively with the many diverse ethnic and cultural groups of today's world. This is one of the main traits that a global citizen should possess. This critical instance case study showed that engagement in a telecollaborative project provided opportunities for students to not only develop the new literacies of effective online communication, but also to develop a cultural awareness of students located on a different continent in a different global hemisphere from their own.

The young children in this study began to identify with and understand the cultural similarities and differences they had as they developed deep connections to their global partners.

Characteristics of care and concern were recognized when Brazilian students were insistently asking the new friends about what had happened to their little travel buddy. At every class they arrived in the computer lab asking the teacher to check the blog for an answer and send messages to the American students. This proves the level of attachment to the mascot they were able to attain, allied to a sense of identity established by this new relationship. In addition to that, they took a lot of care of the visitor, playing with him, patting and holding him as if it was a real friend, which demonstrated the respect they had for their guest.

Curricular Connections

Many subjects from the curricula can be reinforced in a project such as this one. Connections to reading, writing, the arts (e.g. music, dance, artistic expression), and humanities are inherent in all of the learning activities that were developed as part of this exchange. Students engaged regularly in shared reading and writing activities using blogs. They also developed their own individual writing and technical skills through journal entries and the use of software products to create graphic images. Many of the lessons were interdisciplinary in nature and provided opportunities for collaboration between classroom teachers.

In terms of history and social studies in general, this kind of project makes it easier to study the patterns that emerge as humans move, settle, and interact on the planet, how they are arranged, and why they are in particular locations. Additionally, by examining the physical location and human characteristics of particular places, a variety of identities and cultures can be investigated. The use of tools such as maps, globes, charts, graphs and pictures of distinct countries can improve geography classes as students learn about each other's geographic areas. For example, the Brazilian students explored Google Earth® to locate the place their new friends lived. They discovered aspects of the land and the countryside of Kentucky and how it differed from the big city where they lived as well as the close proximity to the Ohio River. This activity provided them a better understanding of why humans settle in certain locations and how human actions modify the physical environment to meet their basic needs.

Technology Integration in the Classroom

The importance that technology has in our lives is unquestionable. The use of ICTs is changing the way people interact with each other, assume relationships, access and obtain information, share information, teach, and learn. Technology integration in the classroom has been gradual but consistent over the past decade. Many schools all around the globe are constantly experimenting with the use of new technologies and testing how they impact students' learning. It is a fact that today's children are born in a technologically infused, global society; many do not know what life is like without the electronic gadgets that consume their attention. Schools and teachers must focus on offering new opportunities "for applying 21st century skills across content areas and for a competency-based approach to learning [in order to enable] innovative learning methods that integrate the use of supportive technologies, inquiry- and problem-based approaches and higher order thinking skills." (P21, 2004, p. 8).

The Travel Buddies Project's main resource for communication and interaction between the participating students is provided through technology, in this case, a weblog or blog. Projects like this encourage students to use technology to communicate in a variety of ways, understanding that technology enhances collaboration to contribute to a learning community. The ability to collaborate as well as acquire an effective range of functional and critical thinking skills related to information, media, and technology are essential characteristics that citizens in the 21st century often need to demonstrate in their schools and workplaces. For that reason, school is the environment where these features must be fostered; otherwise students lose interest, tend to present lower academic achievements, and even evade participation in school because they are required to "power down" in the classroom (Prensky, 2000). This situation describes a reality that many schools and teachers, regardless of the country, are facing today.

Some additional skills that are essential for children to develop for success in today's globalized world were also emphasized during this project. The understanding that acceptable technology etiquette is essential to forge respectful social interactions and good citizenship, as well as examining how individuals interpret messages differently, the role language learning plays on communicating a clear message, how values and points of view are included or excluded, and how media can influence beliefs and behaviors were also perceived while Brazilian and American students interacted through the blog postings. Additionally, the students learned how to effectively apply more appropriate expressions and interpretations in diverse and multicultural environments.

It is important to note that an understanding of the ethical and legal issues surrounding the access and use of media must be discussed with the students continuously. Teachers should demonstrate good practices and uses for technology but also make it clear that the same technology sometimes can be used in bad ways, something that even very young children should be aware of as they become accustomed to using the Internet. In this project, ICTs were used to encourage the integration of communities and its resources beyond school walls, supporting international involvement in learning and intercultural exchange through an online and safe environment.

CONCLUSION

With the onset of new technologies that are a way of life for today's digital natives (Prensky, 2005/2006) and a continually flattening world with Internet access and use continually rising, education needs to refocus instruction to develop culturally responsive, global citizens who embrace diversity and multiple perspectives. This can easily be accomplished through the use of instructional models that provide opportunities for critical thinking, collaboration, and global awareness that are inherent to the information-based, global economy of the 21st century and key characteristics of a global citizen. International telecollaborative projects such as the one described herein allow students to (1) respect cultural differences and work effectively with people from a range of social and cultural backgrounds; (2) respond open-mindedly to different ideas and values; and (3) leverage social and cultural differences to create new ideas and increase both innovation and quality of work (P21, 2004).

During the project, the students perceived that there are many ways to interact with people from all over the world, and they learned how to use technology resources to communicate safely and effectively with their new foreign friends. They learned what a blog is and how the Internet is structured so that a message posted from a computer from their school could be seen by other people anywhere in the world. They were eager to talk to their new friends through the blog comments and get their answers. They were able to meet their buddies virtually through digital pictures that were posted on the blog, as well as share their pictures showing the moments they spent with their mascots along with the descriptions of the activities in which they participated.

Students were presented with distinct realities, not only to differences in language, but also customs, such as a typical school day and how people live in different ways depending on the country. However, the differences were lessened by the fact that regardless of where they lived, countryside or big city, they were kids who shared many of the same interests of knowing about the world, talking to their international peers, and discovering that they like and do similar things.

During this project all of the participating students had access to the same resources in the classroom in order to communicate and learn and all participated in the instructional activities. Most importantly, there was no exclusion of anyone based on ethnicity, culture, socioeconomic status, or other culturally specific characteristics. Telecollaborative projects such as this one should be encouraged, as they provide an opportunity for the participants to develop global citizenship skills that are indispensable for their living in a globalized, diverse, and flattened world.

REFERENCES

Anderson, R. H., Bikson, T. K., Law, S. A., & Mitchlee, B. M. (1995). *Universal access to e-mail: Feasibility and societal implications*. Santa Monica, CA: The Rand Corporation.

Bakhtiari, S., & Shajar, H. (2006). Globalization and education: Challenges and opportunities. *International Business & Economics Research Journal, 5*(2). Retrieved September 19, 2010, from http://www.cluteinstitute- onlinejournals.com/ PDFs/ 2006232.pdf

Banks, J. A. (2003). Educating global citizens in a diverse world. *New Horizons for Learning*. Retrieved February 27, 2010, from http:// www.newhorizons.org/ strategies/ multicultural/ banks2.htm

Banks, J. A. (2004, Summer). Teaching for social justice, diversity, and citizenship in a global world. *The Educational Forum*, , 68.

Boisot, M. (2002). *Knowledge assets: Securing competitive advantage in the information economy*. New York, NY: Oxford University Press.

Button, K., Johnson, M. J., & Furgerson, P. (1996). Interactive writing in a primary classroom. *The Reading Teacher*, *49*(6), 446–454.

Castek, J., Leu, D. J. Jr, Coiro, J., Gort, M., Henry, L. A., & Lima, C. O. (2007). Developing new literacies among multilingual learners in the elementary grades. In Parker, L. (Ed.), *Technology-mediated learning environments for young English learners: Connections in and out of school* (pp. 111–154). Mahwah, NJ: Lawrence Erlbaum Associates.

Cushner, K., & Brennan, S. (2007). The value of learning to teach in another culture. In Cushner, K., & Brennan, S. (Eds.), *Intercultural student teaching: A bridge to global competence*. Lanham, MD: Rowman & Littleton Education.

Davey, L. (1991). The application of case study evaluations. *Practical Assessment, Research & Evaluation, 2*(9). Retrieved September 18, 2010, from http://PAREonline.net/ getvn. asp?v=2&n=9

DiMaggio, P., & Hargittai, E. (2001). *From the digital divide to digital inequality: Studying Internet use as penetration increases*. Retrieved July 2, 2006, from http://www.webuse. umd.edu/ webshop/ resources/ Dimaggio_ Digital_ Divide.pdf

DiMaggio, P., Hargittai, E., Celest, C., & Shafer, S. (2004). Digital inequality: From unequal access to differentiated use. In Neckerman, K. (Ed.), *Social inequality* (pp. 355–400). New York, NY: Russel Sage Foundation.

Friedman, T. (2005). *The world is flat: A brief history of the twenty-first century*. New York, NY: Farrar, Straus & Giroux.

Gates, W. H. (2007, March). Written testimony of William H. Gates, Chairman, Microsoft Corporation: Before the Committee on Health, Education, Labor and Pensions, U.S. Senate. *Business Week*. Retrieved March 14, 2007, from http://www. businessweek.com/ bwdaily/ dnflash/ content/ mar2007/ db20070307_ 617500.htm

Gay, G. (2002). Preparing for culturally responsive teaching. *Journal of Teacher Education*, *53*(2), 106–116. doi:10.1177/0022487102053002003

Glaser, H. R. (2009). Internet governance in Brazil. *International Telecommunication Union News, 3*, 18. Retrieved September 19, 2010, from http:// www.itu.int/ itunews/ manager/ display.asp? lang= en&year=2009&issue=03&ipage=18&ext=html

Hajela, S. K. (2005, September). *Role of information and communication technologies in managing globalization at the national and regional levels*. Paper presented at the International Conference on Strengthening Regional Cooperation for Managing Globalization. Moscow, Russia.

Henry, L. A. (2008). Tertiary level digital divide: A more complex understanding and broadened definition. In P. Kommers & P. Isaías (Eds.), *Proceedings of the IADIS International Conference e-Society 2008* (pp. 24-34). International Association for Development of the Information Society (IADIS).

Heyl, J., & McCarthy, J. (2003, January). *International education and teacher preparation in the US.* Presentation at the National Conference Global Challenges and U.S. Higher Education: National Needs and Policy Implications, Durham, N.C. Duke University.

Hollins, E., & Guzman, M. (2005). Research on preparing teachers for diverse populations. In Cochran-Smith, M., & Zeichner, K. (Eds.), *Studying teacher education: The report of the AERA panel on research in teacher education* (pp. 477–548). Mahwah, NJ: Lawrence Erlbaum Associates.

Internet World Stats. (2011). *Internet usage statistics: The Internet big picture world Internet users and population stats*. Retrieved September 6, 2011, from http://www.internet worldstats. com/ stats.htm

Lapayese, Y. V. (2003). Toward a critical global citizenship education. *Comparative Education Review*, *47*(4), 493–501. doi:10.1086/379495

Lenhart, A., Purcell, K., Smith, A., & Zickuhr, K. (2010). *Social media and mobile internet use among teens and young adults*. Washington, DC: Pew Internet & American Life Project. Retrieved March 16, 2010, from http://pewinternet.org/Reports/2010/Social-Media-and-Young-Adults.aspx

Leu, D. J. Jr, Kinzer, C. K., Coiro, J., & Cammack, D. (2004). Toward a theory of new literacies emerging from the Internet and other information and communication technologies. In Ruddell, R. B., & Unrau, N. (Eds.), *Theoretical models and processes of reading* (5th ed., pp. 1568–1611). Newark, DE: International Reading Association.

Lima, C. O. (2006). *It's not all about access: A comparative study of global citizenship and ICT Use between Brazilian and American students utilizing a social inclusion framework.* Unpublished doctoral dissertation, University of Connecticut, Storrs.

McIntosh, P. (2005). Gender perspectives on educating for global citizenship. In Noddings, N. (Ed.), *Educating citizens for global awareness* (pp. 22–39). New York, NY: Teachers College Press.

Merryfield, M. M. (2002). The difference a global educator can make. *Educational Leadership*, *60*(2), 18–21.

Munck, R. (2005). *Globalization and social exclusion: A transformationalist perspective*. Bloomfield, CT: Kumarian Press, Inc.

New Literacies Research Team. (2007). New literacies, new challenges, and new opportunities. In Sampson, M. B., Szabo, S., Falk-Ross, F., Foote, M. M., & Linder, P. E. (Eds.), *Multiple literacies in the 21st century: The twenty-eighth yearbook of the College Reading Association* (pp. 31–50). Logan, UT: College Reading Association.

Nielsen Company. (2009, June). *How teens use media: A Nielsen report on the myths and realities of teen media trends*. Retrieved January 11, 2010, from http://blog.nielsen.com/ nielsenwire/ reports/ nielsen_ howteensusemedia_june09.pdf

Norris, P. (2001). *Digital divide: Civic engagement, information poverty, and the Internet worldwide*. Cambridge, UK: Cambridge University Press.

Ogbu, J. U. (1992). Understanding cultural diversity and learning. *Educational Researcher*, *21*(8), 5–14.

Partnership for 21st Century Skills (P21). (2004). *A framework for 21st century learning*. Retrieved from http://www.p21.org

Prensky, M. (2000). *Digital game-based learning*. New York, NY: McGraw Hill.

Prensky, M. (2005/2006). Listen to the natives. *Educational Leadership*, *63*(4), 8–13.

Rideout, V. J., Foehr, U. G., & Roberts, D. F. (2010, January). *Generation M2: Media in the lives of 8-to18-year olds*. Menlo Park, CA: The Henry J. Kaiser Family Foundation. Retrieved from http://www.kff.org/ entmedia/ upload/ 8010.pdf

Symantec Corporation. (2009). *Norton online living report 09*. Mountain View, CA: Author. Retrieved from http://www.symantec.com/ about/ news/ release/ article.jsp? prid=20090317_02

Torres, C. A. (1998). Democracy, education, and multiculturalism: Dilemmas of citizenship in a global world. *Comparative Education Review*, *42*(4), 421–447. doi:10.1086/447522

Zhao, Y., & Bryant, F. (2006). Can teacher technology integration training alone lead to high levels of technology integration? A qualitative look at teachers' technology integration after state mandated technology training. *Electronic Journal for the Integration of Technology in Education*, *5*, 62–63.

KEY TERMS AND DEFINITIONS

Cultural Awareness: An understanding of the characteristics, history, beliefs, behaviors, and values of an ethnic or cultural group outside one's own.

Cultural Responsiveness: Demonstrating tolerance for racial, ethnic, cultural, linguistic, and religious diversity.

Global Citizenship: One who shows awareness and concern about the rights and responsibilities of all citizens from a worldview and who understands the interdependency of a globalized society.

Information Communication Technologies: Computer based information technologies that support the transmittal of information, most often via Internet platforms.

New Literacies: Literacy skills and strategies required for Internet-based reading, writing, and communicating.

Telecollaborative Project: A pedagogical-focused project in which collaborators from different geographic areas work together to exchange information using information communication technologies.

Tertiary Level Digital Divide: A differential pattern of Internet access, use, and skill level that is most often related to socioeconomic status.

ENDNOTE

1 Data retrieved in September 2011.

Chapter 9
From Myopia to Global Vision via International Collaboration:
Lessons from Research and Experience

Diane Boehm
Saginaw Valley State University, USA

Lilianna Aniola-Jedrzejek
Poznan University of Technology, Poland

ABSTRACT

One effective strategy to prepare students to be successful participants in a globalized world is the use of online collaborative projects with students from other countries. New technologies and new opportunities for such collaboration may reshape teaching practices in unexpected ways. Three challenges need to be addressed for such projects to be successful: the specific circumstances of the classes, the structure and patterns of team interactions, and the technologies for collaboration. These challenges can be addressed through careful student preparation, well-designed assignments, monitoring of student progress, and a vision of the workplace of the future.

INTRODUCTION

Educating students for successful participation in a globalized world has become a priority for many university faculty members, in the United States as well as across the world. To address this new direction, many educators have taken advantage of evolving technologies. The new opportunities that are made possible by these technologies have also motivated faculty members to explore international connections themselves, since it is difficult to teach convincingly about globalization without having developed a global perspective.

Such global perspectives are urgently needed. As writers such as Thomas Friedman (2005), Lawrence Lessig (2008), and many others have pointed out, most 21st century problems (e.g., energy, climate, food distribution, clean water,

DOI: 10.4018/978-1-61350-077-4.ch009

even copyright issues) can no longer be solved by individuals or even by nations, but must be confronted on a global level. Thus students need to develop problem-solving skills in a global context.

Furthermore, growing globalization has increasingly led to internationally distributed virtual work teams, as companies with multiple international locations seek to hire workers who are able to collaborate across country and cultural boundaries. A significant body of literature has already developed relating to this growing phenomenon, and some of the recent studies (e.g., Gupta, A., Mattarelli, E., Seshasai, S., and Broschak, J., 2009) suggest that there are no significant differences in the efficiency and the quality of outcomes of such globally distributed teams compared to traditional site-based teams. Preparing students for this type of workplace requires educators to develop new approaches.

Over the past seven years, the students we teach in the US and Poland have had first-hand experience with new opportunities and new challenges through the development of international working partnerships. As our students from Saginaw Valley State University (SVSU) in Michigan and Poznan University of Technology (PUT) in Poland have completed online projects each semester, we have experimented with a variety of project designs—and continue to do so. However, as Starke-Meyerring and Andrews (2006) have pointed out, even for experienced instructors, three challenges typically arise in the context of such international collaborations: the specific circumstances of the classes; the structure and patterns of interaction of the teams; and the technologies used for collaboration. Each of these three challenges is addressed below, demonstrating how targeted strategies, such as specific types of student preparation to surmount language and cultural barriers, and explicit assignment design with timetables, accompanied by a tactical monitoring of student work, can develop the problem-solving skills that students will need to succeed in the globalized workplace of the future.

These firsthand experiences have refined our methods and reshaped our teaching practices in unexpected ways, as this chapter will demonstrate. What has been learned from both research and experience will, we hope, provide guidance for those who wish to do the same. More importantly, it will encourage teachers to meet the challenges of such global collaboration and develop their own projects.

LITERATURE REVIEW

In 2003, when the development of online international student collaborations in courses was new, the idea was fairly untested. Early in the history of these authors' collaborative projects, we adapted the model of Chickering and Gamson's "Seven Principles for Good Practice in Undergraduate Education" to the work (Boehm & Aniola-Jedrzejek, L., 2006). The seven principles thereafter developed have continued to guide these authors' online global student collaborations:

1. Develop cross-cultural awareness and mutual understanding of the culture of each group
2. Create a multi-faceted virtual environment
3. Coordinate calendars
4. Require intermediate level of English proficiency
5. Create relevant, engaging collaborative assignments with rubrics and shared understandings for evaluation of student work
6. Establish methods for successful group interactions, including information-sharing and relationship-building
7. Evaluate project outcomes. (pp. 1-2)

As other educators have also developed their own international collaborative projects, multiple models have evolved. Bell and Whatley (2004) have reflected on the complexities of developing collaborations between students who come from different cultures (e.g., British and Dutch), different university timetables, and different assessment models, but conclude that the pedagogical benefits include both language acquisition skills as well as the less tangible outcome of cultural awareness. Their identification of potential cross-cultural barriers echo those identified by Cramton & Hinds (2005) in their analysis of potential pitfalls in the virtual workplace. The long-term effects of international student collaborations have yet to be studied. However, our research (Boehm, Kurthen, & Aniola-Jedrzejek, 2010) to determine whether these types of collaborations had an immediate measurable impact on student ethnocentrism showed that they did not.

Some universities have developed institutional partnerships that allow transnational collaboration to be a central element in their course objectives and activities. As Starke-Meyerring and Andrews (2006) have demonstrated, such institutional relationships that are sustained over time often forge unique partnerships between faculty and students, developing projects with upper-class students in similar disciplines that may span an entire semester or longer. Starke-Meyerring and Andrews offer five recommendations, based on partnerships between McGill University in Montreal and the University of Delaware, that confirm many of the same principles stated above. Starke-Meyerring and Wilson (2008) have further illustrated how the three "pillars" of globally networked learning environments — partnerships, policies, and pedagogies — create the foundation for project success.

As this literature of best practices has emerged in recent years, it has made planning for such partnerships easier, since experienced faculty members have shared their successes and challenges. Bell and Zaitseva (2005) conclude that computer-mediated communication experiences may cause difficulties such as anxiety about cross-cultural communication and the potential for misunderstandings due to the lack of social cues such as gestures, facial expressions, and body language, but that such experiences also offer unique opportunities for learning. Hashimoto and Lehu (2006), pairing French and American teams, discuss the necessity for students to agree on a method to accomplish tasks in spite of individual and cultural differences. Kurthen's best practices for transnational online collaboration (2008) include recommendations to guide educators in five aspects: pedagogical strategy, organizational context, technology use, cultural challenges, and English language skills. Starke-Meyerring, Duin, Palvetzian, & Wilson (2008) have created a 10-part heuristic for partnership development that begins with questions regarding vision and mission and concludes with outcomes (pp. 25-26).

New technologies can address some of the relationship issues, especially the lack of social cues in the online environment. O'Brien and Alfano (2009) have, for a number of years, successfully used videoconferencing to connect students in the US and Sweden. Wikis have also become a useful tool for such collaboration (Bruns, A., & Humphreys, S., 2005). Though a 2008 Information Technology Survey of technology applications at the University of Michigan showed that many faculty members are slow to adopt complex features like wikis and blogs, even if they are available in a Course Management System (Zhu, 2008), wikis are adaptable for many types of collaboration.

Research literature and guidelines for global student collaboration do not mean, however, that such projects are necessarily easy to manage, for they demand extra and often unfamiliar work not only from students, but also from the faculty in the partnership. Perhaps that is why few partnerships have been sustained over time periods as long as the seven years during which we have been collaborating. There is much we have learned since our early collaborations. This chapter's discussion will be organized around the three challenges identified above: the specific circumstances of the classes; the structure and patterns of interaction of the teams; and the technologies used for collaboration. As each of these three challenges is addressedwe will demonstrate how targeted strategies, such as specific types of student preparation, explicit assignment design, accompanied by tactical monitoring of student work, can develop both the intercultural and the technical skills that will prepare students for success in the globalized workplace of the future.

CHALLENGE 1: SPECIFIC CIRCUMSTANCES OF THE CLASSES

Every partnership begins with individual people and the choices they make when developing such collaborations. The benefits of common ground and sustained partnerships between equivalent groups when setting up international partnerships have been described in the literature. However, based on our experiences, even shorter-term asymmetrical partnerships, spanning at most six weeks, such as the type we have developed between quite different student populations, have intrinsic value in developing global awareness in students (see Appendix A for the type of assignment used in our partnership). It is helpful to analyze this challenge of specific class circumstances through three lenses: the contexts for partnership, the language barriers, and academic conventions. Each of these can be addressed when preparing students for the project.

Contexts for Partnership

Since specific class circumstances set the parameters for what is possible, the more students are able to understand each other's contexts, the better their teamwork is likely to be. The partnerships described in this chapter are conducted in widely different course contexts, which can affect everything from the students' commitment to the work to how the projects will be evaluated upon their conclusion. The American students, a majority of whom are likely to be females preparing to be teachers, business persons, or health professionals, are taking required courses either at the freshman or upper class level, depending on the semester; the goals for these courses are to develop writing proficiency. Most of the students bring significant writing experience from their previous schooling. Many are regional first generation college students who have not traveled extensively; most are monolingual. Few have much knowledge of the international workplace.

In the European Union, higher education is in a state of flux, with the 2011 Bologna Agreement deadline to standardize education within the EU fast approaching. PUT students, like those in most European universities, enter the university in their field of study; in this case, generally physics, computer sciences, or mechanical engineering. Like many European students, they are likely to have traveled to other countries; they may even plan to study in other EU universities as part of the Bologna Agreement. They are accustomed to a traditional lecture and test pedagogy. Since writing in a second language is not taught in Polish secondary schools or technical universities, writing in English is likely to be a significant challenge; of the four aspects of language learning (listening/reading/speaking/writing), writing is generally the most difficult, and the last to be developed by non-native speakers of any language. The goal for the PUT students is to develop proficiency in English and in basic writing skills simultaneously in an integrated approach. PUT students, predominantly male, recognize the necessity to become fluent in English if they wish to be successful in the ever more globalized and mobile EU work force, exemplified in the Poznan region by Volkswagen, a large local employer, which has a CEO from South Africa, managers from Germany, and a workforce from the region.

The cultural contrasts between these groups of students extend beyond the educational contrasts. To take just a few varied examples: the population of the US is about 308 million vs. 38 million for Poland. The US has always been a democracy; Poland regained its democracy only 20 years ago, after spending the post-WWII decades as a member of the Soviet Union bloc. The US has great ethnic, racial, religious and cultural diversity, but it is relatively monolingual; Poland is relatively homogeneous ethnically, religiously, and culturally, but most educated persons have competence in one or more foreign languages. In Poland, developing personal and social relationships generally precedes business interactions; in the US, the reverse is often true. Students do not often recognize these differences. Thus, building cultural awareness increases the likelihood of successful outcomes.

Language Barriers

Language barriers, a predictable challenge, go beyond the occasional misunderstood text, as we have learned. Students writing in their native language may lack empathy for the amount of effort their partners must invest in order to communicate using the same language. American students, with limited experience crossing language barriers, may not understand the ways in which first language interference occurs. To illustrate just two of the many linguistic differences in the students' home languages, the Polish language lacks articles (a, an, the) and, unlike English, has flexible word order. The lack of articles is fairly easy to ignore, but word order creates difficulties, since it directly affects meaning-making. Pointing out how the languages differ can help students understand patterns of error that they are likely to encounter, and to recognize the origins of these errors. A related language aspect is that students conducting research are likely to prefer their native language; when they translate information they have found (or use online translation programs, which may exacerbate the problem), the language difficulty can become intensified. It is useful to remind students that access to research sources in two languages can significantly enrich their base of information.

Language barriers underlie an even more significant challenge: the double burden that written language must carry in such partnerships. Students' written language must not only do the work of project development and document construction, but also simultaneously serve as the platform for relationship-building. This requires care not to misread tone and intention. Though American students, in particular, are familiar with the use of language for self-representation in different rhetorical contexts (e.g., Facebook), this self-representation rarely has to bridge cultural boundaries. In our early collaborations, few students had previously given any thought to this second, relationship-building aspect of writing.

Few courses and textbooks give attention to the complexity of this relationship-building aspect of writing in a cross-cultural context. American writing courses generally center around basic rhetorical principles, instructing students to define their purpose, audience, and type of writing. Though language choices and tone are discussed in most American writing classes, this rhetorical analysis generally occurs in the context of audiences with similar ethnic, social class and cultural backgrounds—who share a common understanding of connotations of words, as well as a shared history that makes it easy to recognize allusions and historical or cultural references. In cross-cultural communication, these common understandings no longer apply. Suddenly writing has a new complexity; as students write, they must develop rapport and trust without physical interaction, body language or other non-verbal clues, in order to build the partnerships necessary to complete the assigned task. With international partners, students begin to recognize how difficult it can be to develop working partnerships and shared understandings without a common cultural history when their only medium is words—words in electronic formats without nonverbal cues, at that. They wonder how to write clearly, simply, unambiguously, and directly, honoring cultural differences, without "dumbing down" the language or the concepts they are exploring.

To address this need, the teaching practices in the SVSU writing courses have been adjusted to focus more on communication and teamwork, rather than just on document design and construction, and the assigned reading material has also been adjusted to help students develop relevant insights. For example, Philip Kolin's (2010) "Guidelines for Communication with International Readers" (173-178), used for the SVSU Writing in the Professions course, remind students that they are more likely to be understood if they follow some simple principles: use common, easily understood vocabulary; avoid idiomatic expressions and sports and gambling metaphors; avoid abbreviations, acronyms, or contractions; be aware of different units of measure; keep sentences simple and easy to understand; and pay attention to style and tone.

As these guidelines begin to help students think more strategically about their language choices — and the work their words must do — students generate interesting observations and questions. Easily understood vocabulary may mean something different for a native speaker than for an English language learner. Native speakers may know multiple synonyms for a word, whereas non-native speakers may have few or none. So a non-native speaker may use the word *thin* in describing a malnourished child; a native-speaker might choose words like *scrawny* or even *emaciated,* which suggest much more meaning, but will these words be understood? As well, American English is full of idioms. For example, the students must think about how to describe matters like *hot button issues* without the idiomatic expression. Sports metaphors (*he hit a home run with his new plan*) pervade American English, and literal language (*his new plan was very successful*) may seem uninspiring.

Suddenly the students are asking new types of questions, such as what to do if they do not understand what a partner has written, how many personal details are appropriate in communication, how they can encourage their partners to meet deadlines, and how they should handle misunderstandings. All of these questions are part of the larger central question: how can their language be used to both build relationships and enable their partners to work with them to accomplish a common task?

With support, students do indeed develop their own experience-based answers to such questions. For example, a strategy that can assist American students when confronted with puzzling language is to have them rephrase and test: "*I think you are saying x; am I right*?" In the most recent end-of-semester evaluation, SVSU student comments suggested they had developed this heightened awareness of how to negotiate differences:

- "*It was also interesting to try and use very basic English, since they don't have as wide a vocabulary as we do*."
- "*I learned that working with people around the world is not only possible, but almost easy with today's technology*."
- "*I learned that it is very challenging to work with people from a different country when you can't meet face to face, yet it was a rewarding experience*."

PUT students had interesting comments as well; they valued:

- "*chat with friends from US*,"
- "*possibility of co-operation, meeting new people*,"
- "*learn (sic) new skills while searching for materials*."

Another student indicated he "*developed knowledge of language, some interesting phrases o/a expressions.*" As students acquire strategies to surmount language barriers, they gain informed perspectives and insights they could not have developed any other way.

Academic Conventions

Academic conventions of writing, such as document and citation formats, create yet another difference to resolve. Most American students will have familiarity with standard U.S. formats such as APA or MLA; Polish students, whose previous curriculum may have incorporated little or no academic writing and only limited reading of professional journal articles with extensive reference lists, may be familiar only with the formats accepted at their university. Furthermore, when researching, non-native speakers of English tend to gravitate toward commonplace sources on the Internet (Wikipedia, for example), whereas the expectation for US students is that they will use more scholarly sources when researching.

As the projects develop and students recognize how many different cultural, political, educational and economic perspectives there may be on such topics, shaped by different histories and values, they begin to recognize that their way of thinking is not necessarily the right way, or the only way. This perspective helps students increase their repertoire of problem-solving strategies.

CHALLENGE 2: STRUCTURE AND PATTERNS OF TEAM INTERACTIONS

Not all issues can be systematically addressed through careful student preparation, however. Intercultural learning is always complex, as Harris (2002) explains: "collaborative activities are more difficult to do and to facilitate," because participants must "understand and incorporate plans, procedures and perspectives different from their own. The truly collaborative process usually requires higher-level thinking and interaction" (p. 6). Strategic assignment design and monitoring of student progress can address time limitations and support team-building.

Time Limitations

Assignment design and work schedule play a pivotal role in the project. The single biggest constraint on our collaborations has been time limitations, which make it more difficult to develop a shared group identity to complete the work required. The time span is generally no more than six weeks, since European and American universities may have very different calendars. PUT begins its semesters as SVSU is nearing the halfway point in the semester. Furthermore, semesters may be interrupted by breaks and holidays. The ideal would be to have more time initially for building student relationships. Evaluations at the end of the project inevitably express these sentiments. SVSU students relate: "*I would only change the time constraints, if possible*"; "*I would arrange more time to talk with the Polish students.*" PUT students echo their sentiments: "*I only wish we had more time to do all the things we needed and just to talk freely not only about the project to integrate and know each other better*"; "*We didn't have much time.*" Since timelines are not controllable, the only option is careful planning when designing the assignment, to maximize the time available. Multiple assignment types and processes have been evaluated to deal with this need. A recent assignment (see Appendix A) illustrates use of a joint calendar to address the issue.

Setting the Stage

Given the time constraints, it is important to attempt to get projects off to a quick start. To begin with, the PUT students generally suggest topics that interest them, and the SVSU students join a group based on the topic. Groups are generally composed of 4 to 6 members. Students have consistently done well finding common topics of interest that are relevant and current, though often they begin with a topic that is too broad and that must be narrowed and focused to be handled successfully. In recent projects, students have chosen topics like nuclear energy, higher education, genetically modified foods, and censorship in media. Rather than carefully designing student roles, such as wiki manager, discussion forum manager, or editor, as was done initially, students are now given the responsibility to design their own tasks and roles, a process that has worked at least as well.

Other project models have been tested as well. For example, in one recent project, a tridem was conducted, including the students of a colleague in Italy in the project; the experience supports the literature (Hauck & Lewis, 2007), which suggests that teams composed of students from three countries rather than two are less likely to divide along national lines, and more likely to see each other as individuals, rather than as representatives of a particular "foreign" culture. A tridem, however, complicates both planning and timetables, and thus is not always easy to arrange.

Whatever the model, students may also have had only limited experience—even face to face—with carrying out team/group projects. Now they must do so with the added constraints of electronic communication. Thus, a lot of experiential learning must happen in a short time, as they negotiate unfamiliar work patterns and methods to accomplish tasks at the same time they are managing cultural differences. When designing the timetable for the assignment, therefore, it is helpful to provide more time in class for working on the project, with more hands-on mentoring of the students, offering problem-solving strategies when groups appear to be "stuck."

As much as possible, students should be encouraged to conduct their own problem-solving. Sometimes they are successful, other times not; but often they learn more from missteps that must be resolved than from conventional instruction, and they feel intense satisfaction when, at the end of the project, they are able to give a presentation to their class colleagues, demonstrating that they have surmounted the challenges and are ready to present what they have learned.

Monitoring Team Interactions

Team interactions, of course, determine the success or failure of the project. American students, with their typical "let's get to work" attitude, will often be very direct and straightforward. (For example: "*There are still 2 sections that are not finished on the Wiki. We will do the US regulations, can your group finish up the disadvantages?*") Polish students, already under pressure because their English language skills are less developed, sometimes feel defensive when deadlines are looming and expectations seem overwhelming. They may misread the direct tone of a message and overreact to comments that Americans might simply perceive as straightforward language. Monitoring and intervening when necessary can keep teams moving forward.

To avoid power imbalances, the ideal suggested in the literature is to avoid giving one team an advantage over another in any way (Starke-Meyerring, Duin, Palvetzian, & Wilson, 2008; Starke-Meyerring & Andrews, 2006), because this may give rise to ethnocentricism, the notion that one's own culture is superior to others (Cramton & Hinds, 2005). However, American students have an immediate advantage, since the projects are conducted in English. Unfortunately, this language advantage, coupled with the fact that the SVSU students are less likely to have traveled internationally, may lead the monolingual American students to become impatient with their counterparts' use of English, seeing it as a deficiency rather than the typical patterns of language learners. Groups may also subdivide by language proficiency, with the stronger students blaming the weaker students for project difficulties.

To combat these tendencies, the goal in the assignment design and the monitoring of student progress has been to create what Cramton and Hinds (2005) describe as "structured interdependence." An effective way to do this is to require components that represent each culture. When students are able to build relationships at the psychological/emotional level and begin to recognize their own culturally-bound identities and perspectives, such as how they have been shaped by growing up in a particular culture's values, norms, and beliefs, they learn how to negotiate these differences to achieve a common goal.

CHALLENGE 3: TECHNOLOGIES FOR COLLABORATION

Any online project is subject to a unique set of issues, not only human but also technological (Cramton and Hinds, 2005); the support structure must be easily accessible, encourage team interaction, and support the goals for the assignment. It should also help develop students' proficiency with communication tools of the future. Even access to technology needs to be considered, such as examining the technologies that are available, the expertise that the participants have, and the access to technology available to the students. Both access and tools determine how a project will evolve and how easy it is to monitor student progress.

Technology Access

Access to technology needs to be considered at the outset of any project involving computer-mediated communication. In the projects described herein, all the students possess the requisite expertise, but the PUT students' access to computers is more limited; some have lab access as a class only once a week and may not have easy access outside of class. This can slow down their pace of work.

If a learning management system (LMS) is used, it is likely that one university will need to enroll the students of the other university in its system. Some universities may restrict such access. For expediency, our work is conducted within the SVSU learning management system (Sakai, an open source Learning Management System, dubbed VSpace on our campus); SVSU hosts PUT by enrolling students in its system. This choice privileges the SVSU students even further, because they are already familiar with the system. However, it is also a practical choice, since it means that the groups are able to get off to a faster start, and the SVSU students can "coach" their counterparts when needed.

Choice of Electronic Tools

Equally important is the second question: what specific tools are best suited to such a project? Most learning management systems offer a wide range of electronic tools. Multiple tools can be used for different phases of the project: Resources folders (which can include model texts, guidelines for citation formats, links to web documents, grading rubrics, and the like); Mailtool for email messages; Discussion Forums; Virtual Chat; and the Wiki tool.

Other types of collaboration are possible. As initiatives such as the Stanford University Cross-Cultural Rhetoric Project (O'Brien & Alfano, 2009) illustrate, adding audio and video components enhances real-time communication and reduces the need for written language to carry all of the double burden of relationship-building as well as of project completion. When possible, our project has, in some semesters, used Skype™ Internet phone, which students found especially exciting. Now that Skype™ has added a video component, video-conferencing may become an option no longer limited by cost. NetMeeting™ and other conferencing software could also be excellent for students to gain experience with; these programs have not yet been accessible for both groups of students, however.

As the assignment guidelines demonstrate, the project typically begins with whole-class Discussion Forums, in which students introduce themselves to the class and post photos, representing themselves in the virtual environment. In the most recent semester, the SVSU students had previously introduced themselves and posted a photo at the beginning of the semester. When it came time to begin the collaboration, and the class discussed how to write to be understood by a non-native speaker of English, they revised their initial introductions with their PUT partners in mind. As they were doing so, the students frequently asked the question, "Will they understand this?" Though the question is simple, it is also indicative of a dawning awareness of the need to be consciously aware of their audience when communicating with those from other cultures.

After students form project groups, each group conducts a Virtual Chat to become acquainted, divide up the tasks, and brainstorm strategies for completion of the graded project, the Wiki. The Chat tool remains available throughout the project. Here, even a simple aspect like keyboarding skills may influence the interaction; since American students are taught these skills fairly early in their education, they are likely to be more proficient than their Polish counterparts. This, coupled with their ease of language, means they are likely to post longer and more complex messages.

To carry out the project, each group is given its own identified Discussion Forum to use as a repository for its research, a place to discuss ongoing work, and a way to measure the progress of the group. Though this may seem an unconventional use of a Discussion Forum, it easily allows both interpersonal communication and task-related collaboration. It also gives students a benchmark for completing research, since a certain standard of quality and quantity can be required of their research before opening the group's Wiki. In the projects recently completed, one group, which chose the topic of higher education systems in the US and Poland, set up multiple discussion threads within their Forum to conduct and monitor their work:

- Costs and financial benefits of higher education in the US
- Degree options and dropout rates in the US
- Polish higher education system
- Scholarships in Poland
- Research sources
- PowerPoint presentation
- Condolences (a thread established when the President of Poland and his entire entourage were killed in April 2010 in the plane crash at Katy near Smolensk in Russia).

Another advantage to this use of the Discussion Forum is the fact that all of the project materials and messages are stored in a central location, easily accessible to the participants at all times. The Discussion Forum also makes it easy to add attachments such as drafts, PowerPoint presentations, images, and the like.

Having used several different tools, we have found the Wiki to be particularly appropriate for the graded assignment for several reasons. Wikis are structured for collaboration and automatically save earlier versions of the documents ; the technology is also familiar to most students around the world, because of Wikipedia. Furthermore, Wikis such as Wikipedia have already taught students that nobody "owns" a Wiki; collaboration is its essence, an exemplar of what Lessig (2008) calls "the sharing economy" that has been made possible by nearly universal access to the Internet. The Wiki is likely to be new to both groups of students, so neither has the advantage of significant prior experience. It is multidimensional in format, blending image and text in ways that allow non-native speakers of English to use images such as diagrams, charts, photos and the like to explain complex processes that would likely be difficult to put into words. Furthermore, it allows for the insertion of links to research sources, videos, and other sources of information.

Beyond these technological advantages, the Wiki teaches "technical literacy, content creation in a digital environment, the art of collaboration, consensus building, creating explicit knowledge from tacit understanding, and effectively communicating ideas to other people through networked knowledge environments" (Bruns and Humphreys, 2005); all of these are valuable professional and life skills. The Wiki also supports students' development as writers, as it incorporates all four aspects of revising (adding text, deleting text, reordering text, and rewording text) in multidimensional and nonlinear ways. Because the Wiki keeps a history of the previous versions of each document posted, students may revisit and retrieve information as they draft and redraft, revise, reflect and revise again.

The Wiki also demands logical structuring, with headings to guide the reader; this structure can then easily be adapted to a presentation. And as it is built part by part, rather than on individual contributions, it also focuses on the work of the team, since no names are attached to the sections of the Wiki. The Wiki is always dynamic, always available for further contributions and revisions, and encourages "ongoing meta-analysis on the part of the authors" (Bruns and Humphreys, 2005).

Proficiency with this wide array of electronic tools gives students much flexibility, laying a foundation for them to work and communicate effectively in the electronically-interconnected work teams of the future.

CONCLUSION

International partnerships have motivated educators to recalibrate their teaching in multiple ways. Our end-of-project evaluations often identify student concerns about many of the issues we have discussed: *"more time for the project," "more time for talking to them," "time differences," "different class schedules."* Interestingly, many students both at SVSU and at PUT have had as much difficulty with some of their own team members not completing scheduled tasks, as they had with their international counterparts; one student commented: *"Because you are in a group, you have to pull the weight for the rest of the group, or else you all look unprofessional."*

Each of the three challenges described above can be met in multiple ways. New strategies for student preparation, experiments with assignment design, and use of varied electronic tools to support and monitor students as they develop international teamwork skills are some of the methods that experience has taught us. It is quickly evident to any faculty member interested in online international student collaborations that such initiatives are likely to be an ongoing exercise in problem-solving. That is, of course, true for any and all interesting teaching situations, and makes them particularly intriguing. New teaching practices evolve in each situation, and then evolve yet again in subsequent collaborations.

Educators, students, and global citizen all have much to learn from international partners. A recent team project on nuclear energy, for example, gave insight into the lingering impact of Chernobyl on Eastern Europe, the primary reason Poland has no nuclear energy sources at present. Polish political leaders are reconsidering that decision, even as US leaders, anxious to reduce US dependence on foreign oil, are having similar conversations. Joining expertise worldwide can bring better outcomes for everyone.

For this reason, we remain firmly committed to the value of such learning experiences, both for ourselves and for our students. The cross-cultural benefit, together with the value of experiential learning, has been evident in each collaboration that we have conducted, as student comments reflect: *"It was fun and interesting to get to know the PUT students and learn the differences between our cultures. It was an eye opening experience." "[I learned] how to collaborate with someone you have never met in person." "It was very interesting getting an outside view on global issues." "I loved the fact that I could discuss the topic with the American group as well as get to know them and what they do and do not like."*

As students learn to use a range of electronic tools to work as a team and gain experience with globally distributed teams, they build expertise in computer-mediated communication that equips them for the rapidly changing workplace of the future. This growth alone is reason enough for a teacher to expand his or her professional horizons.

REFERENCES

Bell, F., & Whatley, J. (2004). Trans-national online activities for students—a pragmatic approach. In *ILIA—Innovative Learning in Action, issue one: Internationalising the curriculum* (pp. 13–18). UK: University of Salford.

Bell, F., & Zaitseva, E. (2005). Only connect: Complexities in international student communication. *E-learning, 2*(4), 342–355. doi:10.2304/elea.2005.2.4.341

Boehm, D., & Aniola-Jedrzejek, L. (2006). Seven principles of good practice for virtual international collaboration. In Ferris, S. P., & Godar, S. (Eds.), *Teaching and learning with virtual teams* (pp. 1–31). Hershey, PA: Information Science Publishing. doi:10.4018/978-1-59140-708-9.ch001

Boehm, D., Kurthen, H., & Aniola-Jedrzejek, L. (2010). Do international online collaborative learning projects impact ethnocentrism? *E-Learning and Digital Media, 7*(2). Retrieved from http://www.wwwords.co.uk/ elea

Bruns, A., & Humphreys, S. (2005). *Wikis in teaching and assessment – The M/Cyclopedia project.* Paper presented at The 2005 International Symposium on Wikis, October 16-18, San Diego, CA. Retrieved from http://snurb.info/ files/ Wikis% 20in% 20Teaching% 20and% 20Assessment.pdf

Cramton, C. D., & Hinds, P. (2005). Subgroup dynamics in internationally distributed teams: Ethnocentrism or cross-national learning? In M. Staw & R. M. Kramer (Ed), *Research in organizational behavior, 26,* 231-263. Greenwich, CT: JAI Press.

Friedman, T. L. (2005). *The world is flat: A brief history of the 21st century.* New York, NY: Farrar, Straus and Giroux.

Gupta, A., Mattarelli, E., Seshasai, S., & Broschak, J. (2009). Use of collaborative technologies and knowledge sharing in co-located and distributed teams: Toward the 24-h knowledge factory. *The Journal of Strategic Information Systems, 18,* 147–161. doi:10.1016/j.jsis.2009.07.001

Harris, J. (2002). *Wherefore art thou, telecollaboration?* Eugene, OR: International Society for Technology in Education.

Hashimoto, K., & Lehu, J. (2006). Students International Collaboration Project (SICP): A cross-cultural project using virtual teams to learn communication styles. In Ferris, S. P., & Godar, S. (Eds.), *Teaching and learning with virtual teams* (pp. 221–244). Hershey, PA: Information Science Publishing. doi:10.4018/978-1-59140-708-9.ch010

Hauck, M., & Lewis, T. (2007). The Tridem Project. In O'Dowd, R. (Ed.), *Online intercultural exchange* (pp. 167–192). Clevedon, UK: Multilingual Matters.

Kolin, P. C. (2010). *Writers at work* (9th ed.). Farmington Hills, MI: Cengage Learning.

Kurthen, H. (2008). Transnational online student collaboration: Does it work? In B. E. Dietz & L. H. Ritchey (Eds.), *Scaffolding for student success in learning: Best practices in using instructional strategies* (pp. 202-210). Washington, DC: American Sociological Association (ASA Resource Materials for Teaching).

Lessig, L. (2008). *Remix.* Penguin Press, 2008.

O'Brien, A., & Alfano, C. (2009, December). Connecting students globally through videoconference pedagogy. *Journal of Online Teaching and Learning, 5*(4). Retrieved from http://jolt.merlot/org/vol5no4/obrien_1209.htm

Starke-Meyerring, D., & Andrews, D. (2006). Building a shared virtual learning culture: An international classroom partnership. *Business Communication Quarterly*, *69*(1), 25–49. doi:10.1177/1080569905285543

Starke-Meyerring, D., Duin, A. H., Palvetzian, T., & Wilson, M. (2008). Enabling and sustaining globally networked learning environments. In Starke-Meyerring, D., & Wilson, M. (Eds.), *Designing globally networked learning environments*. Rotterdam, The Netherlands: Sense Publishers.

Starke-Meyerring, D., & Wilson, M. (2008). Learning environments for a globally networked world: Emerging visions. In Starke-Meyerring, D., & Wilson, M. (Eds.), *Designing globally networked learning environments*. Rotterdam, The Netherlands: Sense Publishers.

Zhu, E. (2008). Breaking down barriers to the use of technology for teaching in higher education. In Robertson, D. R. (Ed.), *To improve the academy* (*Vol. 26*). San Francisco, CA: Jossey-Bass.

KEY TERMS AND DEFINITIONS

Cultural Differences: The socially transmitted behavior patterns, arts, beliefs, institutions and thoughts characteristic of a community or population.

Ethnocentrism: The belief that one's own ethnic group is superior to others.

Globalization: The process of making something worldwide in scope.

International Collaboration: Working with others across national boundaries.

Problem-Solving Skills: The abilities to successfully resolve difficult or challenging situations.

Teaching Practices: The activities teachers use to assist students to learn.

Transnational Collaboration: Working with others across national boundaries.

Wiki: An Internet application that allows anyone visiting a website to edit its content.

APPENDIX

Assignment 5: Intercultural Collaborative Project

The workplace of the 21st century will increasingly demand employees who bring a global perspective as well as a multiplicity of abilities to their work, including an understanding of world cultures, an ability to work collaboratively, and the capacity to integrate technology into many facets of their work. To develop this global perspective, students from Saginaw Valley State University (SVSU), USA, and Poznan University of Technology (PUT), Poland, will collaborate online to prepare a final project that will be based on research and an exchange of ideas and information.

The Project is designed to achieve the following objectives in student learning:

- Develop an international perspective on relevant current issues
- Build collaborative skills in multinational groups
- Increase competence in the use of technology in collaborative settings (e.g., wiki)
- Enhance research and writing skills in a collaborative context
- For PUT students: Enhance second language skills through content – CLIL (Content and Language Integrated Learning) approach

The project will be based on a current global issue chosen by the group. (Some examples have been posted to VSpace.)

The project will have the following two outcomes:

1. A Wiki Report (4,000 words minimum, plus References; 3 images minimum) on the topic each group has chosen. Your audience will be your class colleagues, who may have limited knowledge on your topic. Your purpose is to help them understand the significant aspects of this issue. Each person in the group should provide at least 1 piece of authoritative research on the topic. Reports must include appropriate citations for both the Report and the images.
2. An oral group Presentation in each country, using PowerPoint.

Project development:

- PUT students will post Introductions to the VSpace forum; SVSU students will review their posted Introductions for appropriateness for International Readers (see Kolin *Guidelines* 173-178).
- Polish and American students will work in groups (of both nationalities) to complete both the Wiki Report and the Group PPT Presentation.
- Each group will choose a topic on which to conduct research. Use authoritative sources for your research and cite your sources.
- Groups will organize their work so that each person in the group makes a contribution (you may wish to divide up tasks). (Kolin's models for computer-supported collaboration and wikis, 103-107, may be useful.)
- Groups will use VSpace to conduct their collaboration. Each group will have several tools:
    a) a *Discussion Forum* for group members to exchange information and files for subsequent use in the wiki; use this tool also to exchange the PPTs

Table 1.

SVSU	PUT
25 March: Complete posting of all **research** to group Discussion Forum	**25 March**: Same
1 April: * Complete posting of all **images** to group Discussion Forum * Post outline for Wiki Report on wiki	**1 April:** Same
8 April: * Post draft of Wiki Report * Post outline and graphics for PPT presentation	**12 April:** Same. Note: PUT students have Easter break between 2 and 6 April 2010
15 April: Group presentations begin	
20 April: Group presentations conclude; Wiki Report due	**TBA:** Wiki Report due, Group presentations

b) a *Virtual Chat* forum in which your group can meet synchronously to discuss the project
c) SVSU: access to *Skype* in the Student Technology Center
d) a *Wiki* for the group Wiki Report (will be available by 26 March).

The project tasks will be completed according to the Project Calendar for each class (Table 1):

The project, including both the Wiki Report and the Group PPT Presentation, will be evaluated according to the rubrics provided.

Notes on Oral Presentation

A workplace project is often accompanied by an oral presentation in which the investigators present key findings to co-workers or supervisors. Your oral report will be developed collaboratively and presented to a general audience (each class). Each presentation will have a 15-minute time limit.

Prepare a formal, *rehearsed* presentation using PowerPoint (we will discuss best practices for using it); you may also prepare handouts, posters, charts, etc. as relevant for your topic. In keeping with the nature of this assignment, professional attire would be appropriate for the presentation. (SVSU: See text Chapter 16 "Making Successful Presentations at Work" [670-691] to guide you.)

Preparing for Collaboration

1. Revisit/revise or post your VSpace Introduction to make it an effective way for someone from another country to get to know you as a working partner.
2. SVSU: Use Kolin's "Collaborative Writing and Meeting at Work" strategies (Chapter 3, 76-106) to conduct effective collaboration (e.g., organize your group to assure contributions by all).
3. We will request that you evaluate the collaboration when it concludes as a way to enhance future collaborations.

Chapter 10
How Similar or Different Are We?
A Perception of Diversity in Global Virtual Teams

Gaelle Picherit-Duthler
Zayed University, UAE

ABSTRACT

Global virtual teams are becoming vastly popular amongst public and private organizations. This prevalent way of organizing can be used for students to gain experience in working with various cultures using technology. GlobCom is a global communication project that provides students with this opportunity but also presents them with challenges such as cultural barriers. This chapter explores students' perceptions of diversity in their global virtual teams. It provides insight as to how students overcome issues of cultural differences by working in an online environment.

INTRODUCTION

Virtual teams are becoming almost a fixture of organizations throughout the United States, Europe, and beyond, as a response to globalization. Organizations need to have a wider international reach and be present 24/7. By having individuals working together, but from different parts of the world, they can meet their international 24-hour needs.

The field of public relations (PR) is no stranger to this new way of doing business. Many PR firms are expanding globally. For example, as of 2010, Hill & Knowlton has 79 offices in 43 countries; Porter Novelli has 90 offices in 60 countries; Weber Shandwick has 121 offices

DOI: 10.4018/978-1-61350-077-4.ch010

in 76 countries. Similarly, more multinational organizations need to communicate globally by having communication specialists in various geographical locations. These specialists are there to grasp the economic, political and cultural intricacies of their part of the world as PR practices vary in different countries. They work across time, space, and culture to accommodate the needs of these multinational companies.

As PR becomes increasingly international, there is an increasing demand for students to partake and practice international public relations. However, it is rare that students have the opportunity to truly experience and practice at that level. Most of the time, they tend to work on case studies, listen to guest speakers who have traveled the world, and when possible, they can participate in a study abroad program. These activities are limited in their scope. They do not let students practice their skills and apply their knowledge toward this growing field. Organizing students in global virtual teams, however, allows them to practice these skills. In 2003, VanSlyke Turk & Martin (2004) experimented with that model by grouping students from six different universities, and saw the benefits for students to work at the international level. However, this was a one-time project and was not replicated the following year.

One project that has been ongoing since 2003 and that provides students with the opportunity to develop these skills is GlobCom. GlobCom is an international project that has grown to involve eleven universities in eleven countries on five continents. The students are grouped into global virtual teams (GVT), and each virtual team works on the same public relations brief, developing and submitting an international public relations strategy. The students communicate and cooperate with each other in chat rooms, through email, and/or by Voice over Internet Protocol (VoIP), such as Skype. At the end the project, the educators and students meet in a symposium where the top teams present their strategy.

The purpose of this chapter is to describe and analyze the global communication student project in terms of diversity. One of the most common and basic issues in GVT is overcoming the challenges of cultural diversity. Rather than looking for differences amongst the various cultures of the team, this research is designed to understand how students perceive their teams and work through the challenges of communicating with a diverse team via communication technology. Through online interviews, students answered open-ended questions that provide an insight into the ways students dealt with diversity and overcame cultural barriers.

The first part of this chapter will provide a brief review of GVT literature and an introduction of the GlobCom project. The second part of this chapter will analyze students' perceptions of diversity. Finally, the author concludes with recommendations for use of GVT in international education.

BACKGROUND

As virtual teams become more common in the business world, education institutions need to prepare students to excel in that environment. Some academic institutions emulate this by creating virtual teams. Students participate in projects with other students using a variety of communication technologies. Only a few projects incorporate a more international component by including students from various countries to create Global Virtual Teams (GVT).

Global Virtual Teams are of great use in the field of international public relations. They offer the responsiveness and expertise needed in certain situations. Jarvenpaa & Leidner (1999) define a Global Virtual Team as "temporary, culturally diverse, geographically dispersed, electronically communicating work group" (p. 792). This definition focuses on three important defining components. The first one is about whether the teams tend to be permanent or temporary. It does not provide a sense of time in terms of length, whether the team will be together one month or one year. However, GVT are usually formed with team members who have no common history and will not work together in the long term.

This definition also emphasizes that GVT members are located in a variety of different countries, but also need to be culturally diverse. A group of American managers located in different countries and working together on a project together would not be considered a GVT, because the members are too similar culturally. The last component of the definition emphasizes that all the communication is done electronically via technology, thus making the team virtual. There are many challenges for GVT to overcome in order to become effective and productive.

In general, there is a tendency for researchers to compare virtual teams with face-to-face teams (Krebs, Hobman, & Bordia, 2006; Polzer, Crisp, Jarvenpaa, & Kim, 2006; Staples & Zaho, 2006). Trust and leadership are two significant areas of research concerning GVT. Trust is an essential element of effective teamwork (Meyerson, Weick, & Kramer, 1996; Jarvenpaa, Knoll, & Leidner, 1998; Jarvenpaa & Leidner, 1999; Paul & McDaniel, 2004). Leadership is another area of interest with regards to virtual teams (Carte, Chadambaram, & Becker, 2006; Kayworth, Leidner, & Mora-Tavarez, 2000). Virtual teams have also been studied in the education environment, including looking at learning outcomes (Ferris & Godar, 2006; Liu, Magjuka, Lee, 2008; Rutkowski, Vogel, Van Genuichten, & Saunders, 2008).

However, this chapter's focus is to understand how students communicate with different cultures using communication technologies, particularly in terms of diversity, as this is one main component of the GVT definition. Most of the research on diversity has been to compare homogeneous team to non-homogeneous teams (Staples & Zhao, 2006), to compare virtual teams to face-to-face teams (Staples & Zhao, 2006; Krebs, Hobman, Bordia, 2006), and to look at the relationship between cultural differences and technology use (Karpova, Crreia, & Baran, 2009; Lee, 2002; Shachaf & Hara, 2007).

Dafoulas & Macaulay (2001) presents diversity in GVT as an advantage in terms of time zones and geographical dispersion. These become competitive advantages for companies as they can decrease costs. However, diversity is more than just about times zones and geographical location. There are many different types of diversity: 1) demographic diversity (age, sex, race); 2) deep level diversity (attitudes, values, preferences); and 3) functional diversity (knowledge, expertise), resources (Pinjani & Palvia, 2007). The most obvious type of diversity in GVT relates to the national and cultural background of team members thus including diverse nationalities (demographic diversity) and diverse cultural values (deep level diversity). Shachaf (2008) defines cultural diversity as "the heterogeneity of national cultures of team members" (p. 132). Global Virtual Team members are different in terms of national culture, as thus have different beliefs, values, attitudes, competencies, perceptions, experiences, etc. (Hofstede, 1991). However, the issue with cultural diversity is whether these differences are seen as strengths or weaknesses.

Polzer, Crisp, Jarvenpaa, & Kim (2006) review the various arguments supporting diversity as a source of higher performance or diversity as a source of conflict and malfunctioning teams. However, there are many factors affecting the two outcomes such as the types of diversity, stages of a team's life, and organizational contexts. VanSlyke Turk & Martin (2004) observed that diversity stimulated creativity. When students worked together, they started understanding the different cultures better and eliminated the stereotypes. After these barriers were gone, students listened to each other's contributions and ideas. The main advantage of having diversity in a team is to utilize the different opinions and the larger pool of skills. When companies use GVT for international public relations purpose, they need the understanding, the knowledge, and the experience of each country member's perspective. It is the strength of diversity in this context.

When comparing virtual teams to face-to-face (FTF) teams, diversity is found to be more advantageous in virtual teams. Diversity in FTF teams is found to lead to group dissatisfaction and below average performance when compared to virtual teams (Schoenecker, Martell, Michlitsch, 1997). Staples and Zhao (2006) compared heterogeneous and homogeneous teams and virtual and FTF teams, and found that the virtual heterogeneous teams performed better than the FTF heterogeneous teams. Similarly Krebs, Hobman, Bordia (2006) found that virtual teams who were dissimilar in country of birth (i.e. diverse) reported higher level of trust than FTF teams. Peters & Karren (2009) also found that diversity in virtual teams was positively related to performance. More specifically, they found that functional diversity was significantly related to team-rated performance. Interestingly, they also found that skill dissimilarity was positively related with helping behaviors, meaning that team members do not possess the same level of skills, they will tend to help each other.

Overall, most research supports the claim that diversity is advantageous for GVT. The fact that these teams are completely virtual (participants only communicate electronically) is one reason diversity leads to more trust, helping behaviors, and performance. Shachaf (2008) found that communication technology mitigates the disadvantages of diversity. Specifically, the bridging of space and time differences as well as the coding and decoding abilities of certain technologies help overcome the differences in communication. In fact, Carte & Chidambaram (2004) advances a theoretical model proposing that computer-mediated communication helps leverage the advantages of diversity while reducing the disadvantages of diversity. However, their model has yet to be empirically tested.

METHODOLOGY

In order to look into some of the issues of diversity in GVT, this chapter uses a qualitative analysis of students' answers to an open-ended question. Students expressed their opinion from their experience in participating in the GlobCom project. Because the purpose of this chapter is exploratory and descriptive, qualitative methodology is appropriate.

During the Spring Semester 2009, 118 students were emailed four different times a set of open-ended questions. These questions were sent by the GlobCom team coordinator and dealt with a variety of issues, such as leadership, technology, and preparation. One series of questions asked specifically about diversity. All students were asked to reply to the following questions: "How diverse is your team? Do you feel that most students are similar or different? Similar in what ways? Different in what ways?" 55 students answered these questions in small paragraph format. The respondents represented all the various countries involved in the GlobCom project: Australia, Germany, India, Italy, Portugal, Russia, South Africa, Spain, UAE, United Kingdom, and United States.

The students' answers were analyzed using thematic coding. Common themes and trends were identified and analyzed.

Description of GlobCom Project

GlobCom is an international public relations project run by various universities from across the world. This program started in 2003 with only a few universities from only a few countries across the world. In 2010, students from 11 universities of 5 continents formed 10 virtual teams. The students study public relations and/or communication. All the virtual teams are tasked with responding to a public relations brief and to develop a strategic communication campaign for a specific and real client. These students communicate and collaborate with each other using various technologies, including a special chatroom created for the GlobCom Project.

The main reason behind the project was to overcome the lack of experience among public relations professionals with the necessary skills of running international and global accounts. Increasingly, companies demand professionals to have international and intercultural experiences. The GlobCom project provides a platform to train public relations students to be a part of a multicultural team in a real international situation. In order to be successful, these teams must overcome cultural barriers and recognize cultural diversity as a strength.

In general, the GlobCom project is part of the curriculum of the participating university. It can be a special seminar or part of the International Public Relations curriculum. Course credits are granted by all participating universities. In most cases, a lecturer in each university is in charge of the project and responsible for the course.

The teams consist of 14 to 22 students, 1 or 2 students per university on each team. They all work together to solve a global public relations problem, similar to the work of an agency, and the communication is in English. Each global virtual team is led by a Global Team Leader (GTL). This leader is appointed, elected, or volunteered for the position. There are no direct guidelines in terms of how teams should organize. For example, the virtual teams can form working groups based on time zones or topics. All decisions are made within the team by their members.

Each virtual team must prepare a public relations strategic plan for a client, in essence acting like an international public relations agency. In the past, clients have included General Electric Foundational Policy, Kingdom of Swaziland, Gognis GmbH, and Brand Abu Dhabi. The team organizes itself by allocating tasks and work for each team member and by keeping all team members informed. Teams conduct research, create promotional material, write drafts, develop strategies, and submit a detailed response to the PR brief at the semester's end.

RESULTS

The results indicate that, in general, students were overwhelmingly focusing more on what brings them together rather than what sets them apart. The themes are discussed below. Each theme is illustrated by two or three direct quotes from the participants.

Diversity: Force Toward Similarity

The obvious similarity that most virtual team members mentioned was the fact that they were all college age students. Many of them stated that they were all around 20 years old and studying at a university. They can communicate easily as they understand each other.

"The students I have communicated with seem similar to me. Enjoyable college students. It seems we are all trying to balance GlobCom with 100 other things in our lives?" (USA)

"all the students in my team are just normal young students.... which are pretty alike to the Germans I know. Everyone has his problems with university, many have similar spare time activities, have jobs besides their classes at university, many are doing extracurricular activities." (Germany)

The fact that participants could relate to one another based on age is a strong factor in perceiving each other as similar (Krebs, Hobman, & Bordia, 2006). It is not only the age, but also the fact that they are college students and as such have certain lifestyles. College students from all universities face similar experiences, such as lectures, exams, social activities, and friendships. Related to their experience as college students is also the fact that they study the same field.

"All our group members have similar previous knowledge and preconditions" (Germany)

"Most of us have comparable professional experience" (Germany)

Many students acknowledged that their similar educational background provides a strong common ground, similar to the concept of mutual knowledge. Davis & Khazanchi (2007) define mutual knowledge as "common ground, common knowledge, and shared knowledge." They suggest that mutual knowledge should positively affect virtual teams' outcomes. In GlobCom, students have a shared understanding of public relations, and their strength as a team can be seen through their focus on being public relations practitioners and "speaking the same language." When they speak about public relations, they understand each other because of their common background.

"The majority of my team members have a united passion for Public Relations." (Australia)

"It is true that our values and beliefs are so different, but our idea how to make our campaign and the tactics are similar. Most of the students are similar in their thinking about public relations." (Spain)

Most participants' attitudes towards the project appear to be similar as they worked really hard to finish the strategic plan in less than four months. As participants are located in different countries, this common focus actually helps. The fact that GVT are temporary implies that team members have a limited amount of time to accomplish the task. This limitation of time can put pressure on members to focus on the task at hand and not on socializing which is easier done face-to-face than virtually. Interestingly, the lack of commitment for the project can also be what differentiates and fragments the virtual team.

"We all love this project because it's a perfect opportunity to reach all this objectives" (Italy)

"There are two main groups of students in our team. The first is very alike: the students who are very involved in the project, no matter in what year they are, or how much they know, that group has also different cultural background. Then there's the other group: students who show up from time to time and upload a document when they are reminded for the third time" (Germany)

Diversity: Force Toward Differences

Global Virtual Teams, by definition, comprise people from different countries and cultures. These differences were obviously felt by participants, especially when it came to communicating in English. Not everyone had the same skill level in English, and that accentuated the differences within the team, even though it was not a real concern or problem to students. Some students were more fluent in English than others, but this did not lead to many problems.

"I find that the Spanish and Italians struggle a bit with the English, that is where I differ a lot to them." (South Africa)

"The only big difference I notice is spelling. As English is the second language of many, the way sentences are phrases and spelled can be interesting and hard to decipher at times" (UK)

Language is an important issue, and some participants spoke English as a second language. However, students did not describe language as a barrier or an obstacle, just as a difference. Participants also felt differences by the sheer number of different nationalities involved in the project.

We have members from Australia, the UK, the US, India and other countries. I think the great-

est thing of diversity is the timings of our group meetings! Night in one country, early morning in another (India)

Similar to language, countries were just noted as a difference, not as an obstacle. Students expressed more of a concern with the differences in education. Many focused on the fact that, even though they were all studying public relations, there were major differences in perceived knowledge about public relations. Students perceived that concepts about public relations and communication were being taught differently in the various countries.

"The only ways in which I believe we all differ is in the ways in which we have been taught. Some members didn't even know what a PR planning model was and how to carry out a SWOT and PEST/EL analysis" (UK)

"We all have separate educational background, at least in terms of teaching methods and syllabi, there are times when only a few can immediately grasp a particular term or references...these minor hurdles are usually crossed without much difficulty" (India)

Pedagogical issues and knowledge about public relations were perceived as impediments to the process of planning a successful strategic plan, with students expressing more understanding toward participants not knowing English than participants not knowing public relations. They perceived knowledge as an important factor in the success of the project. However, again, students did not express these differences as source of conflict or problems within the GVT. One student from Portugal states: "I don't think that are a big difference between us because we know that we are in different years academics and when someone doesn't understand something someone who understand will make it clear." This statement provides support to Peters & Karren's (2009) claim that skill dissimilarity is positively related with helping behaviors. When there are different skill levels within a team, members will step up and help.

The main source of conflict is one that is common in teamwork. It appears that one universal issue when working with others remains: some people work harder than others. This difference was the most cited and one that students perceived most negatively.

"What is apparent to me is the very different workload of students within our team, and consequently, the resulting disparate involvement in the project." (Germany)

"All of the active people, even if they are more or less prepared, they always want to be part of the group and they bring some element, which are fundamental for the group itself." (Italy)

Slacking students is a universal phenomenon that will create conflict. When students were having team problems, most of the time it was due to the lack of commitment to the project by some of the team members. One last surprising finding was the fact that certain students perceived differences in the team as a reinforcement of cultural stereotypes and cultural barriers. A few students expressed wide cultural differences as "boxes" where you categorized individuals.

"I am like the Indians from a developing world and understand the difficulties and stigma that may be attached to your nationality" (South Africa)

"It's not worth to make prejudices thinking about the same stereotypes but at the end we come from different and long aged cultures and we won't change in some habits. For example the fact that German people are very serious and strictly minded, on the other hand Italians are very funny and social living." (Italy)

Diversity: Strength or Weakness?

The overwhelming sentiment of participants in GVT was that diversity served as a strength for their project. There were four mitigating factors team members identified that helped with diversity: multiculturalism, rule-making, leadership, and communication technology.

Multiculturalism

Students recognized that the diversity of opinions and ideas made their project stronger. Students did not indicate that those differences were a hindrance in their work. It made their project more interesting, because people were approaching it from different perspectives.

"And everybody has different opinions, but that is the strength of our team, because on that way we can learn for each other." (South Africa)

"Interacting with people who live in a different country, speak different language and share different perspectives than I am is just an amazing experience that I am going through by interacting with people from 11 different countries. Each member from our team have different ideas and we share them to have better results" (UAE)

The diversity of ideas and opinions was seen as the core component of their team and their project. Participants felt the benefits of these multicultural teams. They increased their cultural knowledge and discovered that different was not always bad.

Students reported that differences are normal, and therefore do not seem important or noticeable.

"I think that we are young and for this reason I feel similar and I don't notice and I don't want to notice differences between us." (Italy)

"I think there are not too many differences between countries…some countries are more punctual, accurate, others less, but…I suppose it is the normality" (Italy)

These cultural differences appeared to be an accepted norm for students working on an international project. These students expected to find cultural differences and as a result, are prepared to deal with them. Students accept those differences, and can focus on the project and not focus on the difficulties. Accepting cultural differences is a main component of living in multicultural world. Some participants mentioned their attitude towards the world and their acceptance of being part of a multicultural world as a way of being similar.

"We all value multiculturalism and we all share an interest in wanting to learn about different countries" (Australia)

"My team is not so diverse, the people who work really good in the project are well open minded to every idea of the others, and accept ideas of everybody" (Italy)

These comments indicate that some students are already open to new cultures, and thus seek situations that will expose them to people similar to them. Being open-minded is one way to overcome cultural differences and is an indicator of intercultural communication competency. Penbek, Yurdakul, & Cerit (2009) found that student with international experience are more open minded and respectful of different cultures. Overall, students indicated that they are accepting of cultural differences and ready for multiculturalism.

Rule-Making

Students accommodated for cultural differences by creating mechanisms that allowed for the differences. The mechanisms created are similar to the concepts of rule making in groups by Walther, Buz, & Bazarova (2005). The researchers found the presence of rules in virtual teams leads to a greater level of trust and liking. Students reported that teams made everything possible to ensure that everyone knew what was expected of them but also that everyone had a chance to participate.

"Everybody do their respective tasks in their own way but at the end of the day we all reach a consensus and decide on what's best for the team (India)

"All of us realize that we have a common goal and an objective that we have to reach together hence even if there are differences in opinion, we try to resolve them amicably" (India)

The development and emergence of rules was organic. Yoon & Johnson (2008) found this procedure to be a normal development pattern of virtual teams. However, some of these procedures were not always organic and developmental. Occasionally the rules were dictated by their team leaders.

Leadership

Even though students were not asked specifically about leadership in their GVT, many group members identified leaders as an important factor of success. The main communication conduit ran through the global team leader.

"More than anything else depend on our group leaders to take the final call" (India)

"For the convenience of the team, our team leader has chalked out a pattern according to which all of us work" (India)

These comments confirm the importance of leadership in GVT. Sivunen (2006) points out four techniques that a leader can use to strengthen members' identification with the team: catering to the member, giving positive feedback, outlining common goals, and working the meetings. As illustrated above, the students mentioned a few of these tactics.

Communication Technology

Finally, as Carte & Chidambaram (2004) suggested, similarities were mitigated by communication technology. Students mentioned that technology was a mediating factor in their perception of similarity and differences.

"I am sure that all the people in my team are very different but when we work online it doesn't really seem that way." (India)

"The team has a common goal which is to do the communication plan for the client and because we are all working on a virtual team I still can not find many differences." (Portugal)

"The meetings couldn't be verbal they had to be written on MSN; I feel that with a verbal conversation it would be far easier to communication and that team differences and similarities would have been further highlighted." (UK)

These students' comments provide insight as to why diverse virtual teams can outperform FTF teams. Communication researchers tend to view virtual teams as less effective because of the lack of nonverbal communication. However, as Carte and Chidambaram (2004) suggest, communication technologies are a "bundle of capabilities" and can enhance the virtual team.

CONCLUSION

All GlobCom teams completed their project and performed very well at the conclusion of the course. Surprisingly, they never complained that the task was too hard or too complex, and they never complained about technological problems. Through these online interviews, students

Table 1. Diversity as force towards similarity and differences

Forces toward similarity	Forces toward differences
Age	English language skills
Lifestyle	Cultural stereotypes
Educational background	Nationalities
Shared knowledge	Lack of shared knowledge
Involvement in the project	Lack of involvement in the project

provided interesting insights as to how they perceived diversity in GVT. The majority of the students focused more on the similarities than the differences. The main themes are listed in Table 1.

Students identified their knowledge for PR as a force towards similarity, but the lack of knowledge had the opposite effect as a force towards differences. If students felt that other team members lacked the same type of knowledge, they felt more differences within their team. This conclusion is consistent with Davis & Khazanchi's (2007) discovery about how mutual knowledge can affect virtual team performance. When teams have a shared knowledge, they tend to perform better. Thus, when creating GVT, educators need to consider not only the knowledge the students possess, but also the type of educational system students come from.

Regardless of whether they felt different or similar, students explained how they overcame these barriers. There were four mitigating factors that played an important role in making diversity an asset to their GVT's success. First, students recognized and learned the value of diversity. Teams expressed not only an interest in listening to others' opinions and ideas, but valuing these opinions and ideas. Second, students self-managed their teams successfully. Rules were established, teams scheduled meetings, established work procedures, and so on. These self-imposed mechanisms were important and established guidelines for everyone, regardless of the country or nationality. Third, students recognized the importance of a leader. Teams can create their own rules and guidelines, but they still need a good leader to provide more authority to the team. The last factor was that the communication technology mitigated the differences. Many researchers tend to look for cultural differences in use of technology (Karpova, Correia, & Baran, 2009). The fact is that these technologies de-emphasize cultural differences as they eliminate certain cues, such as physical appearances and accents. This claim provides more legitimacy to a concept called "perceived similarity." Zellmer-Bruhn, Maloney, Bhappu, & Salvador (2008) argue that perceptions of diversity are as important as the actual diversity. The fact that certain students did not seem to notice the differences amongst their team members indicates that their perceptions of the team would lean toward being more similar than different.

The overwhelming outcome of the GlobCom project is that students do not perceive cultural diversity as a barrier. Students bring different perspectives from their own culture, but focus on the commonalities among members. Participants in the GVT did not express some of the problems that are put forth in other studies, such as lack of nonverbal cues. In fact, students stated the opposite. These students experienced the closest simulation to practicing international public relations. This experience will be incredibly valuable for employers. Courses at higher education institution can integrate the use of Global Virtual Teams in their curriculum as technology is more widely implemented and international collaboration among faculty increases. With less expensive technology and international cooperation increasing, projects such as GlobCom are becoming more common. The resources required for virtual teams are minimal: faculty members willing to create an international project and internet access.

The use of global virtual teams in education encourages students to work in an environment of international cooperation and enrich their academic knowledge of diversity. For teaching international public relations, this type of experience facilitates the understanding of one of their main challenges of "think global, act local."

REFERENCES

Carte, T. A., & Chidambaram, L. (2004). A capabilities-based theory of technology deployment in diverse teams: Leapfrogging the pitfalls of diversity and leveraging its potential with collaborative technology. *Journal of the Association for Information Systems*, *5*(11/12), 448–471.

Carte, T. A., Chidambaram, L., & Becker, A. (2006). Emergent leadership in self-managed virtual teams. *Group Decision and Negotiation*, *15*(4), 323–343..doi:10.1007/s10726-006-9045-7

Dafoulas, G., & Macaulay, L. (2001). Investigating cultural differences in virtual software teams. *The Electronic Journal on Information Systems in Developing Countries*, *7*(4), 1–14.

Davis, A., & Khazanchi, D. (2007). Does mutual knowledge affect virtual team performance? Theoretical analysis and anecdotal evidence. *American Journal of Business*, *22*(2), 57–65. Retrieved from http://www.jstor.org. doi:10.1108/19355181200700010

Ferris, S. P., & Godar, S. (2006). *Teaching and learning with virtual teams*. Hershey, PA: Information Science.

Hofstede, G. (1991). *Cultures and organizations*. New York, NY: McGraw-Hill.

Jarvenpaa, S. L., Knoll, K., & Leidner, D. E. (1998). Is anybody out there? Antecedents of trust in global virtual teams. *Journal of Management Information Systems*, *14*(4), 29–64.

Jarvenpaa, S. L., & Leidner, D. E. (1999). Communication and trust in global virtual teams. *Organization Science*, *10*, 791–815. doi:10.1287/orsc.10.6.791

Karpova, E., Correia, A. P., & Baran, E. (2009). Learn to use and use to learn: Technology in virtual collaboration experience. *The Internet and Higher Education*, *12*(1), 45–52..doi:10.1016/j.iheduc.2008.10.006

Kayworth, T. R., Leidner, D. E., & Mora-Tavarez, M. (2000). *Leadership effectiveness in global virtual teams*. Working Paper, Baylor University.

Krebs, S. A., Hobman, E. V., & Bordia, P. (2006). Virtual teams and group member dissimilarity: Consequences for the development of trust. *Small Group Research*, *37*(6), 721–741. Retrieved from http://online.sagepub.com. doi:10.1177/1046496406294886

Lee, O. (2002). Cultural differences in e-mail use of virtual teams: A critical social theory perspective. *Cyberpsychology & Behavior*, *5*(3), 227–232. doi:10.1089/109493102760147222

Liu, X., Magjuka, R., & Lee, S. H. (2008). The effects of cognitive thinking styles, trust, conflict management on online students' learning and virtual team performance. *British Journal of Educational Technology*, *39*(5), 829–846.. doi:10.1111/j.1467-8535.2007.00775.x

Meyerson, D., Weick, K. E., & Kramer, R. M. (1996). Swift trust and temporary groups. R. M. Kramer, & T. R. Tyler (Eds), *Trust in organizations: Frontiers of theory and research* (pp. 166-195). Thousand Oaks, CA: Sage Publications.

Paul, D. L., & McDaniel, R. R. (2004). A field study of the effect of interpersonal trust on virtual collaborative relationship performance. *Management Information Systems Quarterly*, *28*(2), 183–227.

Penbek, S., Yurdakul, D., & Cerit, A. G. (July 2009). *Intercultural communication competence. A study about the intercultural sensitivity of university students based on their education and international experiences*. Paper presented at the European and Mediterranean Conference on Information Systems. Izmir, Turkey.

Peters, L., & Karren, R. J. (2009). An examination of the roles of trust and functional diversity on virtual team performance ratings. *Group & Organization Management*, *34*(4), 479–504. Retrieved from http://online.sagepub.com. doi:10.1177/1059601107312170

Pinjani, P., & Palvia, P. C. (2007). *Diversity in global virtual teams: Results from a field survey.* Abstract from the International Conference on Advances in Interdisciplinary Statistics and Combinatorics, Greensboro, NC.

Polzer, J. T., Crisp, C. B., Jarvenpaa, S. L., & Kim, J. W. (2006). Extending the faultline model to geographically dispersed teams: How colocated subgroups can impair group functioning. *Academy of Management Journal*, *49*(4), 679–692. doi:10.5465/AMJ.2006.22083024

Rutkowski, A. F., Vogel, D., Van Genuichten, M., & Saunders, C. (2008). Communication in virtual teams: Ten years of experience in education. *IEEE Transactions on Professional Communication*, *51*(3), 302–312. doi:10.1109/TPC.2008.2001252

Schoenecker, T. S., Martell, K. D., & Michlitsch, J. F. (1997). Diversity, performance, and satisfaction in student group projects: An empirical study. *Research in Higher Education*, *38*(4), 479–495. doi:10.1023/A:1024966627400

Shachaf, P. (2008). Cultural diversity and information and communication technology impacts on global virtual teams: An exploratory study. *Information & Management*, *45*(2), 131–142.. doi:10.1016/j.im.2007.12.003

Shachaf, P., & Hara, H. (2007). Behavioral complexity theory of media selection: A proposed theory for global virtual teams. *Journal of Information Science*, *33*(1), 63–74. doi:10.1177/0165551506068145

Sivunen, A. (2006). Strengthening identification with the team in virtual teams: The leaders' perspective. *Group Decision and Negotiation*, *15*(4), 345–366..doi:10.1007/s10726-006-9046-6

Staples, D. S., & Zhao, L. (2006). The effects of cultural diversity in virtual teams versus face-to-face teams. *Group Decision and Negotiation*, *15*(4), 389–406..doi:10.1007/s10726-006-9042-x

VanSlyke Turk, J., & Martin, E. F. (2004). Opportunities and challenges for intercultural virtual teams: Human factors in project Connect and their implications for using virtual teams in public relations. *Proceedings of the 7th Annual International Public Relations Conference*. Miami, FL. Retrieved from http://instituteforpr.org

Walther, J. B., Bunz, U., & Bazarova, N. N. (2005). The rules of virtual groups. *Proceeding of the 38th Hawaii International Conference on System Sciences* (pp. 1-10). Waikoloa, HI.

Yoon, S. W., & Johnson, S. D. (2008). Phases and patterns of group development in virtual learning teams. *Education Technology Development*, *56*, 595–618..doi:10.1007/s11423-007-9078-x

Zellmer-Bruhn, M. E., Maloney, M. M., Bhappy, A. D., & Savaldor, R. (2008). When and how do differences matter? An exploration of perceived similarity in teams. *Organizational Behavior and Human Decision Processes*, *107*(1), 41–59.. doi:10.1016/j.obhdp.2008.01.004

KEY TERMS AND DEFINITIONS

Diversity: Recognizing and accepting individual differences based on culture.

Intercultural Communication: Understanding how people from different cultures communicate and perceive the world.

GlobCom: International student project that require global virtual teams to create a public relations strategic plan for a client.

Global Virtual Team: Group of dispersed individuals from various cultures who communicate primarily through electronic communication.

Multiculturalism: Belief that multiple cultures are valuable to an organization or society.

Mutual Knowledge: Knowledge communicated to team members and that is known to be shared.

Public Relations: Communication activities that support an organization's goal of creating and maintaining relationships with various publics.

Chapter 11
A Framework for Networked Experiments in Global E-Science:
Perspectives for E-Learning in Global Contexts

Diego Liberati
Italian National Research Council, Italy, Italian National Nuclear Physics Institute, Italy & Politecnico di Milano University, Italy

ABSTRACT

This chapter presents a framework that creates, uses, and communicates information whose organizational dynamics allow individuals to perform a distributed cooperative enterprise in public educational environments. The approach presented here assumes Web services (possibly offered over a grid) are the enacting paradigm used to formalize educational interactions as cooperative services on various computational nodes of a network. By examining a case study involving a well known micro-array experiment in the growing field of bioinformatics, this chapter will detail how specific classes of interactions can be mapped into a service-oriented model that can be implemented in a variety of e-learning contexts. This framework illustrated by this case study allows for a sophisticated degree of e-learning that can be applied to a range of local or international contexts.

INTRODUCTION

The topic of e-learning is often associated with a sophisticated and often manned teaching platform used to allow remote users to access educational information that enhances an individual's knowledge about a topic. In this chapter, a broader view of e-learning is taken by examining how knowledge exchanges can be distributed across a network of notes. In particular, the chapter presents a framework educators can use to facilitate co-operation and discovery associated with research and teaching practices.

DOI: 10.4018/978-1-61350-077-4.ch011

In today's world, both researchers and educators face problems linked to the virtualization of resources and to the orchestration of services in the heterogeneous and distributed context of e-science (De Roure, Gil, & Hendler, 2004). In this sense, distributed network e-science approaches, like the one proposed in this chapter, could provide tremendous benefits to e-learning by improving an institution's or a programs' ability to recruit a higher portion of the human capital (i.e., researchers and educators) who are not yet fully involved in scientific research and teaching. Additionally, such approaches can help educational institutions in other nations – particularly developing nations – overcome the subtle form of digital divide affecting educational practices in those nations by providing researchers, educators, and students with access to instrumental information associated with effective science education.

SCIENCE, NETWORKS, AND COLLABORATION

Today, effective science research and education are increasingly a question of having a critical mass of skilled people, often with complementary backgrounds, who can collaborate to become a kind of unique global organism in pursuit of a substantial common goal. Historically, scientifically specialized workforces have involved concentrating the right group of individuals in unique sites where special scientific instruments are available. Today, however, the availability of information and communication technologies enables researchers to interact as part of a distributed network of individuals living in different locations, but collaborating through information communication technologies (ICTs) to engage in a variety of projects.

In the current global context, many of the problems faced in the present renaissance of biology do not need a unique, big site, while other disciplines, like geology, are intrinsically distributed for data collection. Such differences can have important effects on or consequences for science education. Within this context, important educational benefits can arise from the use of possible networks of experimenters who are not necessarily located in the same physical space, but who can use different technologies to collaborate on the same major experiments in a way that involves all of their complementary competencies.

In the cases of bioinformatics paradigm of e-science examined in this chapter, a proper virtual organization is required to manage the flow of information over a given network. A networked organization such as this might be logically defined through workflows acting over Web services, possibly exploiting a grid context. The technology addressed in this chapter allows the most competent scientist to use such a networking approach to design project workflows even if the individual is not at a very high level of competence in ICTs (which are, in a sense, logically embedded). Rather, the proposed context offers participants a special opportunity to electronically learn complementary skills, with respect to the ones already possessed. The idea is that participating individuals, through their distributed collaboration on a project, can constitute a kind of learning tool related to the whole process on which they are working. In this way, a distributed approach to doing scientific research could provide educators in a variety of fields with a framework for developing effective and distributed e-learning platforms.

COLLABORATION AND E-SCIENCE

In recent years, the concept of virtual organization has been developed as a result of the grid computing paradigm (Foster, Kesselman, Nick, & Tuecke, 2002; Foster & Kesselman, 2004) that is used as a general conceptual model to coordinating activities. Such virtual organization is a set of individuals and institutions that all have direct access to services, knowledge, tools, data, software, computers, and possible other resources in a heterogeneous dynamic way. The aim of such organizations is to use ICTs to achieve a common goal through collaboration. The basis of virtual organization is the virtualization of resources. The underlying idea is to create and associate a generic interface that allows collaborators to access offsite technologies via remote access in order to make use of technologies and expertise located in other areas. Through this approach, individuals do not need to be co-located in order to collaborate on the same experiment.

Within the context of scientific collaboration, a distributed scientific experiment can be defined on the basis of aggregating, sharing, using, and re-using virtualized resources. Such resources may be human beings, data, ICT components, or computational power. They can also include services in a heterogeneous and distributed context, according to the service-oriented architecture paradigm (Crawford, 2003). Some of these services could be pre-existent and made available by some organizations within the pool, while others have to be virtualized in a suitable way and shared within the scientific community. For example, a collaborator in one location might have access to a particular piece of technology needed to conduct the first part of an experiment, while a collaborator in another location might possess a different technology needed to perform the next part of the overall experiment. By collaborating virtually via ICTs, these individuals can pool their resources and complete the overall experiment without needing to travel.

"E-science" is the term currently applied to the use of advanced computing technologies to support scientists in such collaborative research undertakings (De Roure, Gil, & Hendler, 2004). Because of their need for high-performance computing resources, as well as cooperative ICTs like those found on the Web, many scientists are increasingly drawn to grid computing and to the Web as providing the infrastructures needed to support data management and analysis across organizations. High-performance computing and communication technologies are thus enabling computational scientists, or e-scientists, to study and better understand complex systems, like neurophysiology (Baraldi, Manginelli, Maieron, Liberati, & Porro, 2007), systems biology (Milotti et al., 2008), agrobiotechnology (Liberati 2008) pathophysiology (Liberati 2009) and pain (Storchi et al., 2009).

These online technologies also allow for new forms of collaboration and learning over large distances and the ability to process, share, disseminate instructional and educational information across vast distances. Global-scale experimental networking initiatives have been developed in recent years: the aim is to provide advanced cyber-infrastructures for e-scientists through the collaborative development of networking tools, advanced grid services, and data-intensive applications (Newman, Ellisman, & Orcutt, 2003). Such technologies could easily be grafted onto or into existing science education curricula at a variety of universities to provide students with access to a richer and a broader range of educational experiences within the sciences. Grids provide basic facilities for robust computation and efficient resource management, transfer, and sharing, and they support distributed computation – all of which could be used to augment educational practices.

Additionally, new Web developments have also been undertaken to address the need to support scientific collaboration by addressing multidisciplinary distributed science research at the end-user level. Such projects have enabled semantic integration through transparent document sharing and metadata annotations – all of which could be made available to students in order to enrich their educational experiences beyond the confines of the on-site classroom. Because both grid computing and Web services deal with interoperability, from the e-science and e-learning perspectives, both are useful, for on its own, neither technology would be able to achieve the full e-science vision. The integration of the two, often called semantic grid, creates a key framework that can provide a myriad of rich new e-leaning opportunities for students in the sciences as well as in a variety of other disciplines.

DISTRIBUTED PROCESSES AND MUTUAL OBJECTIVES

The aim of the e-science approach is to enable a set of remote scientific (but possibly also commercial) co-workers to design the workflow of distributed experiments. The purpose for such an approach is to allow the execution of experiments over a set of distributed, cooperative nodes with each node in the overall network providing and using a set of services. This approach creates a kind of network-based synergy in which the whole is greater or is able to accomplish more than any of its individual parts.

Typical the services involved in this system are mathematical tools, specific applications and scientific methods, database access services, and even whole existing portions of other experiments that are collectively viewed as parts of the flow of the present experiment process. According to this approach, virtualization issues of resources (i.e., relying on resources that are not physically present in all locations) might be easily achieved using Web service technologies (Alonso, Casati, Kuno, & Machiraju, 2004). Coordinating overall processes through this system may, in turn, be achieved by interpreting each single virtualized resource as a component of a distributed cooperative process, modelled in terms of workflows (Van der Aalst & Van Hee, 2002). That is, the services offered by each collaborator or available at each location contribute to the whole of the related experiment. The general process of a problem-solving strategy in a virtual organization can thus be seen as the whole of all the single collaborative procedures (Travica, 2005). Each procedure might be seen as the flow of input data through a computing core and done to produce output data. The desired orchestration (Peltz, 2003) of such single procedures constitutes the aggregate workflow related to the particular scientific (or any other business) activity. Again, each part in the distributed process moves toward achieving a mutual goal of interest to all involved.

Within such a paradigm, collaborations are the joined efforts made by the involved organizations, and they are undertaken to achieve a common goal through some services described with a workflow, recalling other formalisms for specifying execution procedures like Petri nets. The single procedures, in general, can represent tasks performed by human actors (e.g., using a specific technology to perform a particular kind of analysis), with various and possibly repetitive interactions with the system. The geographic, logical, and institutional location of the procedures is often distributed over more than one organization. (From an educational perspective, this approach means that networked organizations can provide students with access to technologies and expertise not locally available at their home institution.) In such systems, the definition of the workflow is typically executed by a human actor, possibly partially assisted by an expert system, competent within the application domain.

Such a workflow designer is in charge of selecting and composing the distributed resources of the experiment. This individual, moreover, facilitates such processes without caring about the implementation and technical details related to the physical distribution of the experiment workflow parts. Among the advantages of such an approach, a few are worth mentioning:

- **Efficiency:** The workflow definition procedures should help discover and localize the resources, hence relieving the workflow designer from technical and repetitive tasks and thus letting him better contribute to the creation of best practices that can be evaluated, compared, and shared with other people.

- **Reproducibility:** The workflow can be precisely repeated automatically, also by third parties possibly belonging to external organizations, even on instances of different data and parameters.
- **Reuse and automatic enhancement of knowledge:** The produced outputs are potential new inputs to some of the subprocesses: a virtuous cycle is set up, incrementally augmenting knowledge about the experiment, the domain, and also its execution needs or requirements, like speed enhancement or better allocation of experiment resources.
- **Traceability:** The workflow is monitored within an environment where the data sources and tools can be traced and checked.

As discussed, the application domain of this chapter is bioinformatics, or more generally e-science (Hey & Trefethen, 2003). These concepts from e-science, however, can be generalized to generic range of online educational undertakings. In the e-science context, where the concept of collaboration can be considered stronger than in other areas, the term "virtual organization" is often replaced by the term "collaborative environment." This distinction is made to denote a collaborative laboratory (Bosin, Dessi, Fugini, Liberati, & Pes, 2006a). Such an approach, however, can be applied to laboratory-based classes or other context-based educational environments in a way that can enrich and expand e-learning practices.

Based on this premise of a collaborative environment, the approach discussed here goes in the direction of developing support methods and tools for virtual laboratories—that is, virtual organizations where cooperation to execute a scientific experiment is supported- and scientific e-learning. Within this context, Web services and the grid enable technologies that can support the simulation and execution of different classes of experiments—from visualization (like browsing and search interfaces) to model identification through clustering and rules generation—in application fields such as drug discovery, micro-array data analysis (Garatti, Bittanti, Liberati, & Maffezzoli, 2007), or molecular docking. The benefits such an approach can have to the teaching of science at a variety of levels is impressive. By applying Web services and the grid, an experiment or a simulation can be executed in a cooperative way on various computational nodes of a network that get aggregated dynamically for the specific experiment in a virtual laboratory. In this way, a student's educational experiences can transcend the limitations of his or her local, physical environment. Such approaches also allow knowledge to be exchanged among researchers beyond the walls of a given laboratory, and thus introduce a range of perspectives and expertise to the process. (Again, this increase access to expertise only enriches the educational experiences of students involved in such networked undertakings.) Upon the conclusion of the experiment, the virtual laboratory can be released and the cooperation network closed.

For such distributed educational opportunities to be effective, a correct design and set up of the experiment workflow, visualization methods, and information retrieval tools (e.g., for searching similar protocols or descriptive datasheets for chemical reactors) must be established. Such a design supports the cooperating scientists (and science educators) who participate in joint experiments where specialized tools (e.g., data mining or database tools) or computational power (e.g., for protein analysis based on their spatial structure) are available only at specific nodes or participating sites in the overall network of contributors (e.g., science educators) participating in the experiment.

Within the context of such collaborative undertakings, the visualization part of the process needs special care particularly in relation to considering "friendly" (i.e., easy-to- use even by individuals who are computer illiterate) interfaces and graphical simulations enable improved comprehension of currently textual explanations. (This chapter does not address the various aspects of the visualisation issue related to this process, for such an examination is a topic best examined in its own, dedicated work.) Improving processes to allow even intermediate results to be examined is of paramount importance in order to allow the scientist to concentrate on a given research problem from many different perspectives. The focus of the remaining discussion will thus focus more on processing data than to visualizing them.

THE ARCHITECTURE OF PROJECTS

To illustrate how such a distributed approach operates, this chapter uses the example of a distributed experiment involving the process of DNA microarrays clustering based on previously examined techniques (see Bosin, Dessi, Liberati, & Pes, 2006b; Ferrari-Trecate, Muselli, Liberati, & Morari, 2003; Garatti et al., 2007; Muselli & Liberati, 2002). While this example is used to illustrate how such a networked approach might be used within the context of scientific research, readers should be able to view this process as one that could also be used to teach students about such research. Thus, the example is one that can be applied simultaneously to research and education and can be used to acculturate students to new approaches to conducting scientific research.

Within this context, the architecture of the work is as follows. The term "organization" represents the members of a virtual laboratory (universities, hospitals, research centers, enterprises, single researchers) who have decided to cooperate to run a specific experiment. We can distinguish the following sites:

- Experiment definition and launch site
- Data storage site
- Computation site
- Results visualization site
- Experiment invocation site

The decomposition in modules and the high interoperability of the elements favours reuse both inside and outside the organizations and at different levels of granularity. The granularity and the terms of the contract about collaborations taking place outside the virtual laboratory – be it used for research or educational purposes – are negotiable and adjustable depending upon specific demands and according to the usual variables of complexity, performance, and costs.

Experiment Definition and Launch Site

In this process, the workflow designer is generally someone associated with the leading organization (the organization taking the lead on the project). That designer creates the workflow model that characterizes the distributed experiment or educational undertaking by selecting the needed resources (either local or available on the Web) and specifies their choreography – or the order in which they will take place – in the workflow of the overall experiment or educational exercise. The results of the experiment will usually be delivered at that same site. The results are accessible both to a human actor and an automatic system that might have to provide adequate storage in local archives for analysis and for future and/or remote reuse.

The main actor operating in this area is the domain expert, such as the researcher or scientist for a research project or the lead professor or instructor for an educational collaboration. Usually, that main actor is not deeply skilled with technical aspects of information and communication technologies; as a consequence, the proposed solution allows this actor to skip technical problems in such field. The tool used to support this actor in specifying the experiment workflow is a workflow editor (e.g., an open source workflow management system like Taverna – see http://taverna.sourceforge.net/) that allows participating scientists or educators to define and execute their workflows. Such a system would also allow participants to analyze the deriving outputs through the use of Web service discovery, selection, and link that can be executed through a graphical support.

Once the definition of the experiment – be it for research or educational purposes – has been completed, an instance of a workflow model is created and produces results. Such a model can be created in the Simple Conceptual Unified Flow Language (Scufl) format or XML file. Scufl (Oinn, 2004) is a workflow description language, more known in the commercial and software engineering environments (Andrews et al., 2003), similar to the Business Process Execution Language (BPEL).

Data Storage Site

Different organization can use various approaches to store and make available, according to the Web service standard way of operating (Booth et al., 2004), all the data necessary for the given experiment. These data resources can also be drawn from various sites that could belong to different organizations participating in the overall project. In such cases, several instances of the experiment have to be launched at the same time in order to achieve the best parallelisms and resource allocation policy for the project. According to this approach, data can be found in any location on the Web provided that such data are identified through a specific URL.

Computation Site

The computation associated with such processes is usually done on selected machines that have the computational power adequate for achieving the goal of the overall process. In our example, the processing is done using the Matlab tool (see http://www.mathworks.com) that is modelled as a Web service implementing, among other things, the clustering process described by Garatti et al. (2007); the Bayesian networks described by Bosin et al. (2006b); the logical networks described by Muselli and Liberati (2002); the piece-wise affine identification described by Ferrari-Trecate et al. (2003); the neural networks described by Drago, Setti, Licitra, and Liberati (2002); and the deconvolution described by Sartorio, De Nicolao, and Liberati (2002). The computation can be partitioned in functional sub-modules.

Single functions of the processing are isolated, made independent, and published – possibly assigning each part of the overall procedure to a physically different sites. This way, modularity and reuse are facilitated both inside the organization (with respect to hardware service and licensing, load balancing, security and parallelisms) and outside it (with respect to collaboration, knowledge sharing, and possible distribution of provided services, including external e-learning, under possible payment). Analogously, workflow computing processes of external organizations can possibly be included in order to achieve collaboration, knowledge sharing, and externalisation of procedures. From an educational perspective, such an approach can greatly reduce the costs associated with doing all parts of such a process locally as well as make such processes an option in educational environments where the needed technology might not be available locally.

At this site, additive modular potentialities are available through interfaces that can be linked to to existing services, such as Web service definition language interfaces (Christensen, Curbera, Meredith, & Weerawarana, 2003), with possibly additional information regarding the modalities of execution of the experiment. If, for example, data localization is at the same node that will be involved in processing, the download of the dataset will probably require less time. There is the possibility to add a module that caches the last used datasets. Such an approach means that if the experiment – be it for research or educational purposes – needs to be reset, datasets will be already formatted and locally ready. (Consider the implications such an approach has for the efficiency with which institutions can engage in educational activities related to not only the sciences, but a range of subjects in which extensive data collection and processing occurs).

Result Visualization Site

The site where the experiment results can be displayed and analysed is usually the same as the experiment launch site (i.e., where the overall process began). However, because data are identified through a URL, they can also be visualized on other sites belonging to other organizations (e.g., other educational institutions). The visualization will be the same on all the participating nodes and will be independent of the platform used at the various sites due to interoperability issues provided by the support environment. The results are provided in XML to all participants, and these results do not need appropriate rendering tools, because results are presented through a standard Web page.

Experiment Invocation Site

The whole experiment is itself viewed as a service that carries an added value for the virtual laboratory; hence, it is potentially interesting for external organizations, which can invoke it from within a complex Web service. Through the additional myGrid tools (see http://www.mygrid.org.uk/), such an environment allows scientists and educators to save the experiment, or parts thereof, to a repository of reusable experiments from where it can be discovered and reused inside the consortium and within the whole scientific community for subsequent experiments. From an educational perspective, this option allows instructors to reuse parts or all of a previously run experiment to teach the same concepts or approaches to different groups of students over time.

MODULES

In the framework discussed here, the implementation of some components of the virtual laboratory support system, three macro-modules, are worth mentioning. Each of these components, in turn, can be implemented as a Web service and placed in different physical nodes. These components are

- Data extraction module. Since a variety of data sources (like on patients, on medical cares, on health protocols) are currently available online, often on public sites, the data extraction module executes a pre-processing of the data discovered on the sites in order to take different dataset categories into a single kind of internal representation, useful to start the data clustering procedures. Other types of pre-processing can be added to this module.
- Experiment engine. This module is in charge of setting a link with the Matlab software (or in general the computational language) and running the classification algorithms, like, in the proposed example, clustering, principal component analysis, and Bayesian networks.
- Visualization module. The visualization module deals with problems related to the most suitable visualization mode of the experiment results.

The decomposition of the virtual laboratory support architecture thus provides the well known benefits of modular systems. These benefits include flexible composition, addition/deletion, reuse of modules, and isolation of faulty modules – all of which can be of great benefit to educators who wish to examine the same topics across different classes. In particular, these benefits come from the ability to substitute, reuse, and update single functional parts without affecting other modules – a factor that can add stability to educational processes and pedagogical practices.

Experiment Engine

In such distributed projects, the experiment engine interfaces with the Matlab environment and executes the code used for the clustering procedure. The core of the elaboration is a Matlab function. This is made active by Java code through a suitable interface called JMatLink (see http://jmatlink.sourceforge.net/). The whole Java code is exposed as Web service. Single functions can be isolated as independent Web services and allocated on different sites according to a reuse principle. The creation and deployment of Web services can involve Apache Axis that has been used and installed on the Web container Apache Tomcat (see http://tomcat.apache.org/).

Data Extraction Module

Within the context of the example experiment examined here, nearly all DNA microarrays data is available in a few standard formats. Each can be further distinguished in different variants. A wide choice of data is available, for example, from the Broad Institute (see http://www.broad.mit.edu/cgi-bin/cancer/datasets.cgi). Data extraction is done through a Java module that can express these different kinds of formats into a sole representation. The module, exposed as a Web service, can be adopted and invoked also by other clustering processes of other organizations as an independent element placed backwards in the respective supply chain.

Visualization Module

The visualization module deals with problems related to the most suitable visualization mode of the experiment results. The results can be presented as graphs and textual descriptions and can be put in an HTML page available to any Web browser. Also, the visualisation module is a virtual resource and, as a consequence, can be included in other experiments as an independent module in the workflow of other organizations. As a result, this module needs to be specially customized with respect to e-learning targets.

ENHANCING FLEXIBILITY IN RESEARCH AND TEACHING

Starting from the architecture presented here, an evolution can promote the flexible execution of an experiment for research or educational purposes. In particular, the use of agents (Wooldridge, 2002) can allow for dynamic selection, at run time, of the services that execute the activities of an experiment.

The improved architecture can make use of two kinds of agents: supervisor agents and resource agents. A supervisor agent is in charge of supervising the overall execution of the experiment and belongs to the organization in which the execution of the experiment has been launched. For each activity of the experiment, such an agent discovers from a telephone directory service the set of agents able to provide the needed service(s). These resource agents represent and manage the actual services, modelled as Web services and described with various properties. These properties include the quality of services parameters—like time, cost, availability, robustness, and security—needed to select the Web services. For example, the activity assignment can be done according to the contract net protocol (Smith, 1980) in which a supervisor announces an activity to some resource and awards it to the best offering one. A given organization within the related the collaborative network or researchers or educators can have more supervisors, and more supervisors means more experiments can be running at the same time and more resources can be made available.

In this scenario, the workflow designer does not need to be aware of the physical structure of the network that will be used to execute the experiment in a distributed collaborative way. For example, the workflow designer does not need to specify that the initial activity of the workflow will be executed by a service from a specific organization (e.g., a specific university). Rather, the workflow designer has only to specify the kind of the initial activity. The supervisor agent will then find the most appropriate service to execute the experiment according to quality of service parameters. Note that the service from a given organization will execute the activity if it is the only available service for the activity or if it is considered by the supervisor to be the best service (i.e., requiring minimum time and with a high degree of security). Setting up this improved architecture requires participants to address a number of issues, including the dynamic negotiation of contracts between agents and the workflow engine used between agents and Web services.

DISCUSSION AND CONCLUDING REMARKS

This chapter has described an approach to the realization of virtual laboratories for e-science and has indicated how this approach can be extended to e-learning. The architecture of a prototype has been illustrated in order to show how a distributed process can be modelled and executed in the bioinformatics field. In so doing, the chapter has noted how this process can be applied to e-learning environments in both the sciences and in other fields. The technologies of Web services in synergy with the Taverna tool have adequately answered to the pre-fixed goals: flexibility, reuse, interoperability of the components, possibilities of parallelisms, simplicity of the workflow creation for actors with poor ICT knowledge, and a certain general robustness. When the grid service (Foster & Kesselman, 2004) reachs a better stability on the market, it will be possible to build a prototype that implements the presented work by using this technology also, as well as the emerging clouds paradigm. The goals of such an undertaking will be the ones that the grid services take from Web service: quality of service, security, load balancing, self-configuring, self-healing, and self-optimizing support to semantics.

The approach presented in this chapter is general, and it could be extended to wider areas of e-learning. The evolution of such approaches will occur in the virtual marketplaces of services, where the modalities of distribution are characterised by ways of payment—for subscription or for amount consumed—for single transactions. In this way, the proposed approach could constitute a distributed way to build e-learning infrastructures. Such processes, however, bring with them a range of associated concerns educators will need to address when using such approaches. These include issues of privacy and data security, considering methods used to select trusted nodes within the cooperation network and meeting the learning requirements of the different educational organizations participating in such projects.

ACKNOWLEDGMENT

Thanks to the research groups of the Università di Cagliari: Professor Nicoletta Dessì, Dr. Andrea Bosin, and Dr. Barbara Pes; and of Politecnico di Milano: Professor Maria Grazia Fugini and Dr. Francesco Amigoni, for common ideas and work. The prototype implementation by Alberto Trovato is also acknowledged.

REFERENCES

Alonso, G., Casati, F., Kuno, H., & Machiraju, V. (2004). *Web services: Concepts, architectures, and applications*. Wiesbaden, Germany: Springer.

Andrews, T., Curbera, F., Dholaki, H., Goland, Y., Klein, J., Leymann, F., et al. (2003). *BPEL4WS: Business Process Execution Language for Web Services,* version 1.1.

Baraldi, P., Manginelli, A. A., Maieron, M., Liberati, D., & Porro, C. A. (2007). An ARX model-based approach to trial by trial identification of fMRI-BOLD responses. *NeuroImage*, *37*(1), 189–201. doi:10.1016/j.neuroimage.2007.02.045

Booth, D., Haas, H., McCabe, F., Newcomer, E. L, Champion, M. F., & Orchard, D. (Eds.). (2004, February 11). Web services architecture. *W3C Working Group Note.* Retrieved February 15, 2011, from http://www.w3.org/ TR/ ws-arch/

Bosin, A., Dessì, N., Fugini, M. G., Liberati, D., & Pes, B. (2006a). Applying enterprise models to design cooperative scientific environments. [Wiesbaden, Germany: Springer.]. *Lecture Notes in Computer Science*, *3812*, 281–292. doi:10.1007/11678564_25

Bosin, A., Dessì, N., Liberati, D., & Pes, B. (2006b). Learning Bayesian classifiers from gene-expression microarray data. [Wiesbaden, Germany: Springer.]. *Lecture Notes in Computer Science*, *3849*, 297–304. doi:10.1007/11676935_37

Christensen, E., Curbera, F., Meredith, G., & Weerawarana, S. (2003). Web service definition language (WSDL) 1.1. *W3C Note.* Retrieved February 15, 2011, from http://www.w3.org/ TR/ 2001/ NOTE- wsdl- 20010315

Crawford, D. (Ed.). (2003). Special issue on service oriented architectures. *Communications of the ACM*, *46*(10).

De Roure, D., Gil, Y., & Hendler, J. A. (Eds.). (2004). Special issue on e-science. *IEEE Intelligent Systems*, *19*(1).

Drago, G. P., Setti, E., Licitra, L., & Liberati, D. (2002). Forecasting the performance status of head and neck cancer patient treatment by an interval arithmetic pruned perception. *IEEE Transactions on Bio-Medical Engineering*, *49*(8), 782–787. doi:10.1109/TBME.2002.800788

Ferrari-Trecate, G., Muselli, M., Liberati, D., & Morari, M. (2003). A clustering technique for the identification of piecewise affine systems. *Automatica*, *39*(2), 205–217. doi:10.1016/S0005-1098(02)00224-8

Foster, I., & Kesselman, C. (Eds.). (2004). *The Grid 2: Blueprint for a new computing infrastructure* (2nd ed.). San Francisco, CA: Morgan Kaufmann.

Foster, I., Kesselman, C., Nick, J., & Tuecke, S. (2002). *The physiology of the Grid: An open Grid services architecture for distributed system integration.* Globus. Retrieved February 15, 2011, from http://www.globus.org/ alliance/ publications/ papers/ ogsa.pdf

Garatti, S., Bittanti, S., Liberati, D., & Maffezzoli, A. (2007). An unsupervised clustering approach for leukemia classification based on DNA microarrays data. *Intelligent Data Analysis*, *11*(2), 175–188.

Hey, T., & Trefethen, A. (2003). E-science and its implications. *Philosophical Transactions of the Royal Society London A. Mathematical, Physical, and Engineering Science*, *361*, 1809–1825.

Liberati, D. (2008). Plasmopara viticola germination virulence forecasting via Piece-Wise Affine identification for hybrid systems. *Nonlinear Analysis. Hybrid Systems*, *2*, 1217–1221. doi:10.1016/j.nahs.2008.09.014

Liberati, D. (2009). Biomedical applications of Piece-Wise Affine identification for hybrid systems. *Annals of Biomedical Engineering*, *37*(9), 1871–1876. doi:10.1007/s10439-009-9750-x

Milotti, E., Chignola, R., Dalla Pellegrina, C., Del Fabbro, A., Farina. M., & Liberati, D. (2008). VBL: Virtual biophysics lab. *Il Nuovo Cimento*, *31 C*(1), 109-118.

Muselli, M., & Liberati, D. (2002). Binary rule generation via Hamming clustering. *IEEE Transactions on Knowledge and Data Engineering*, *14*(6), 1258–1268. doi:10.1109/TKDE.2002.1047766

Newman, H. B., Ellisman, M. H., & Orcutt, J. A. (2003). Data-intensive e-science frontier research. *Communications of the ACM*, *46*(11), 68–77. doi:10.1145/948383.948411

Oinn, T. (2004). *XScufl language reference*. European Bioinformatics Institute. Retrieved February 20, 2011, from http://www.ebi.ac.uk/ ~tmo/ mygrid/ XScuflSpecification.html

Peltz, C. (2003). Web services orchestration and choreography. *Computer*, *36*(10), 46–52. doi:10.1109/MC.2003.1236471

Sartorio, A., De Nicolao, G., & Liberati, D. (2002). An improved computational method to assess pituitary responsiveness to secretagogue stimuli. *European Journal of Endocrinology*, *147*(3), 323–332. doi:10.1530/eje.0.1470323

Smith, R. G. (1980). The contract net protocol: High-level communication and control in a distributed problem solver. *IEEE Transactions on Computers*, *29*(12), 1104–1113. doi:10.1109/ TC.1980.1675516

Storchi, R., Biella, G., Liberati, D., & Baselli, G. (2009). Extraction and characterization of essential discharge patterns from multisite recordings of spiking ongoing activity. *PLoS ONE*, *4*(1), e4299. doi:10.1371/journal.pone.0004299

Travica, B. (2005). Virtual organization and electronic commerce. *ACM SIGMIS Database*, *36*(3), 45–68. doi:10.1145/1080390.1080395

Van der Aalst, W., & Van Hee, K. M. (2002). *Workflow management: Models, methods, and systems*. Cambridge, MA: MIT Press.

Wooldridge, M. J. (2002). *An introduction to multiagent systems*. New York, NY: Wiley.

KEY TERMS AND DEFINITIONS

Bio-Informatics: The application of the ICT tools to advanced biological problems.

Transcriptomics and Proteomic: Involving huge amounts of data.

Cooperative Information Systems: Independent, federated information systems that can either autonomously execute locally or cooperate for some tasks towards a common organizational goal.

E-Science: Modality of performing experiments in silico in a cooperative way by resorting to information and communication technology (ICT).

Grid Computing: Distributed computation over a grid of nodes dynamically allocated to the process in execution.

Interoperability: Possibility of performing computation in a distributed heterogeneous environment without altering the technological and specification structure at each involved node.

Web Services: Software paradigm enabling peer- to-peer computation in distributed environments based on the concept of "service" as an autonomous piece of code published in the network.

Workflow: Stream of information within the network related to the accomplishment of every single orchestrated task.

Chapter 12
Speech Codes Theory as a Framework for Analyzing Communication in Online Educational Settings

Tabitha Hart
University of Washington, USA

ABSTRACT

Knowing how best to assess and evaluate the communication that takes place in online educational settings can be a challenge, especially when the features of educational platforms continue to develop in their complexity. This chapter will discuss Speech Codes Theory, which is grounded in the Ethnography of Communication, as a theoretical and methodological framework for conducting qualitative, interpretive research. It will show how Speech Codes Theory can potentially be used to analyze and understand communication in a range of online educational settings.

INTRODUCTION

New communication technologies such as Web 2.0 and Voice-over-internet-protocol (VoIP) [1] continue to rapidly increase the potential for effective teaching, training, and learning in online environments. These technologies help students and teachers from different continents, different time zones, and different cultural settings, to connect with one another. They link students and teachers in meaningful ways, making it possible to chat in real time and share knowledge and information without ever meeting face-to-face. Through these media, students and teachers can exchange information both synchronously and distantly, and these powerful communication technologies can

DOI: 10.4018/978-1-61350-077-4.ch012

facilitate learning between people with diverse cultural identities who are physically located half a world away from each other.

As this edited volume argues, there is an ever-increasing need for instructors, students, trainers, and other professionals involved in education and training to better understand and better address teaching and learning in online environments. Online educational settings must be evaluated, and their challenges and opportunities identified, to make sense of the communication that is going on within them. Educators must continue to explore how to adapt their teaching approaches and communication methods to online environments (St.Amant, 2002). Towards that end, this chapter will discuss the theoretical and methodological framework of Speech Codes Theory (SCT), which is grounded in the Ethnography of Communication (EC), as a means of studying, evaluating, and making sense of communication in online educational settings. It will begin by showing the relationship between ethnography in the traditional sense and the Ethnography of Communication (two distinct but related approaches). Next, it will give an overview of Speech Codes Theory. After that, it will provide an overview of some extant EC/SCT work on online communication, identifying gaps in the field. Finally, it will highlight potential questions for research into communication in online educational settings using the EC/SCT framework.

ETHNOGRAPHY AND THE ETHNOGRAPHY OF COMMUNICATION

It is useful to precede a description of Speech Codes Theory with a brief introduction to the Ethnography of Communication, in which it is grounded. The Ethnography of Communication is distinct from, but closely related to traditional ethnography. While ethnography is commonly equated with ethnographic methods, it is best understood as its own branch of anthropological research, traditionally associated with the following features. It is geared towards the study of human behavior and culture, and seeks to "[reveal that culture] through discerning patterns of socially shared behavior" (Wolcott, 1999, p. 67). Ethnographies are naturally driven by research questions that are fitting to an ethnographic approach, such as "descriptive questions as to how, and underlying questions… as to meanings imputed to action" (Wolcott, 1999, p. 69). Since ethnographies are intended to produce highly contextualized accounts of human behavior and culture, they necessarily involve immersion in a setting, i.e. the continuous and attentive presence of a researcher in a place of study (Gordon, Holland, & Lahelma, 2001; Smith, 2001; Wellin & Fine, 2001; Wolcott, 1999). While in that place, the researcher may use a variety of ethnographic methods to collect data, such as observation, participant observation, and interviews. The data collected are primarily qualitative, but may be quantitative as well (Gordon, et al., 2001). In either case, "the researcher [is] a major instrument of research" (Gordon, et al., 2001, p. 188; see also Wolcott, 1999) in the sense that a researcher's analysis is based on experiences, observations, and interactions in the field. An ethnographic analysis produces an ethnographic account, which is not only a highly detailed description but also an interpretation of cultural processes, "out of which cultural patterning can be discerned" (Wolcott, 1999, p. 68). Many successful ethnographies of educational settings have been produced, a partial account of which may be found in Gorden et al. (2001).

The Ethnography of Communication, developed by Dell Hymes (1962, 1972, 1977), combines "ethnography, the description and analysis of culture, with linguistics, the description and analysis of language" to show "relationships between language and culture" (Keating, 2001, p. 285). True to the epistemology of ethnography, the EC approach contextualizes a study of communication by including detailed information on what happens in, around, and through speech, and does not simply look at speaking alone – divorced from context – as an object of study (Philipsen & Coutu, 2005). This is so because the EC approach sees speech and human behavior as intertwined. Together, speech and human behavior merit studies on "the situations and uses, the patterns and functions, of speaking as an activity in its own right" (Hymes, 1962, p. 101). The ethnographer of communication seeks to discover the structure inherent to the context of participants' socio-cultural worlds, believing that there are patterns and rules (socio-cultural ones) shaping communication. These patterns and rules will guide, for example, what speech interlocutors consider to be appropriate in what settings and when, and will inform what speech (and its many local varieties) signifies to speakers, and so on.

Significantly, Hymes calls for ethnographers of communication to examine not only socio-cultural structure, but also "pragmatic meaning" (1962, p. 104), i.e. meaning in practice, or everyday, real-life meanings attached to speech. Here again the EC approach stresses the importance of context, since accounts of pragmatic meaning must necessarily look at the larger situations (of activity, of human relationships, of shared histories and experiences) in which speaking takes place. For example, an utterance itself has meaning, but contextual factors play a role in how an utterance is understood. A statement such as "You'd better complete that assignment on time or face the consequences," may have intrinsic meaning (in what Hymes terms its "form") but its meaning also depends on the relationships between the speakers, the situation in which they find themselves, their shared experiences, their ideas about how they may speak to each other under what circumstances, and so on. Again, the EC approach emphasizes the pressing need for descriptions of speaking as well as the relationship between the speech and its contextual factors. It allows one to better understand the connection between "social structure and linguistic form," (general), as well as "the ways in which speakers associate particular modes of speaking, topics or message forms, with particular settings and activities" (specific) (Hymes, 1972, p. 36).

The EC approach has a heuristic value in that it helps researchers tease apart how elements of speaking differ from one group to another, thus allowing for a comparative approach that can be useful not only for understanding ways of speaking but also for developing theories about them. Like ethnography, the Ethnography of Communication is thus not simply a descriptive endeavor, but a method of generating theories about human behavior. Through describing and interpreting speech in context, we may work through its subtleties and complexities in order to understand and make predictions about the social world. A rigorous examination of speaking-in-context can produce not merely descriptions of what is (in terms of human behavior), but also informed projections about what may be. EC accounts can serve to "give rise to a comparative study of the cross-cultural variations in a major mode of human behavior" (Hymes, 1962, p. 102). In other words, a rigorous body of descriptive speaking-in-context research will help scholars to compare different socio-cultural systems. In this way, EC work may help with "prediction and inference about behavior" (Hymes, 1962, p. 114), including speakers' motives, responses, and other actions.

Just as ethnographers have a distinct "way of seeing" (Wolcott, 1999), so do ethnographers of communication. Ethnographers of communication are engaged with "the systematic, comparative knowledge of phenomena and systems…" (Hymes, 1977, p. 170). They do not merely describe context, but demystify it, making informed inferences about it, analyzing it rigorously and systematically while still open to new discoveries, rather than trying to fit data into a predetermined model. Identifying patterns and discovering structure that is present but not obvious, requires skill and the proper mindset to make sense of it. Finally, the end goals of the Ethnography of

Communication are to produce insight on this discoverable structure of speaking and context, and not (as with linguistics) to merely make sense of "language organization."

In terms of doing Ethnography of Communication, Hymes explicitly states that "the concern is, first of all, with the attitudes and knowledge of the members of the community" (1972, p. 36). Knowledge and truth are located in the social world and in the research informants (i.e. what is the significance of speaking as interlocutors themselves understand it?) rather than in a mechanical, non-social system (syntax, grammar, etc.) In other words, the search is for meaning that is co-created between speakers in the social world. Ethnographers of Communication focus on groups, not languages or dialects, as the unit of analysis, and often study "speech communities," which "differ significantly [from one another] in ways of speaking, in patterns of repertoire and switching, in the roles and meanings of speech. They indicate differences with regard to beliefs, values, reference groups, norms, and the like..." (1972, p. 42). The idea of studying speech communities again highlights the difference between language/dialect (a focus of linguistics) and ways of speaking (associated with beliefs, relationships, traditions, social life – the focus of the ethnographer of speaking).

As with ethnography, presence in the field (typically through fieldwork) is an important part of the EC approach (Keating, 2001; Saville-Troike, 1982). Specifically, in order to learn about the structure in speaking, an ethnographer of communication would typically look at naturally occurring speech in the settings in which it occurs, considering how contextual factors such as the features of the settings, the relationships between participants, the goals of the speech event, and norms and rules pertaining to the event, were implicated in or constitutive of the communication. The order in which speech acts, or the components of speech acts, occur, as well as their tone or manner might also be considered (See Hymes, 1962 for a complete description of his SPEAKING model.).

The situated, highly contextualized, richly descriptive ethnographic approach applies equally to online as well as offline settings. In fact, there is already substantial historical precedence for using the Ethnography of Communication methodology to study traditional offline educational settings. See, for example, Keating's (2001) excellent summary of the Ethnography of Communication, which details key studies that have used EC methods to look at, among other phenomena, how educational frameworks impact student performance and achievement, particularly among minority children. Other helpful resources on EC studies are Duff (2002), Gordon et al. (2001), and Philipsen & Carbaugh (1986b). Duff's work, in particular, is a good model for how to organize and carry out EC research in a traditional school setting. While there are few published studies that use EC methods to look specifically at online education, the general use of ethnography to study online communities and communication has been widely embraced. See, for example, Goodfellow & Lamy (2009), Hine (2000), Mann & Stewart (2000), and Miller & Slater (2001).

SPEECH CODES THEORY: A METHOD FOR STUDYING CMC IN EDUCATION

In this chapter, I am proposing the theoretical and methodological framework of Philipsen's Speech Codes Theory (Philipsen, 1997; Philipsen, Coutu, & Covarrubias, 2005), which is grounded in the Ethnography of Communication, as a useful heuristic for studying computer-mediated communication (CMC) in online teaching, training, and learning settings. As the Ethnography of Communication presupposes, speaking can reveal a great deal about people, their histories, ways of living, and notions of self and society. It is just this belief on which Speech Codes Theory is founded.

SCT provides ethnographers of communication with a framework as well as conceptual and methodological tools for exploring and making sense of situated communication and interaction, and for explicating the connection between communication and culture (Carbaugh, 1995, 2005; Philipsen, 1992, 1997; Philipsen & Coutu, 2005).

As already noted, ethnographers of communication subscribe to a belief in deep, inextricable "relationships between language and culture" (Keating, 2001, p. 285; cf. Philipsen & Coutu, 2005). Here, "language" is "all forms of speech, writing, song, speech-derived whistling, drumming, horn calling, gesturing, etc." (Keating, 2001, p. 287), and "speaking" is "the use of language, in all its modes and including those manifestations and derivations of language for which speaking can be a surrogate term... includ[ing], but not necessarily limited to, systems of body movement, gestural expression, music, graphic communication, and drum and chanting systems." (Philipsen & Coutu, 2005, p. 355). The relationships between language, speaking, and culture are "[suggestive of] the possibility that there are, in any given place and time, locally distinctive means for, and ways of organizing, communicative conduct, and that these ways implicate a culturally distinctive system of meanings pertaining to communicative conduct itself" (Philipsen & Coutu, 2005, p. 355).

Speech codes make up "a system of socially constructed symbols and meanings, premises, and rules, pertaining to communicative conduct" (Philipsen, 1997, p. 126). That is, speech codes are part of a larger culture (historically transmitted, socially constructed); they provide code-users with names for communicative or social phenomena (symbols and meanings); they specify what communicative or social phenomena go together (premises); and they specify what should/shouldn't be done (rules). Speech codes are a resource that speakers can draw on in three important ways. First, they comprise "a rhetorical resource" that aids speakers in constructing persuasive speech appropriate to the situation at hand, useful in achieving the social goals mentioned above. Second, speech codes are "a system of interpretive resources" that help speakers decode (or encode) the meanings of social interaction. Third, they form an "identificative resource" that "answer[s] questions about why [speakers] exist and where they fit in a scheme of sense and meaning..." (for all 3 points see Philipsen, 1992, p. 16).

Underlying SCT are three important assumptions about speaking: that it is structured, distinctive, and social (Philipsen, 1992). First, to say that speaking is structured means that there are patterns in when and how to speak, and to whom. Speaking is organized, consistent and systematic; ways of speaking are not random or haphazard, but have a "systemic order" to them (Carbaugh, 1995, p. 273; Philipsen, 1992, pp. 9-10). Furthermore, because speaking is structured, it can be discovered, described, and analyzed by ethnographers of communication. Second, speaking is distinctive, i.e. linked to its social, historical, and cultural contexts and unique from setting to setting. Rules and meanings, the beliefs about speech and the norms regulating it are not uniform across groups, thus understanding a given group's way of speaking is to understand something unique about them and their culture (Philipsen, 1992, pp. 12-13). Because this speaking is distinctive, SCT provides a very useful framework for analyzing and then comparing different groups' speech codes. Finally, to say that speaking is social implies that it is a means of accomplishing goals in the social world, such as indexing membership in particular groups, or reinstantiating beliefs about identity. As such, speaking "shapes and constitutes social life" (Philipsen, 1992, p. 13). Researchers and educators may ask what social goals interactions accomplish, what ideas of self and other they involve, and what means of carrying them out are socially sanctioned. Because speaking is social, the discovery, description, and analysis of it can generate insights on the "culturally distinctive psychology, sociology, and rhetoric" of particular communities (Philipsen, et al., 2005, p. 61). In other words, to understand a group's way of speaking is to understand their social life – notions of personhood, the social categories by which members define one another, how people are linked in social units, what they perceive strategic communication use to be, and so on (Philipsen, 1992; Philipsen & Coutu, 2005).

Speech codes reveal the strong connections between communication and culture. Whether in employing them (as speakers do) or in identifying and articulating them (as ethnographers of communication do) speech codes "[mark] off a universe of meaning and [supply] a system of interpretive resources" that may be "rhetorical, interpretive, and identificative" (Philipsen, 1992, p. 16) in nature. They are a useful heuristic for analyzing culture and communication, as well as for cross-cultural (or cross-group) comparisons.

Speech codes exist in all speech communities and, because they are part of a larger structure of speaking, they may be identified and described (in some fashion) by local speakers and ethnographers alike. To identify a community's speech codes, an ethnographer of communication first goes into the field to observe the communicative conduct of members of a speech community, being careful to explicate this as members themselves enact and see it (Carbaugh, 2005; Hymes, 1977; Philipsen, 1992; Philipsen, et al., 2005). The ethnographer focuses her attention on one or more of these key elements of a speech code: ideas of personhood and the social categories used to define people in the community; ideas of social relations and the ways in which people are connected to one another in the community; ideas of how people in the community may use communication strategically to achieve desired outcomes; metacommunicative vocabularies (words, phrases, and/or expressions about communication and communicative conduct); and symbols, meanings, premises and rules that shape or regulate communicative conduct in that community. (Philipsen, 1992, 1997; Philipsen, et al., 2005) While learning about and makes sense of the community's communicative conduct, the ethnographer tries to synthesize the findings into a systematic explanation of how this community operates. In so doing, a name is given to the "system of resources that these participants use to [enact, name, interpret, and judge communicative conduct]" (Philipsen, et al., 2005, p. 57). This name is a speech code. Classic examples of speech codes are the Nacirema code of dignity and the Teamsterville code of honor (Philipsen, 1975, 1992).

In terms of how "culture" fits in with speech codes, it should be noted that the SCT perspective does not equate culture with nationality, ethnicity, class, or religion. From this view, people do not act in a particular way because they are Canadian, or Asian, or working class, or any other ethnicity, class, or religion (Philipsen, 1997). Rather, culture is defined as a code or a system, "a socially constructed and historically transmitted pattern of symbols, meanings, premises, and rules" (Philipsen, 1992, p. 7). This system is rooted in traditions and developed through social interaction; it endures but also shifts and changes over time (Carbaugh, 1995). While culture influences how people communicate, it is not monolithic. It has some degree of force on people's behavior, but people may choose to keep, modify, or flout cultural norms (Carbaugh, 1995; Philipsen, 1992; Philipsen, et al., 2005). The important thing here is that the ethnographer of communication starts and ends with practices, and not nationalities, ethnicities or other features associated with a culture. As Carbaugh puts it, "to conceptualize culture, then, as a system of expression, is to emphasize that one explores how a symbol or form (like the choice of last name upon marriage) functions within a larger communicative situation; what the symbol or symbolic form is like and unlike in this system; on what various occasions it is used and to what ends; what are its limits of expression; and what ideas and ideologies go along with it or are refracted by it?" (1995, p. 285). Rather than starting with nationalities or cultural dimensions and correlating them with behaviors, SCT starts with practices (actions, behaviors, knowledge), assuming that their analysis will reveal a great deal about the larger codes, or systems, or cultures of the people who engage in them.

APPLYING SPEECH CODES THEORY TO STUDIES OF ONLINE SETTINGS

Using Speech Codes Theory, the speech and the lived experience of speech communities in online educational settings for evidence of a code or codes in operation there can be explored. For researchers interested in questions of determining the effectiveness of online communication, or its challenges and opportunities, or its implications, SCT can be a very useful tool. SCT can help researchers to discover what norms, rules, and expectations shape interlocutors' interactions with one another, and this can be of great assistance in diagnosing why and how communication in particular situations succeeds or fails.

Since its inception, the first order of business with the Ethnography of Communication has been to generate hundreds of detailed studies that successfully analyzed local ways of speaking/codes of communicative conduct (see both Philipsen & Carbaugh, 1986a; & Philipsen & Coutu, 2005 for extensive bibliographies; see also S. O. Murray, 1993, p. 331-332) There is also precedence for looking at speech codes in such technology-mediated environments (Hanna & De Nooy, 2004; Keating & Mirus, 2003; Murray, 1988; Wick, 1997). Additionally, hundreds of new media studies have explored the notion of "community" in online environments (Baym, 2000 is one popular exemplar) and have successfully shown that members of these groups certainly do have shared ideas about how to conduct themselves in and through shared speech. Internet spaces have been proven to be social spaces (Baym, 2006; Danet, Rudenberg-Wright, & Rosenbaum-Tamari, 1997; Donath, 1999; Miller & Slater, 2001; O'Brien, 1999; Sterne, 1999; Stone, 1995; Turkle, 1995) where online community may unfold through text-only communication, both within and across cultural groups (cf. Baym, 2000; Bretag, 2006; Cassell & Tversky, 2005). That is, online groups constitute their own social world, and most certainly involve social rules, structures, and norms that influence how people interact and the meanings assigned during that interaction. As previously described, the EC approach used in combination with SCT can help researchers discover the rules for speaking/communicating in online educational environments.

Hanna & De Nooy (2004), for example, did an EC/SCT-style study in which they compared posts on French news discussion boards to posts on British news discussion boards. Through an analysis of participants' key symbolic terms (like "debate" "forum" and "talk"), Hanna & De Nooy showed that message posters from the two groups had quite different approaches to communication in their online communities. On the British sites, users expected "conversation" modeled on offline talk, and informal talk and digressions were accepted. On the French sites users characterized their posts as formal "debate" and discouraged digressions. In the SCT framework, symbolic terms express something crucial about the experience of membership in any given community, so one effective strategy for discovering a community's speech code is to explore the meaning of its symbols. (See Philipsen, 1992, chapter 4.) Thus, while Hanna & De Nooy did not explicitly utilize speech codes theory in their analysis, their examination of prominent symbols and meanings pertaining to communicative conduct provides us with a useful example of how research can be conducted using the SCT approach.

In addition to the symbols mentioned above, Hanna & De Nooy look at premises and rules shaping their participants' communicative conduct. Exploring a community's premises and rules is another an effective way of discovering its speech code(s). In the speech codes theory framework, premises "express beliefs of existence (what is) and of value (what is good and bad)" (Philipsen, 1992, p. 8) in regards to speech and communicative conduct. In other words, a community's premises convey assumptions about what is right or wrong, helpful or unhelpful, positive or negative about speaking. Rules, which are closely interrelated with premises, are "prescription[s], for how to act, under specified circumstances, which [have] (some degree of) force in a particular social group." (Philipsen, 1992, p. 7) Put differently, a community's rules (whether spoken or unspoken) are in place to help guide and shape members' communicative behavior, including how to act, how to feel, and how to evaluate and make sense of speech under particular circumstances. By exploring symbolic terms, premises and rules, Hanna & De Nooy's work reveals cultural aspects to Internet-based communication by showing how the individuals taking part in their particular discussion board communities were indeed guided by "codes" of communicative conduct (cf. Philipsen, 1997; Philipsen, et al., 2005) specific to the sites they posted on.

For researchers looking at online educational settings, identifying an online community's rules of communicative conduct can be a powerful step towards understanding participants' communication successes and failure. It can also help researchers understand how (or if) these rules, structures and norms are transferred over from face-to-face (FTF) interaction, how (or if) these rules are created over time, or how codes of communicative conduct compare across online groups.

In terms of identifying speech codes, SCT recognizes that there may be multiple speech codes at play in a given community (Philipsen, et al., 2005), a claim well supported by Coutu (2000). Given this, speech codes theorists must necessarily be able to delineate these codes from one another, describing (in some fashion) where one code begins and another ends. This also raises the question as to where and how codes might overlap. There are cases where speech codes are clear-cut (Coutu, 2000) but it is likely not so straightforward in other situations where multiple and even contradictory codes co-exist (cf. Swidler, 2001). Related to this is the question of how individuals or groups who hold multiple and conflicting codes determine what code (of several, or many) to employ in any given situation. This is mentioned in, for example, in Carbaugh (2005), in his treatment of conflicting codes on public speaking. It is also addressed in Swidler (2001), whose work examines how people make use of cultural resources on love, and the complicated ways in which they do so. In Swidler's work, informants draw from quite conflicting codes in order to make sense of, explain, and/or justify their life worlds. It is a complicated process, particularly when people have competing frames of reference as to the activity, communication, or interaction in question (cf. Manning, 2008).

Understanding how context plays into online communication in educational settings continues to be a challenge for researchers, but is one that may successfully be addressed by using Speech Codes Theory. As described, in the Ethnography of Communication framework, discovering a group's speech codes actually requires contextualizing the particular and the local in studies of speaking. Again, this is because of "[the Ethnography of Communication's concern] with discovering and describing, rather than taking for granted, the means of communication that are used in a given speech community. Means are not... considered independently of use in the life of a particular social group" (Philipsen & Coutu, 2005, p. 368, see also Keating, 2001, p. 285-286) because the EC/SCT framework rejects the traditional sociolinguistic/conversation analytic focus on pure talk for a more rounded, comprehensive analysis of talk-in-context. Researchers using EC/SCT to look at communication in online educational settings would therefore be sensitive to how the speech codes in a particular environment were connected with contextual factors such as particular educational traditions, teacher/student roles, and/or participants' identities as technology users and consumers.

For example, Hewling (2005) looked at text-only discussion board message exchanges shared among a group of English Language Learners of different cultural backgrounds. As with Hanna & De Nooy, Hewling's study shows that the meaning and significance attached to message composition and posting varies among participants. However, although Hewling speculates that this meaning and significance may well be tied to multiple contextual factors, such as the context of the message, the perceived role (i.e. student, teacher) of the person posting it, and even classroom norms in participants' home or local educational settings, she does not present data to support this speculation. Rather, she suggests the value of more nuanced approaches to studying online communication, and the importance of going beyond posted messages when analyzing what users put into and take out of their online talk. With its emphasis on holism and accounting for contextual factors, EC and SCT can be used to help researchers identify how context plays into online communication in educational settings.

Just as social factors influence communication and speech codes, so too do technological ones (Barley, 1986; Danet, et al., 1997; Fischer, 1992; Keating & Mirus, 2003). Researchers utilizing SCT might therefore analyze how students and teachers negotiate their speech codes to suit the affordances and constraints of the online space where their interactions take place. A researcher might look at the ways in which participants adhere to, breach, or enforce the rules in that environment, as well as what aspects of the environment influence their choices to do so. Keating and Mirus' study, for example, examined how American Sign Language (ASL) users drew on both text and live video streaming to chat with one another online, illustrating how communication tools themselves influenced users' communication, and how communication norms and conventions may be adapted to the affordances and constraints of the tools supporting them.

One method of discovering a speech community's codes is to look for patterns of communicative conduct in how people speak and when, what topics are covered, or what sequence talk falls in. All of these factors are bound with ideas about and rules pertaining to communicative competence. Briefly, communicative competence involves "what speakers need to know to communicate appropriately in a particular speech community, and how this competence is acquired" (Keating, 2001, p. 287). Rules tell interlocutors how to act and what to do in particular places and times. To have communicative competence in a community is to know what rules (e.g. who is permitted to speak at what times and how; what form speech is expected to take; what words, tone, and speed should be used; what content speech may have; and other similar considerations) are appropriate there. To understand a speech community's rules is thus a way of revealing operative codes. Though rules have force, they may be broken, and so "are subject to all the whims of social life, including their legislation, transgression, remediation, and negotiation" (Carbaugh, 1995, p. 273). One potential question for those examining communication in online educational settings is therefore what speech codes (i.e. rules of engagement or cultural codes of communication, cf. Philipsen, 1992, 1997; Philipsen, et al., 2005) are negotiated, developed, and/or drawn upon in the environment under study? On a more micro level, one might look at what rules of conversation are established in the online interactions, as well as how participants negotiate turn-taking, or decide how much talk is appropriate from whom? One could look at who typically controls the flow of conversation and how are such conversations are begun and ended.

It is important to note that in keeping with Speech Codes Theory, no a priori codes are used. Researchers using the Ethnography of Speaking and SCT typically collect information on speaking-in-context without preconceived hypotheses as to what they will find. This is consistent with the ethnographic approach, since forming hypotheses in advance of discovery can potentially blind researchers to important contextual information in the field or setting of study.

CONCLUSION

While there are many potential approaches to analyzing teaching and learning in online educational environments, the theoretical and methodological framework of Speech Codes theory is a particularly strong one. It provides researchers with a particularly powerful heuristic for evaluating online educational settings, identifying their challenges and opportunities, and making sense of the communication that takes place there.

REFERENCES

Barley, S. R. (1986). Technology as an occasion for structuring: Evidence from observations of CT scanners and the social order of radiology departments. *Administrative Science Quarterly*, *31*(1), 78–108. doi:10.2307/2392767

Baym, N. K. (2000). *Tune in, log on: Soaps, fandom, and online community*. Thousand Oaks, CA: Sage.

Baym, N. K. (2006). Interpersonal life online. In Lievrouw, L. A., & Livingstone, S. (Eds.), *The handbook of new media* (pp. 35–54). London, UK: Sage.

Bretag, T. (2006). Developing 'third space' interculturality by using computer-mediated communication. *Journal of Computer-Mediated Communication*, *11*(4), 981–1011. doi:10.1111/j.1083-6101.2006.00304.x

Carbaugh, D. (1995). The ethnographic communication theory of Philipsen and associates. In Cushman, D. P., & Kovacic, B. (Eds.), *Watershed research traditions in human communication theory* (pp. 269–297). Albany, NY: State University of New York Press.

Carbaugh, D. (2005). *Cutures in conversation*. Mahwah, NJ: Lawrence Erlbaum Associates.

Cassell, J., & Tversky, D. (2005). The language of online intercultural community formation. *Journal of Computer-Mediated Communication*, *10*(2).

Coutu, L. (2000). Communication codes of rationality and spirituality in the discourse of and about Robert S. McNamara's "In retrospect". *Research on Language and Social Interaction*, *33*(2), 179–211. doi:10.1207/S15327973RLSI3302_3

Danet, B., Rudenberg-Wright, L., & Rosenbaum-Tamari, Y. (1997). Hmmm... Where's that smoke coming from? Writing, play, and performance on Internet relay chat. *Journal of Computer-Mediated Communication*, *2*(4).

Donath, J. S. (1999). Identity and deception in the virtual community. In Smith, M., & Kollock, P. (Eds.), *Communities in cyberspace* (pp. 29–59). New York, NY: Routledge.

Duff, P. A. (2002). The discursive co-construction of knowledge, identity, and difference: An ethnography of communication in the high school mainstream. *Applied Linguistics*, *23*(3), 289–322. doi:10.1093/applin/23.3.289

Fischer, C. S. (1992). *America calling: A social history of the telephone to to 1940*. Berkeley, CA: University of California Press.

Goode, B. (2002). Voice over Internet Protocol (VoIP). *Proceedings of the IEEE*, *90*(9), 1495–1517. doi:10.1109/JPROC.2002.802005

Goodfellow, R., & Lamy, M. N. (Eds.). (2009). *Learning cultures in online education*. London, UK: Continuum.

Gordon, T., Holland, J., & Lahelma, E. (2001). Ethnographic research in educational settings. In Atkinson, P., Coffey, A., Delamont, S., Lofland, J., & Lofland, L. (Eds.), *Handbook of ethnography* (pp. 188–203). London, UK: Sage.

Hanna, B. E., & De Nooy, J. (2004). Negotiating cross-cultural difference in electronic discussion. *Multilingua, 23*, 257–281. doi:10.1515/mult.2004.012

Hewling, A. (2005). Culture in the online class: Using message analysis to look beyond nationality-based frames of reference. *Journal of Computer-Mediated Communication, 11*(1), 337–356. doi:10.1111/j.1083-6101.2006.tb00316.x

Hine, C. (2000). *Virtual ethnography*. London, UK: Sage.

Hymes, D. (1962). The ethnography of speaking. In Gladwin, T., & Sturevant, W. C. (Eds.), *Anthropology and human behavior* (pp. 13–53). Washington, DC: Anthropological Society of Washington.

Hymes, D. (1972). Models of the interaction of language and social life. In Gumperz, J. J., & Hymes, D. (Eds.), *Directions in sociolinguistics: The ethnography of communication* (pp. 35–71). New York, NY: Holt, Rinehart and Winston.

Hymes, D. (1977). Qualitative/quantitative research methodologies in education: A linguistic perspective. *Anthropology & Education Quarterly, 8*(3), 165–176. doi:10.1525/aeq.1977.8.3 05x1511c

Keating, E. (2001). The ethnography of communication. In Atkinson, P., Coffey, A., Delamont, S., Lofland, J., & Lofland, L. (Eds.), *Handbook of ethnography* (pp. 285–300). London, UK: Sage.

Keating, E., & Mirus, G. (2003). American Sign Language in virtual space: Interactions between deaf users of computer-mediated video communication and the impact of technology on language practices. *Language in Society, 32*, 693–714. doi:10.1017/S0047404503325047

Mann, C., & Stewart, F. (2000). *Internet communication and qualitative research: A handbook for researching online*. London, UK: Sage.

Manning, P. (2008). Barista rants about stupid customers at Starbucks: What imaginary conversations can teach us about real ones. *Language & Communication, 28*, 101–126. doi:10.1016/j.langcom.2008.02.004

Miller, D., & Slater, D. (2001). *The Internet: An ethnographic approach*. Oxford, UK: Berg.

Murray, D. E. (1988). The context of oral and written language: A framework for mode and medium switching. *Language in Society, 17*(3), 351–373. doi:10.1017/S004740450001294X

Murray, S. O. (1993). *Theory groups and the study of language in North America: A social history*. Amsterdam, The Netherlands: John Benjamins Publishing.

O'Brien, J. (1999). Writing in the body: Gender (re)production in online interaction. In Smith, M., & Kollock, P. (Eds.), *Communities in cyberspace* (pp. 76–106). New York, NY: Routledge.

Philipsen, G. (1975). Speaking "like a man" in Teamsterville: Culture patterns of role enactment in an urban neighborhood. *The Quarterly Journal of Speech, 61*(1), 13–23. doi:10.1080/00335637509383264

Philipsen, G. (1992). *Speaking culturally: Explorations in social communication*. Albany, NY: State University of New York Press.

Philipsen, G. (1997). A theory of speech codes. In Philipsen, G., & Albrecht, T. L. (Eds.), *Developing communication theories* (pp. 119–156). New York, NY: State University of New York Press.

Philipsen, G., & Carbaugh, D. (1986a). A bibliography of fieldwork in the ethnography of communication. *Language in Society*, *15*(2), 387–397. doi:10.1017/S0047404500011829

Philipsen, G., & Carbaugh, D. (1986b). A bibliography of fieldwork in the ethnography of communication. *Language in Society, 15*, 387–398. doi:10.1017/S0047404500011829

Philipsen, G., & Coutu, L. M. (2005). The ethnography of speaking. In Fitch, K. L., & Sanders, R. E. (Eds.), *Handbook of language and social interaction* (pp. 355–379). Mahwah, NJ: Lawrence Erlbaum Associates.

Philipsen, G., Coutu, L. M., & Covarrubias, P. (2005). Speech codes theory: Restatement, revisions, and response to criticisms. In Gudykunst, W. (Ed.), *Theorizing about intercultural communication* (pp. 55–68). Thousand Oaks, CA: Sage.

Saville-Troike, M. (1982). *The ethnography of communication: An introduction*. Baltimore, MD: University Park Press.

Smith, V. (2001). Ethnographies of work and the work of ethnographers. In Atkinson, P., Coffey, A., Delamont, S., Lofland, J., & Lofland, L. (Eds.), *Handbook of ethnography* (pp. 220–233). London, UK: Sage.

St.Amant, K. (2002). Integrating intercultural online learning experiences into the computer classroom. *Technical Communication Quarterly*, *11*(3), 289–315. doi:10.1207/s15427625tcq1103_4

Sterne, J. (1999). Thinking the Internet: Cultural studies versus the millenium. In Jones, S. (Ed.), *Doing Internet research: Critical issues and methods for examining the Net* (pp. 257–288). Thousand Oaks, CA: Sage.

Stone, A. R. (1995). *The war of desire and technology at the close of the mechanical age*. Cambridge, MA: The MIT Press.

Swidler, A. (2001). *Talk of love: How culture matters*. Chicago, IL: University of Chicago Press.

Turkle, S. (1995). *Life on the screen: Identity in the age of the Internet*. New York, NY: Simon & Schuster.

Valdes, R. (2001). How VoIP works. *HowStuffWorks*. Retrieved from http://communication.howstuffworks.com/ ip-telephony.htm

Wellin, C., & Fine, G. A. (2001). Ethnography as work: Career socialization, settings and problems. In Atkinson, P., Coffey, A., Delamont, S., Lofland, J., & Lofland, L. (Eds.), *Handbook of ethnography* (pp. 323–338). London, UK: Sage.

Wick, N. (1997). Speech community in the virtual world: The case of one listserv. Unpublished Doctoral dissertation. University of Washington.

Wikipedia. (2007). *Voice over IP*. Retrieved from http://en.wikipedia.org/ wiki/ Voice_over_IP

Wolcott, H. F. (1999). *Ethnography: A way of seeing*. Walnut Creek, CA: AltaMira Press.

KEY TERMS AND DEFINITIONS

Communicative Competence: The state of being a skilled or able communicator in a particular situation or context. It requires being able to understand and also formulate communication, and typically involves social knowledge of communicative rules, norms, beliefs, etc.

Culture: A socio-historical construct that involves beliefs, values, ways of expressing information, local meanings, and rules for how to conduct oneself.

Ethnography: A branch of study that involves the description and analysis of human social life.

Ethnography of Communication: An approach to the study of situated human communication.

Speech Code: A construct designated by a researcher to describe and explain the ways of speaking of a particular group in particular contexts.

Speech Codes Theory: A theoretical and methodological framework, grounded in the Ethnography of Communication, that helps researchers to understand situated speaking and communication.

Speech Community: A group of people who share a speech code.

ENDNOTE

1 VoIP, which is defined as "the routing of voice conversations over the Internet or through any other IP-based network," (Wikipedia, 2007) can be understood as internet-enabled telephony. There are a number of variants of VoIP, including those which have analog or digital telephone hardware as their endpoint (cf. Goode, 2002; Valdes, 2001). A third variant has computers as its endpoints. Skype is an example of a popular web-based communication platform that uses VoIP.

Chapter 13
Students' Evaluation of Online Discussion:
An Ethnographic Construction of Learning Contexts

Yun Xia
Rider University, USA

ABSTRACT

This chapter reports on a study in which different learning contexts in an online discussion were constructed, and the effects of these contexts on learning were explored. From a qualitative analysis of data, six themes emerged and were organized to four learning contexts in online discussion: (1) social cultural context; (2) face-to-face classroom context; (3) online context; (4) technological context. The learning contexts in online discussion were then classified into two categories: emotive context and cognitive context.

INTRODUCTION

Scholars and educators increasingly believe that the use of online educational technologies has been changing practices in higher education (Twigg, 2003). Because the implementation of online learning technologies varies from class to class, studies of online learning technologies should go beyond online interactions to broader learning contexts (Kern, Wave, & Warschauer, 2004; Warschauer, 2000). Any research that examines such ideas of broadening learning contexts, thus, has a great deal to offer a growing number of educators and educational administrators. Such research can also serve as an important foundation for future

DOI: 10.4018/978-1-61350-077-4.ch013

examinations into what factors can affect learning in online contexts.

This chapter presents a study that investigated students' perceptions of learning contexts in online discussion and the effect of the perceptions on student evaluation of online learning. To examine these issues, the author focused on how students constructed learning online contexts based upon their perspectives related to learning in the classroom. The results of this research revealed four learning contexts and their effects on learning outcomes. In examining these items, this chapter contributes to our understanding of online education by expanding the focus on online interactions to the broader social, cultural, and technological contexts of online learning.

AN OVERVIEW OF ONLINE LEARNING

Mixed Research Results of Online Learning

The literature on online learning contains mixed research results relating to the topic. On the one hand, a number of studies found positive learning outcomes resulting from the inclusion of various online learning technologies into the educational process. Some studies, for example, reported on the introduction of online discussion to supplement traditional classroom learning. These works also reported increased student involvement, better learning results, and higher learning grades resulting from the uses of online media (Connolly, MacArthur, Stansfield, & McLellan, 2007; Dede, L'Bahy, & Whitehouse, 2002; Sauers & Walker, 2004; Twigg, 2003). Other research compared online discussion with face-to-face discussion and discovered that students who participated in online discussions achieved a higher level of skill development, increased participation, better collaborative thinking, and more ideas generated than peers participating in exclusively face-to-face discussions (Bronack, Riedl, & Tashner, 2006; Davies & Graff, 2005; Kramarski & Mizrachi, 2006; Rodriguez, Ortiz, & Dvorsky, 2006). Additional studies have noted that the use of online discussions in classes can result in students reporting an increased sense of satisfaction and feeling a greater sense of community (Aitken & Shedletsky, 2002; Connolly et al., 2007).

Other research, however, has reported different findings related to online learning. In some cases, researchers found that students in online classes scored lower on course tests and final exams than did their counterparts in conventional versions of the same class (Waschull, 2001). Moreover, students in online classes seemed more likely to fail the class than did their peers enrolled in on-site versions of the same course (Waschull, 2001).

In yet other instances, researchers have found similar levels of learning success when comparing online and on-site instruction relating to the same course or subject. For example, in one study of online and classroom courses in three community colleges, Benson et al. (2005) found no differences in students' motivation, satisfaction, and learning outcomes between online and face-to-face classes. Clark and Jones (2001), in turn, found no real difference in students' ability to develop public speaking skills when they compared face-to-face and online versions of the same public speaking course. Similarly, after comparing online and face-to-face versions of the same accounting class, Basile and D'Aquila (2002) found no significant differences related to students' attitudes toward the course. More recently, Benoit, Benoit, Milyo, and Hansen (2006) examined the differences between traditional face-to-face and web-assisted instruction. After comparing the grades for speech assignments and end-of-semester overall grades, they found no differences between learning – as indicated by grades as a sign of success – for online and face-to-face students.[1]

Learning Contexts and Students' Perspectives in Online Learning

The use of online learning technologies, of course, varies from class to class and can result in different learning outcomes. From a two-year ethnographic study of online learning in four college language and writing classes, Warschauer (2000) concluded that the implementations of online teaching and learning were more shaped by social-cultural contexts. For this reason, Kern and Warschauer (2000) urged more studies to describe and to evaluate the practices and the uses of online instruction in different specific contexts.

More recently, Kern, Wave, and Warschauer (2004) summarized studies in online learning pedagogy as focusing on the texts resulting from online interactions vs. the actual learning processes themselves. For this reason, they have called for a new direction as the "second wave" that would push beyond online interaction and investigate broader contexts related to online teaching and learning. This new direction also required a shift from quantitative methods of examining online education to qualitative methods that could account for the context created by online classrooms. Such a shift could also allow researchers to focus on particular uses of computer-assisted instruction, specific social contexts, and classroom cultures associated with teaching and learning online.

Today, when scholars run extensive studies of online leaning technologies, they often fail to focus on students' perspectives in relation to their research. Some scholars, however, offer conceptual frameworks of online teaching and learning. These frameworks generally synthesize theories from education, communication, psychology, and information management in a way that includes students as one component (Brandon & Hollingshead, 1999; Witmer, 1998). Some studies, in turn, have used students as subjects in experiments in an attempt to better compare face-to-face and online teaching and learning (Basile & D'Aquila, 2002; Connolly et al., 2007). Other researchers have conducted case studies designed to explore online interaction patterns with questionnaires, surveys, and observations (An & Frick, 2006; Heckman & Annabi, 2005; Schrire, 2004).

These reports and studies, however, did not use students' perspectives as the focus for examining online learning. As a result, the students' perspectives remain missing from such research, yet these perspectives are essential to understanding such processes. This oversight in the previous research literature prompted the author of this chapter to raise the following research questions about students' perspectives on learning contexts in online discussion:

RQ1: How do students perceive learning contexts in online discussion?
RQ2: How do learning contexts in online discussion from students' perspectives affect their learning evaluation?

The author then conducted research designed to address these questions – and thus fill a key gap in the literature on this topic. The remainder of this chapter presents an overview of this research project and the results of that project.

METHODS

To address the afore-mentioned research questions, the author employed ethnography as the primary means of gathering data. Ethnographies have been used widely in education to provide useful insights into numerous pedagogical problems including educational evaluation (e.g., Caracellil & Greene, 1997). An ethnographic method includes fieldwork, selection of a place and a people to study, entry to a community of interest, participant observation, and interviewing (Fetterman, 1998). All of these factors made this approach an effective way for addressing the afore-mentioned research questions because the ethnographic approach enables researchers to understand and describe a social and cultural scene from an insider's perspective. Each step of the ethnographic process used for this research project is described in the following design of the research.

Research Design

Ethnographic research involves immersion in a group or culture with maintenance of a professional distance (Fetterman, 1998). It is an approach that requires researchers to interact with people for long periods of time and to do so in natural settings. For this study, the researcher immersed himself into the culture of a particular class that used a combination of face-to-face and online discussion to examine ideas. To do so, the researcher attended every class with students and participated in students' online discussions in a four-month regular semester. A graduate telecommunication class in digital communication at a large Midwest university was selected as the place and people to study. This class was selected because participation in online discussion is required as a complement to face-to-face class interactions. In order to gain entry to the class, the researcher contacted the professor/instructor for that course and was introduced to the class by the professor. The researcher then provided participants (i.e., students in the class) with an overview of the study, and this overview included a discussion of the goal, the purpose, and the process of the study.

The class was comprised of 34 students, twenty-eight of whom were Master's students in interactive multimedia design and six of whom were doctoral students in telecommunication. During the study, common first names were employed as pseudonyms for the participants.

The class used the discussion board of WebCT as the mechanism for engaging in online discussions. Each week, the professor posted six to ten discussion questions online and required all students to participate with at least five postings as a supplement to classroom discussion. Altogether, 1,838 postings were collected from the online discussion board.

During the semester (15 weeks in the spring of 2002), the researcher attended each class and took observation notes about classroom interactions. The researcher also read, downloaded, and organized online postings on a daily basis during the semester. Finally, all participants were invited for an in-depth individual interview at the end of the semester.

Of the 34 members of the class, 29 agreed to participate in these interviews. Each interview was from 30 to 40 minutes long, and all interviews were taped and transcribed. In order to maintain the participants' meanings, the researcher chose non-directive interviews in which the participants talked in their own terms and controlled most of the interview conversations. During each interview, specific questions were structured according to different participants or different responses from the participants.

Data Analysis

The researcher used Hammersley and Atkinson's (1995) process to analyze the qualitative data resulting from this research process. The first step in the analysis process involved careful reading of the online postings and interview transcripts. During the reading, special attention was paid to the communicative vocabulary that indicated the participants' perceptions of learning contexts in online discussion (e.g., "extension of classroom discussion," "multimedia interaction," and "multitasking communication"). Special attention was also paid to vocabulary that indicated how learning contexts affected participants' learning evaluation (e.g., "pleasing," "learning faster," and "understand better"). With the communicative vocabulary as the guideline, segments of the online postings and interview transcripts were reassembled and regrouped according to identified patterns (Fetterman, 1998). Analytical categories that were central to the study were generated from

Table 1. The students' perceptions of the learning contexts in the online discussion

Social Cultural Context: Digital Generation
Face-to-Face Classroom Context
Online Communication Context
• Multimedia interaction
• Flexible communication
• Massive information sharing
Technological Context

the communicative vocabulary and the patterns from the data. Examples and stories from the data were then used to explain and illustrate these analytical categories.

FINDINGS

In the analysis, six themes emerged and were classified to four broad categories that constituted learning contexts (see Table 1). For each of the 6 themes, exemplars were used to illustrate the participants' perceptions of the learning contexts in the online discussion and their learning evaluations. Some perceptions and learning evaluations were experienced by individuals or some groups, but others were more universal in the class.

Social Cultural Context: Digital Generation

This context was the culture that the participants brought with them to the classroom. When participants talked about the learning contexts in the online discussion, they mentioned the cultural changes that new communication technologies have brought to their lives. The participants called themselves a "digital generation." One student (David) explained the meaning of "digital generation" by taking out his cell phone and iPod. He placed these items on the table in front of him, and explained, "When I walk on campus, I listen to MP3 music. When I sit in classroom, I text message my friends. When I work in the computer lab, I surf on the Internet." With these "digital gadgets," David said he was in "digital operation" all the time.

More than half of the participants (18 out of 34) admitted that they had at least one of these digital devices with them all the time. These devices allowed the information to flow from one media format to another. In discussing this process, participant Lee said he could hook a Palm Pilot to a computer to transfer people's telephone numbers (information) from one device to the other. Lee could then plug his iPod into that computer to download music he had copied onto the computer from CDs. Lee also noted he could email photos to his friends after he had uploaded them from the camera on his cell phone to his computer. In this way, individuals were able to move data from one device to another almost reflexively.

Many participants agreed that with "digital operation," all communication media formats converged. This convergence resulted in constant digital communication through an "endless digital web." With a cell phone, for example, participants could constantly reach others and be reached by others. With Instant Messager (IM), participants could constantly talk with others. With a Palm, participants could constantly check their next appointment or meeting with others.

When asked about how much time, on average, participants spent on digital communication (e.g., cell phone calling or the Internet surfing) each day, two-thirds of the participants (25 out of 34) admitted to more than six hours of such communication. Participant Jill explained, "There's no difference between online and offline. For us, we're online all the time. By online, I don't mean on the Internet. Like, I mean you're engaged in some sort of communication through those digital devices 24/7."

When participants evaluated their learning in this context, they agreed that the culture of the "digital generation" shaped their attitudes and feelings about learning in the online discussion. The online discussion was one format into which their endless digital web had transformed. Through this digital web, the participants felt "at home" and "comfortable" in the online discussion as a natural extension of their outside-classroom digital culture. Participant Dane said, "For me, I feel so good going online and posting messages. They are no longer class requirements. Whenever I have time, I sit before a computer; start to listen to music from my iPod, open up IM [Instant Messager], at the same time, read postings on the board and type in some messages. That's what I do anyway."

Thus, the participants felt "extremely happy" and "enthusiastic" that online discussion as a part of their digital culture could allow them to engage in a constant learning process. The constant learning process was also reflected in the combination of entertainment and learning. Participant Seth explained, "I am glad that finally learning and entertainment can happen at the same time. I was text messaging my friend about virtual reality in the movie Matrix. Then, I asked him about the second orality question in multimedia communication that I read from the textbook. I switched to our discussion board and finished my posting for the class. I switched back and continued our talk. They are all one."

Face-to-Face Classroom Context

In their discussions of learning contexts, the participants compared online interactions with the face-to-face classroom discussions. First, the participants agreed that the online discussion could not be separated from the face-to-face classroom discussion. The participants perceived the online discussion was an extension instead of a substitution of the face-to-face classroom discussion. Participant Lucy posted a message:

Well Jay, of course nothing replaces face-to-face interaction (which is why online classes are not as good as regular ones). But what you have to remember is that we live in a society where it's all about convenience. And I think that this idea of online discussion is in a sense about convenience. It extends classroom discussion beyond time and place limits. It is convenient.

Lucy liked online discussion because of the convenience they offered students.

Second, the participants used face-to-face classroom discussion as a reference for evaluating online discussion. Participant Jay explained, "If I'm happy offline in class, I will be happy online. If I am not happy in class, I won't be happy online." One reason for this perception was the idea that the participants could know people better in a face-to-face interaction. After knowing people, the participants felt more comfortable participating in online discussion with those same individuals. In one participant's words, online discussions were something that "you can't touch and you can't feel."

In this "virtual" environment, knowing people in a class added contextual or background cues that made online interactions "a little bit real" and "easier." Additionally, participants believed in the "personal and natural communication" format in a face-to-face classroom: "You can hear. You can watch. You can touch. You can even smell. You can have interactions with the professor face-to-face. It is so real. It's so natural. It is so personal."

In their evaluations of their online learning, the participants seemed to believe the face-to-face classroom set up the emotional tone for the related online discussions. On the one hand, knowing people in a face-to-face classroom helped participants choose the group that they felt comfortable with in the online discussion. Thus, in different forums of the online discussion, participants were grouped together according to different comfort levels that they had with each other in the classroom. Participant Dana, for example, liked to read and respond to others with whom she knew best: "I know Jin is an international student from our talk in class. I know she is doing research on international communication and speech freedom. I always watch out for her postings because I like her. I am curious about how she is doing in class."

On the other hand, some participants complained about the face-to-face discussion and its negative effects on the online class discussion. Participant Alex, for example, posted a message complaining that the class discussion was poor: "When do the students get to talk in the class? Every class is just a steamroller of imparted knowledge, without scholarly discussion. I tell you. I am extremely unhappy in class. Thus, I don't feel comfortable online. I don't trust people online." Alex felt uncomfortable and could not "trust people" in the online discussion because of his bad experience of face-to-face classroom interactions. In his end-of-semester interview, Alex admitted that the face-to-face classroom interaction made him angry during the online discussion. As a result, Alex claimed he would never again take any classes that used online discussions.

Online Communication Context

The online communication context was where the participants interacted with each other in the online discussion. During these discussions, participants talked about multimedia interaction, flexible communication, and massive information sharing in this context.

Multimedia Interaction

All participants recognized the "multimedia interaction" in the context of online discussion. The participants agreed that multimedia interactions referred to communication messages with graphic, audio, video, and text. The participants defined "multimedia" as any postings online that contained more than text, including postings with hyperlinks that pointed to graphic elements. The kinds of materials posted, in turn, varied from individual to individual.

At the beginning of the semester, some participants attached their personal photos to a self-introduction posting to the class. Other persons posted graphics to illustrate their ideas. In order to answer a posting about web server and client communication on the Internet, participant Shawn posted a graphic illustration with two icons of computers that one sent a web address request and the other sent a web interface back.

In some cases, participants included hypertext links pointing to audio and video elements that exemplified certain ideas being discussed. Participant Jay, for example, posted a link referencing a short video clip about the history of media technology. The video clip showed the process of different media converged on the basis of digital technologies.

In the evaluation of learning in the context of multimedia interaction, the participants claimed that multimedia interactions created a "real," "dramatic," and "interesting" environment in the online discussion, making them feel comfortable and satisfied in learning. The participants used "dry," "virtual," "flat," and "indirect" to describe text-based online interactions that could turn them away. The participants liked multimedia interactions with audio, video, graphics, and animations. In one participant's words, multimedia interactions had a sense of "real" personal communication that made online learning more "pleasing."

The participants seemed to attribute their preference for multimedia interactions to the potential of simulating "the real-life world." Participant Caroline said:

Look, the real life world is like multimedia. You listen to speeches. You watch pictures. You watch people walking and cars moving. You are living in a multimedia world out there. It is comfortable and natural to be with multimedia messages since they simulate the outside world.

Participant Caroline also thought that the best online learning technologies should simulate the real-life world. She even predicated that the future of online discussion should be in virtual reality technologies that made people feel more comfortable because online discussions could be an actual extension of face-to-face classroom interactions.

On the other hand, the participants agreed that they could learn information more quickly and comprehend it better through multimedia interactions. The participants claimed that they were a "generation of visual communication": "We grew up in front of TV. Now, we live on the Internet. We're online 4 or 5 hours a day. We are very visual in learning."

Seventeen of the participants mentioned they could learn more quickly and comprehend better by watching and reading. Participant Wei, for example, could not understand how TCP/IP worked from layer to layer until she watched a short animation referenced by one posting in the online discussion. The animation showed the process of how an English message was broken into small chunks, the small chunks changed to binary codes, the codes gained address header and got transferred along different routes through routers, and were finally re-assembled back into the original English message at the final destination. Wei explained her learning in multimedia by noting, "You read, you watch, and sometimes you listen. You are receiving all stuff. You got to get faster and understand more in the end."

Flexible Communication

The participants agreed that the online discussion allowed them to have a "flexible communication." They noticed the online discussion did not require physical proximity. Rather, the participants in different places could have interactions and collaborations. The participants could "talk" or "discuss" with each other when they were in a computer lab, in a library, and at their home. Additionally, participants appreciated the independence allowed in terms of the time when one could participate in the online discussion. That is, the asynchronous nature of the online discussion meant participants did not have to meet at particular time to engage in a given discussion.

This flexibility, moreover, was not only about how the participants communicate, but it also involved what they chose to communicate. That is, participants could choose "different individual threads" in their online discussion. Participant Dana used the term "sharing of interests" to describe how she chose discussions online: "It is impossible to read all the messages. You know, there are about 2,000 messages there. I post messages of my interests and respond to messages of my interests. Of course, anyone in the class who is interested in a specific topic or one message can jump in. It is very flexible"

In their evaluations of learning, the participants related the flexible communication in the online discussion to self-controlled learning that helped knowledge building. With time and place flexibility, online discussion allowed self-paced and self-controlled learning that could better involved the participants with different communication habits and learning styles.

Eventually, with flexible communication, the participants could learn more easily and could acquire knowledge more quickly. In addition, this flexible communication allowed the participants to have "multiple channels" of knowledge building. The participants could follow, in some participants' words, "many-to-many modules of learning." That is, the participants could simultaneously join multiple groups according to their various interests at a given time. Participant James, for example, said: "In the online discussion, you're posting messages about mobile telephony while reading messages about FCC's regulations about the Internet. You're also thinking about responding to the message about wireless laptop in the classroom. You're doing multiple groups at the same time. Other people are probably doing the same thing at that moment." With the "many-to-many module of learning," the participants thus built knowledge from multiple sources.

Massive Information Sharing

All participants felt strongly about massive information sharing in the online discussion. At the beginning of the semester, one self-introduction message in the online discussion board could summarize how the students viewed information sharing in the online discussion: "Hi, I'm Jay. I'm an IMMP student now. Let's share our information with each other and learn more knowledge of mass media technologies through our online discussions."

Because the professor supervising the class did not specify what messages the students should post as long as they were about communication technologies, participants could choose a wide range of topics in their postings. Thus, the students were "exposed to messages of various kinds." In summary, there were seven categories of messages in the online discussion:

- Questions and answers about the classroom discussions (34%)
- Communication technology research articles (26%)
- Discussions of textbook readings (15%)
- News of communication technologies (13%)
- Personal experience of class materials (8%)
- Evaluations of the online discussion (8%).
- Personal comments about communication technologies (5%)

The participants also talked about "multitasking communication" that allowed them to share information massively. Some participants admitted that they might have more than 10 windows of different programs open at the same time on one computer. Within this context, it was very easy to get information to flow among different programs so that it could be shared with other people. Participant Joe explained, "I can use messages from the chat with my friends in the research paper. I can also copy and paste messages from the paper to Google similar ones on the Internet. Whenever I have questions in writing my papers, I text message my friends. I also pass around interesting articles from the Web in the discussion board."

In the evaluation of their learning through massive information sharing, the participants agreed that the online discussion extended the face-to-face classroom meeting and helped their understanding and synthesis of ideas. Some participants were unclear of the discussions in the classroom and expected for clarifications. Not surprisingly, the biggest category of messages in the discussion board was questions and answers about the classroom discussions. Some participants posted concepts that they did not understand in their reading of the textbook. When the participants posted messages, they extended their class discussions and could clarify their understandings.

In addition, the participants mentioned that the massive sharing information helped synthesize new ideas for their research projects. Such information sharing saved time for information collection. One participant explained the process in the following way: "We have 30 something students in the class. Each person has to post five messages. You multiply it by 30. That's huge."

Interestingly, the participants did not seem to think that other people did work for them through such sharing of information. Rather, the participants seemed to believe that, as graduate students, they should create insightful ideas through such exchanges. The ultimate goal of massive information sharing was, in turn, to "synthesize good ideas for research projects."

Participant John offered an example about how his idea about mobile telephony in the final project was developed from numerous postings about the Internet technology. John read the postings about digital conversion of information signals on the Internet that incorporated old media formats, such as typographic text, photographic images, radio, and television. He found the same process in mobile telephony and concluded that mobile telephony would be the next new media after the Internet. John used the term "synergy" to describe the information sharing and creation of ideas in learning.

Technological Context

The technological properties became one important learning context in the participants' learning experience. As one participant said, communication technologies facilitated online discussions. As a result, it was impossible to be unaware of the technological context in online learning.

This awareness of the technology took many forms. Some participants had to create a login name multiple times before they could use the discussion board. One participant complained that he had to "spend half an hour each time just to set up stuff before doing anything." In other instances, some participants confused their WebCT passwords with other passwords, such as library or Instant Messager passwords. All participants expressed the expectation that educational technologies should not need logins and passwords.

Some participants also expressed strong feelings about the poor design of the interface used for the discussion board. Participant Catherine, for example, complained that the discussion board window was too small to post messages directly. Thus, she had to use notepad or Word. Then, she switched to WebCT and pasted messages. She had to switch back to notepad or Word if she wanted to add more messages. In one participant's words, the poor interface in the discussion board did not "facilitate the interactions at all."

Using the online discussion board was more than writing and posting messages. The participants, in turn, had to read others' messages, respond to others' messages, and sometimes quote others' messages. The participants then had to switch back and forth between their messages and others' messages.

When evaluating their learning in this context, the participants held a negative feeling for the discussion board offered through WebCT. The login problems, poor interface, and uneasy transition between reading and writing made the participants "very frustrated and uncomfortable" in the online discussion. Participant Jena felt very disappointed when she used "double barriers" to evaluate her learning experience with online discussion: "For me, writing is kind of indirect. It is uncomfortable. The poor technology in WebCT turns me away even further. So, it's double barriers." Jena thought that students in the online discussion were always "two steps" away from actual learning.

The participants also expressed their expectation about the technology that they would feel comfortable with in online learning. Participant Roger used "transparency" to define the role of technology in online discussions:

The best online discussion technology is transparent. I mean invisible online discussion technology. When you use the online discussion board, you feel so natural and comfortable that you do not realize you are using some technology. The painful experience of login and password make you realize you are dealing with machines, stupid machines. The poor interface keeps reminding you are struggling and freaking out, sometimes. It is frustrating.

According to Roger, online discussion technology should be a "transparent facilitator" instead of "visible mediator." As a "transparent facilitator," online discussion should make students comfortable so that they could focus on the content instead of computer technological properties.

DISCUSSION

Summary of the Study

In this study, the participants constructed four different learning contexts in the online discussion for the class. With the social cultural context, the students were in digital operation all the time. The students felt "at home" and "comfortable" in the online discussion that was a part of their digital culture. The face-to-face classroom context set up the emotional tone for the online discussion. Knowing people in a face-to-face interaction helped the students choose groups with which they felt comfortable. The poor face-to-face classroom discussions, however, did make some students feel "frustrated" when interacting through online discussions.

The online communication context for the class had three themes. First, with multimedia interactions the students added audio, video, graphics, and hypertext links in online postings. On the one hand, the multimedia context was closer to the real-life world and made the students feel "pleasing" and "interesting" in the online discussion. On the other hand, multimedia interactions enabled students to learn faster and understand better. Second, the students were free of a time and place limitation in the online discussion with flexible communications. The students had self-controlled and many-to-many module of learning that helped knowledge building. Third, the students extended their classroom discussions with massive information sharing. The students could also understand the class materials better and synthesized new ideas from information sharing. The technological context indicated the significance of technological properties in the online discussion. For example, many students were uncomfortable with and frustrated by the poor design of the technologies associated with the discussion board of WebCT.

Interpretation of the Findings

Bloom's (1956) discussion of cognitive domain and affective domain of learning has sparked decades of efforts to measure cognitive and affective learning. The valid measurement of the students' learning in the online discussions is beyond the scope of the current study. However, Bloom's taxonomy can help explain the students' perception of the learning contexts and their learning evaluation. In this study, the students constructed four learning contexts in the online discussion. Those contexts included:

- Social Cultural Context: Digital Generation
- Face-to-Face Classroom Context
- Online Communication Context
- Technological Context

Those contexts affected the students' evaluation of learning in different ways. With some contexts, the students had attitudes, beliefs, and feelings, such as "enthusiastic," "happy," "comfortable," "easy," "frustrated," or "angry," in their learning evaluation. With some contexts, the students thought that they could "understand better," "learn faster," "create new ideas," and "synthesize new ideas."

In Bloom's definition of affective learning, the students' emotional responses toward learning environment and class content are important elements. In Bloom's taxonomy of cognitive learning, the three categories, knowledge, comprehension, and synthesis, can also explain the students' evaluation of learning in the online discussion. The following model (Figure 1) summarizes the students' perception of the learning contexts in the online discussion according to Bloom's definition of affective learning and taxonomy of cognitive learning.

Within this model, all contexts are a part of the learning context in online discussions. Any one of the contexts or the components of contexts can lead to the evaluation of learning. The emotive

Figure 1. Model summarizing students' perception of the learning contexts in the online discussion

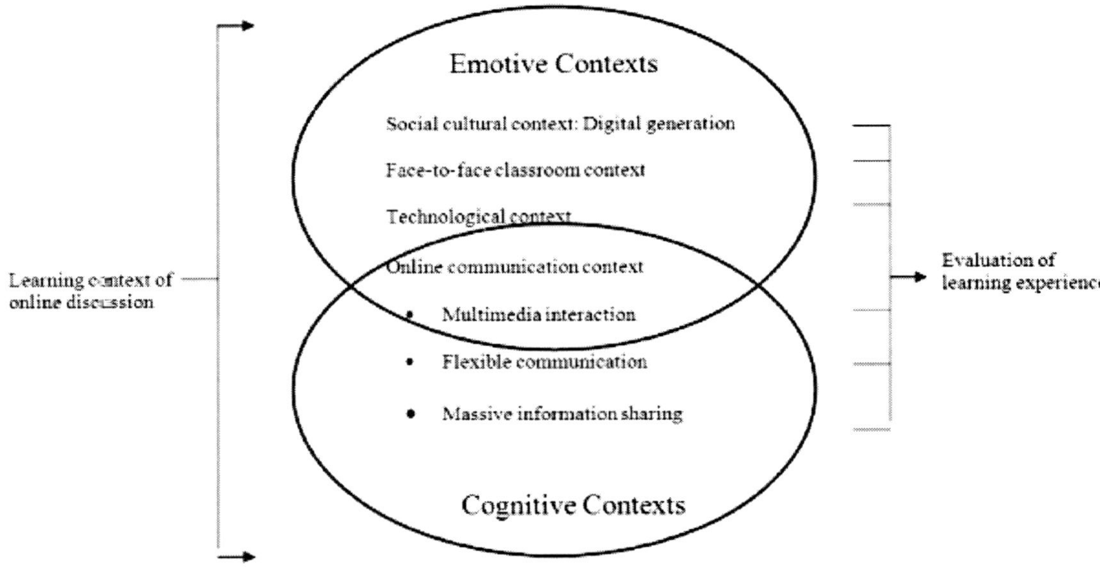

context includes the social cultural context, face-to-face classroom context, technological context, and multimedia interaction in the online communication context that leads to students' emotional evaluations of learning.

With the social cultural context, students take online discussions as a continuation of their "digital web" outside the classroom that makes online discussions a part of their digital culture. Students feel "at home" or "comfortable" in online learning. The face-to-face classroom context sets up an emotional tone for online discussions. In this study, the poor face-to-face classroom discussion made some students "frustrated," "angry," and mistrustful in online discussions.

Some research indicates that computer-mediated communication can create a better learning environment than traditional face-to-face classrooms (Sauers & Walker, 2004; Twigg, 2003). The findings presented in this chapter indicate the significant role of face-to-face classroom interactions as a part of the online learning context and as a mechanism that shapes students' feelings about online learning. The findings about technological context reveal that the best online discussion technology should be invisible in the communication process. In the study presented here, students wished that technological properties were "transparent" and thus they could be more comfortable in learning. Multimedia interactions create a context that simulates the real-life world with graphics, audio, video, and animation. With multimedia interactions, students can feel natural and pleasing in online discussions.

The cognitive context includes all themes of the online communication context – themes that lead to students' cognitive evaluation of learning. The cognitive properties of multimedia interactions are based on the dual coding theory of the human mind (Baddeley, 1999). According to this approach, the human mind processes information in visual and aural channels, and multimedia environments make use of the dual channels for the full capacity of active information processing. These notions of information processing explain why students can learn more quickly and understand better by both watching and reading.

The cognitive properties of multimedia interactions constitute both the knowledge and comprehension domain of cognitive learning (Bloom, 1956). Flexible communication in online context has long been a widely accepted feature of online communication in the related research literature (Bordia, 1997; Kiesler, 2007; McGrath & Hollingshead, 1994; Walther & Parks, 2002). Flexible communication also leads to self-controlled and many-to-many module of learning that helps knowledge building (Vrasidas & McIsaac, 2000).

As a theme in the online communication context, massive information sharing can extend classroom discussions and can help students understand concepts and seek answers for questions that constitute comprehension and knowledge building in cognitive learning (Bloom, 1956). Massive information sharing also allows students to create new ideas from synthesis of different sources. The students do so by putting together elements and parts to form a new pattern or structure (Bloom, 1956). Massive information sharing thus makes it possible for synergetic creations of ideas. In a synergetic interaction, messages are connected to one another through information sharing. Advanced cognition can, in turn, be found in synergistic interactions through computer conferencing (Schrire, 2004).

Theoretical and Pedagogical Implications

Brandon and Hollingshead (1999) synthesized a descriptive model of computer-supported collaborative learning theory and research. According to Brandon and Hollingshead's model, collaborative learning is based on social construction of knowledge. The cognitive theories for collaborative learning emphasize on the development of cognitive structures through social interactions. In the model, social interactions and social environments are restricted to an online communication process. However, students come to classes with a broader social cultural background. The findings presented in this chapter expand this model to a broader social cultural context. In particular, this study's findings about the social cultural context in which students interact and learn calls attention to the significance of this broader context in online learning. In the new millennium where students call themselves the "digital generation," what students do on computers outside of the classroom affects the learning that will take place inside that same classroom.

With the emotive context in the online discussion, the findings in the study add an affective dimension to cognitive theories of multimedia learning. As computer-networking technologies are in the fourth decade, multimedia has gone beyond a concept to full recognition. However, there are few studies about the application of multimedia in computer assisted teaching and learning. The research of Richard E. Mayer and his colleagues is one of the exceptions.

Based on cognitive psychology and information processing models, Mayer and his colleagues (Mayer, 2001; Mayer, Dow, & Mayer, 2003; Mayer, Hegarty, Mayer, & Campbell 2005; Mayer, Sobko, & Mautone, 2003) developed a cognitive theory of multimedia. The cognitive theory of multimedia learning assumes that humans process information in dual channels for visual/pictorial and auditory/verbal. That is, multimedia learning is compatible with dual channel communication. However, these ideas do not cover students' emotional attitudes toward online learning environment.

The findings presented in this chapter reveal important insights about the emotive context and the significance of affective learning in multimedia. Students might not sit before computers if they are not happy or comfortable no matter how compatible multimedia is with information processing channels in the human mind. As a result, these findings point to a potential theoretical foundation for multimedia learning: emotive dimension. In this study, when students wished online discussion technologies could be "transparent" or "invisible" and simulate the "real-life world," multimedia could satisfy the affective need when students sat before computers.

The finding about the technological context also offers a new pedagogical perceptive on the debate between technological determinism and social constructivism in educational use of technologies. Technology determinism advocates that technology develops independently of human influence and leads to various social cultural changes (Marx & Smith, 1994). As more scholars cast doubt on technological determinism, an alternative argument arises from social constructivism and advocates that negotiations and struggles among people shape the use and view of technologies (Heckman & Annabi, 2005; Kling, 2000; Fulk, Schmitz, & Ryu 1995).

The findings presented in this chapter indicate that technological properties as a learning context can affect the use, perception, and evaluation of online discussions. When students struggle with technological problems, technological properties might work as a poor emotive context that leads to negative learning evaluations. The pedagogy of online discussion must therefore consider technological properties as an important learning context.

The findings presented here on the face-to-face classroom context emphasize the importance of face-to-face discussions in the incorporation of online discussions. Face-to face classrooms can affect students' emotional attitude in online discussions. In the study, students all agreed that the online discussion could not be separated from the face-to-face discussion. With a better face-to-face interaction, students can feel comfortable in a better emotive context of online discussion. On the other hand, the cognitive context in online discussions offers what face-to-face discussions do not have: flexible communication, massive information sharing, and multimedia interaction. In summary, face-to-face discussions create an emotive context in which students feel easy and comfortable in online learning.

Future Research

This study was conducted on one graduate level telecommunication class. Its indication for other classes therefore needs further exploration. In this study, learning evaluations were in the format of students' self reporting via interviews. More methods can and should be used to evaluate students' learning.

This study also constructed four different learning contexts. They are by no means an exhaustive list. Further research can investigate how students' backgrounds affect their learning in online discussions. Further research can also examine how institutional context influences student learning and should include examinations of faculty attitudes toward new technology, administrative policy and support, and technology access issues.

Discussing different learning contexts individually helps to describe themes, patterns, and ideas clearly. However, such discussions do not represent the interconnections among these elements. Apparently, the technological context can affect multimedia interactions in the online communication context. More research is needed to unpack those interconnections among different learning contexts in online discussion.

CONCLUSION

The implementation of communication technologies in teaching and learning varies from class to class. In this study, different learning contexts were constructed and their effects on learning were explored. The finding about the cognitive context provides a holistic perspective about learning in online discussions. More important is the finding about students' emotional attitudes toward online discussion that should urge the research and incorporation of communication technologies in the classroom at a broader emotive context. Students' feelings from various emotive contexts can be equally as important as cognitive contexts in the evaluation of learning experience in online discussion.

REFERENCES

Aitken, J. E., & Shedletsky, L. J. (2002). Using electronic discussion to teach communication courses. *Communication Education*, *51*(3), 325–331. doi:10.1080/03634520216512

An, Y.-J., & Frick, T. (2006). Student perceptions of asynchronous computer-mediated communication in face-to-face courses. *Journal of Computer-Mediated Communication*, *11*(2). Retrieved from http://jcmc.indiana.edu/ vol11/ issue2/ an.html. doi:10.1111/j.1083-6101.2006.00023.x

Baddeley, A. D. (1999). *Human memory*. Boston, MA: Allyn & Bacon.

Basile, A., & D'Aquila, J. M. (2002). An experimental analysis of computer–mediated instruction and student attitudes in a principles of financial accounting course. *Journal of Education for Business*, *77*(3), 137–143. doi:10.1080/08832320209599062

Benoit, P. J., Benoit, W. L., Milyo, J., & Hansen, G. J. (2006). *The effects of traditional vs. Web–assisted instruction on student learning satisfaction*. Columbia, MO: University of Missouri Press.

Benson, A. D., Johnson, S. D., Taylor, G. D., Treat, T., Shinkareva, O. N., & Duncan, J. (2005). Achievement in online and campus–based career and technical education (CTE) courses. *Community College Journal of Research and Practice*, *29*, 369–394. doi:10.1080/10668920590921589

Bloom, B. S. (1956). *Taxonomy of educational objectives*. New York, NY: David Mckay Company.

Bordia, P. (1997). Face-to-face versus compute-mediated communication: A synthesis of the experimental literature. *Journal of Business Communication*, *34*, 98–118. doi:10.1177/002194369703400106

Brandon, D. P., & Hollingshead, A. B. (1999). Collaborative learning and computer-supported groups. *Communication Education*, *48*, 109–126. doi:10.1080/03634529909379159

Bronack, S., Riedl, R., & Tashner, J. (2006). Learning in the zone: A social constructivist framework for distance education in a 3D virtual world. *Interactive Learning Environments*, *14*(3), 219–232. doi:10.1080/10494820600909157

Caracelli, V. J., & Greene, J. C. (1997). Crafting mixed-method evaluation designs. *New Directions for Evaluation*, (74): 19–32. doi:10.1002/ev.1069

Clark, R. A., & Jones, D. (2001). A comparison of traditional and online formats in a public speaking course. *Communication Education*, *50*, 109–124. doi:10.1080/03634520109379238

Connolly, T. M., MacArthur, E., Stansfield, M. H., & McLellan, E. (2007). A quasi-experimental study of three online learning courses in computing. *Journal of Computers and Education*, *49*(2), 345–359. doi:10.1016/j.compedu.2005.09.001

Davies, J., & Graff, M. (2005). Performance in e–learning: Online participation and student grades. *British Journal of Educational Technology*, *36*(4), 657–663. doi:10.1111/j.1467-8535.2005.00542.x

Dede, C., L'Bahy, T. B., & Whitehouse, P. (2002). Designing and studying learning experiences that use multiple interactive media to bridge distance and time. In Vrasidas, C., & Glass, G. V. (Eds.), *Distance education and distributed learning* (pp. 1–30). London, UK: Eurospan.

Fetterman, D. M. (1998). *Ethnography: Step by step*. Thousand Oaks, CA: Sage.

Fulk, J., Schmitz, J., & Ryu, D. (1995). Cognitive elements in the social construction of technology. *Management Communication Quarterly*, *8*, 259–288. doi:10.1177/0893318995008003001

Hammersley, M., & Atkinson, P. (1995). *Ethnography: Principles in practices*. New York, NY: Routledge.

Heckman, R., & Annabi, H. (2005). A content analytic comparison of learning processes in online and face-to-face case study discussions. *Journal of Computer-Mediated Communication*, *10*(2). Retrieved from http://jcmc.indiana.edu/ vol10/ issue2/ heckman.html.

Kern, R. G., Ware, P., & Warschauer, M. (2004). Crossing frontiers: New directions in online pedagogy and research. *Annual Review of Applied Linguistics*, *24*, 243–260. doi:10.1017/ S0267190504000091

Kern, R. G., & Warschauer, M. (2000). Theory and practice of network-based language teaching. In Warschauer, M., & Kern, R. (Eds.), *Network-based language teaching: Concepts and practice* (pp. 41–58). Cambridge, UK: Cambridge University Press.

Kiesler, S. (2007). Network nation: Human communication via computer. In Erickson, T., & McDonald, D. W. (Eds.), *HCI Remixed: Reflections on notable HCI papers* (pp. 69–75). Cambridge, MA: MIT Press.

Kling, R. (2000). Learning about information technologies and social change: The contribution of social informatics. *The Information Society*, *16*(3), 217–232. doi:10.1080/01972240050133661

Kramarski, B., & Mizrachi, N. (2006). Online discussion and self–regulated learning: Effects of instructional methods on mathematical literacy. *The Journal of Educational Research*, *99*(4), 218–230. doi:10.3200/JOER.99.4.218-231

Marx, L., & Smith, M. R. (Eds.). (1994). *Does technology drive history? The dilemma of technological determinism*. Cambridge, MA: MIT Press.

Mayer, R. E. (2001). *Multimedia learning*. New York, NY: Cambridge University Press.

Mayer, R. E., Dow, G. T., & Mayer, S. A. (2003). Multimedia learning in an interactive self-explaining environment: What works in the design of agent-based microworlds? *Journal of Educational Psychology*, *95*, 806–812. doi:10.1037/0022-0663.95.4.806

Mayer, R. E., Hegarty, M., Mayer, S., & Campbell, J. (2005). When static media promote active learning: Annotated illustrations versus narrated animations in multimedia learning. *Journal of Experimental Psychology. Applied*, *11*, 256–265. doi:10.1037/1076-898X.11.4.256

Mayer, R. E., Sobko, K., & Mautone, P. D. (2003). Social cues in multimedia learning: Role of speaker's voice. *Journal of Educational Psychology*, *95*, 419–425. doi:10.1037/0022-0663.95.2.419

McGrath, J. E., & Hollingshead, A. B. (1994). *Groups interacting with technology*. Thousand Oaks, CA: Sage.

Rodriguez, J., Ortiz, I., & Dvorsky, E. (2006). Introducing evolution using online activities in a non-major biology course. *Journal of College Science Teaching*, *35*(6), 31–35.

Sauers, D., & Walker, R. C. (2004). A comparison of traditional and technology–assisted instructional methods in the business communication classroom. *Business Communication Quarterly*, *67*(4), 430–442. doi:10.1177/1080569904271030

Schrire, S. (2004). Interaction and cognition in asynchronous computer conferencing. *Instructional Science*, *32*, 475–502. doi:10.1007/s11251-004-2518-7

Twigg, C. A. (2003). Improving quality and reducing costs: Designs for effective learning. *Change*, *35*, 22–29. doi:10.1080/00091380309604107

Vrasidas, C., & McIsaac, M. S. (2000). Principles of pedagogy and evaluation for web-based learning. *Educational Media International*, *37*(2), 105–111. doi:10.1080/095239800410405

Walther, J. B., & Parks, M. R. (2002). Cues filtered out, cues filtered in: Computer-mediated communication and relationships. In Knapp, M. L., & Daly, J. A. (Eds.), *Handbook of interpersonal communication* (pp. 529–563). Thousand Oaks, CA: Sage.

Warschauer, M. (2000). Online learning in second language classrooms: An ethnographic study. In Warschauer, M., & Kern, R. (Eds.), *Network-based language teaching: Concepts and practice* (pp. 41–58). Cambridge, UK: Cambridge University Press.

Waschull, S. B. (2001). The online delivery of psychology courses: Attribution, performance, and evaluation. *Teaching of Psychology*, *28*, 143–147. doi:10.1207/S15328023TOP2802_15

Witmer, D. F. (1998). Introduction to computer-mediated communication: A master syllabus for teaching communication technology. *Communication Education*, *47*, 162–173. doi:10.1080/03634529809379120

KEY TERMS AND DEFINITIONS

Ethnography: a research method in the field social sciences. It is used to collect empirical data through participant observation, interviewing or questionnaire from the fieldwork. It can be used to describe the nature of the people it attempts to study.

Learning Contexts: The classroom contexts where students' learning takes place.

Learning Effectiveness: Learning outcomes as the final result from learning process.

Online Learning: Learning process through different forms of communications mediated by the Internet.

Multimedia Learning: Learning process involving communications through graphic, audio, video, animation, and text.

Students' Perspectives: Students' own interpretations of meanings and messages in their perceptions.

ENDNOTE

1 While grades are not necessarily an effective marker of student learning, they are widely used by educators as an indicator of how well students grasped the material covered in a given course.

Section 3
Case Studies

Chapter 14
The Agile Teaching Library:
Models for Integrating Information Literacy in Online Learning Experiences

Mariela Gunn
Oakland University, USA

Elizabeth W. Kraemer
Oakland University, USA

ABSTRACT

This chapter discusses several models of integrating information literacy instruction into computer-mediated learning processes on university campuses with an eye towards the sustainability of each model, its advantages and limitations, as well as its demands on both personnel and the institution as a whole. We strive to provide professionals with the insights to make informed decisions suitable for their context in terms of pedagogical outcomes, organizational resources, and technology infrastructure.

INTRODUCTION

Information literacy (IL) instruction refers to teaching students how to design a meaningful research process for themselves that meets the information needs of a specific project or assignment. The skills necessary for the effective, competent, and ethical use of research materials have been codified as information literacy standards by the Association of College and Research Libraries (ACRL). These standards include the skills of individuals to "recognize when information is needed and have the ability to locate, evaluate, and use effectively the needed information" (American Library Association [ALA], 1989). The definition of information literacy is relevant well beyond the academic context, because it encompasses lifelong learning, as well as any other personal research pursuits. Higher education, however, offers a research-intensive context for students, making universities the perfect environment for teaching information literacy skills, thus

DOI: 10.4018/978-1-61350-077-4.ch014

providing students with the conceptual tools needed to achieve academic excellence and to develop research habits that will enable self-directed learning throughout their lives.

ACRL last reviewed the information literacy standards in 2000, and the standards are currently being used as the foundation of information literacy programs in academic libraries throughout the United States. They have also been endorsed by the American Association for Higher Education and the Council of Independent Colleges (ACRL, 2010). Educational accrediting bodies, too, have come to recognize the fundamental and inter-disciplinary nature of information literacy skills. For example, the Middle States Commission on Higher Education, the Western Association of Schools and Colleges, and the Southern Association of Colleges and Schools have all incorporated information literacy as a key component of student learning (Yang, 2009, p. 684). Given the central role of information literacy to student learning across the disciplines, special attention ought to be brought to the manner in which academic librarians incorporate such instruction in the ever-increasing number of courses offered online and within hybrid courses that offer varying degrees of computer-mediated instruction. A conceptualization of various integration models that together comprise an agile information literacy program can aid librarians, subject faculty, and administrators in navigating the pedagogical and organizational choices involved in creating a rich, student-centered learning experience in relation to information literacy skills. Each model constitutes but one facet of an information literacy program that enables students to become better researchers and life-long learners.

Traditionally, information literacy instruction has happened within university classes at the request of a specific subject faculty member, who invites a librarian to teach a research session relating to a course assignment. The librarian's teaching objectives have focused on a few key information literacy competencies, such as generating appropriate keywords for a topic and finding relevant information in the library catalog and databases. Rapid technology changes, along with an abundance of information available both freely on the web and in databases subscribed to by the library, have highlighted the need to find ways of teaching all information literacy competencies, including the evaluation of resources and their ethical use. In effect, changes in the research context have shifted librarians' focus from teaching tools and strategies to teaching the overall research process and the critical thinking skills needed to manage it. Integrating information literacy instruction into subject courses necessitates approaches that go beyond the traditional one-class visit.

Classroom-based instruction uses traditional pedagogies to teach a computer-mediated research process that encompasses digital versions of materials and diverse interfaces for accessing both physical and digital collections. The explosion of online publications and digitized versions of print publications increase the need for information literacy skills. Computer-mediated research introduces the need to understand how the print world is structured and to then apply critical judgments to digital publications based on that knowledge. Students required to use peer-reviewed scholarship in their research assignments, for instance, might be faced with online texts that look visually similar to many free websites, but which in fact constitute the online versions of articles published in peer-reviewed, print-based journals. Computer-mediated communication (CMC) and its attendant technologies have transformed the research context by offering immediacy of online access combined with the added burden to be all the more discerning and knowledgeable in selecting appropriate sources of information.

In addition, web-based information literacy instruction poses further pedagogical challenges in teaching the complexities of the research process, not only by introducing CMC technologies into teaching itself, but also by positioning library faculty as embedded visitors within online learning environments. As the one-time classroom visit is replaced by a consistent virtual presence within the learning environment, library faculty members have an opportunity to re-envision information literacy instruction and to enhance its integration into the course-driven learning process of students through the deliberate use of CMC technologies.

Research has shown that web-based instruction – which, by its very nature, relies on computer-mediated communication – in any discipline is most effective when it integrates hands-on, interactive exploration of concepts rather than relying simply on the traditional 'sage on the stage' lecture or presentation (Ali, 2003). Because information literacy instruction is especially suited to scenario-based lessons in which students are expected to interact with a tutorial or database in order to accomplish an assignment, it is reasonable to offer more library instruction in online environments. A review of the library literature demonstrates repeatedly that many college and university libraries are doing just that, relying on some form of online instruction to help ease the demands that a growing emphasis on information literacy places on instruction librarians—many of whom are already over-scheduled in the classroom (Kraemer, 2003). Online library instruction benefits learners by providing new opportunities to engage students in the research process through the integration of CMC technologies such as video, audio, and interactive content—all of which can be customized to address a specific subject area or audience. In addition, the flexibility of web-based instruction gives students control over the timing and speed of content delivery, an advantage of CMC which is simply not possible in face to face communication, such as a session in which a librarian is addressing an entire class (Bi, 2000). The effectiveness of online library instruction has been confirmed by numerous studies, which have shown that student performance in a pre- and post-test assessment situation improves after receiving information literacy instruction, regardless of whether that instruction is delivered all-online, face-to-face, or in a hybrid fashion (Kraemer, Lombardo, & Lepkowski, 2007).

The objective of this chapter is to discuss several different models of integrating information literacy instruction into the overall learning process on campus with an eye towards the sustainability of each model, its advantages and limitations, as well as its demands on both personnel and the institution as a whole. Rather than recommending a specific set of best practices, this chapter strives to provide professionals with the insights to make informed decisions that are suitable for their contexts in terms of pedagogical outcomes, organizational resources, and technology infrastructure. In the following sections, four integration models are presented and analyzed: program-wide partnerships, credit-based information literacy courses, online learning modules, and system-based integration into learning management systems (LMS) on campus. Each integration model embraces CMC technologies in instruction, and each is considered as a facet of a broader, agile information literacy program through a comparison of common implementation concerns. It is recognized that each information literacy program is unique as each strives to address the specific needs of its unique campus community. Yet an overall analysis of program components can inform faculty and administrators as they consider the place of information literacy on their respective campuses, implement each component, and balance the use of their organizational and technological resources.

MODELS FOR INTEGRATING INFORMATION LITERACY INSTRUCTION

Program-Wide Partnerships

It is common practice for librarians to be involved in subject-based instruction through what are informally known in the profession as "one-shots"

– one-time in-class sessions, or guest lectures, by librarians, focusing on subject specific or assignment specific research tools and skills. Such a guest lecture will include demonstrations of relevant databases and, ideally, hands-on research time for students to apply the skills covered by the librarian. Courses taught entirely online also can benefit from course-integrated library instruction in a variety of ways, including synchronous web-casting sessions, online tutorials and quizzes, and asynchronous librarian participation in the course via class news forums or discussions boards. Providing information literacy instruction within the context of a specific area of study allows students to see direct connections and immediate application of critical thinking techniques and also "increases the likelihood that students will see the transferability of IL concepts and skills" across research activities and disciplines (Hunt & Birks, 2004, p. 33). One significant problem with the one-shot approach, however, is that it often lacks consistency and continuity. For example, if there are multiple sections of a course being taught by different teachers, some classes may request a library instruction session while others may not, or an instructor may involve a librarian one semester but not have time the next semester that he or she teaches the class. Computer-mediated instruction has the potential to alleviate the problem of time, regardless of the format of the class. Information literacy that is communicated asynchronously via computer can be assigned by an instructor as homework, thus alleviating the problem of a guest lecture taking time lecture time away from the regular instructor. This method can also save time for the librarian; recorded lectures can be replayed for various classes, without requiring the librarian to give the lecture in person each time it is needed.

In an article describing the importance of information literacy instruction to the university experience, author Edward Owusu-Ansah notes this very issue and suggests that students would profit even more from an information literacy program that is integrated into the curriculum, one that "avoided haphazard distributions of engagements and ensured an organized plan for engaging the entire student population at one time or the other of their college attendance" (2004, p. 4). At Oakland University (OU), students benefit from just such an integrative arrangement, thanks to the collaborative relationship between Kresge Library and the Department of Writing and Rhetoric. Through this partnership, every on-campus section of Writing 160 Composition II (WRT 160) receives in-class library instruction supplemented by online instructional content and testing in the university's learning management system, Moodle. This program-wide partnership is intended to introduce OU students to research concepts and tools near the beginning of their college careers—the majority of those enrolled in WRT 160 are in their first year— thereby laying the foundation for advanced research as they progress in their majors.

Advantages

As noted in the literature, integrated instructional programs thrive through the shared cooperation of librarians, classroom faculty, and administrators (Grafstein, 2002, p. 198); therefore, a factor not to be overlooked in a program-wide approach to information literacy instruction is the need to sustain a close relationship with the collaborating department. Though library instruction is a mandatory part of WRT 160 at Oakland through a long-standing arrangement, teacher support makes the program significantly more effective; for instance, when the course instructors promote in the classroom the importance of information literacy—and, even better, assign a portion of the overall course grade to the successful completion of the library modules—their students tend to be more attentive and productive during the library sessions. Therefore, information pertaining to the library instruction component of WRT 160 is regularly distributed in order to keep the classroom faculty involved and updated. Here, too, CMC facilitates the process: in order to maintain open lines of communication, the library's Coordinator of Information Literacy has been granted access to the Writing and Rhetoric Department's online news forum, allowing for the posting of library announcements and requests for input from the Writing faculty. In addition, representation at unit faculty meetings or professional development retreats is a key method of communicating in-person the library's plans, goals, and expectations for the partnership, as well as those of the collaborating department. Through such meetings ideas can be exchanged, new initiatives can be announced, and relationships between librarians and classroom faculty are strengthened.

Of utmost importance to the success of such an integrated program of IL instruction is the development of specific student learning outcomes for the library instruction portion of the course. Outcomes provide focus for the program as a whole and direction for the in-class sessions, and also allow for assessment of library content to ensure that students are learning the research concepts and skills expected of them. Regular reviews of the established information literacy learning outcomes are called for, to confirm that the standards are up-to-date and that they retain their relevance; for example, the library faculty at Oakland University recently underwent a semester-long review of the WRT 160 IL outcomes and found that an almost complete overhaul was needed due to obsolescence of certain tools and concepts once central to library instruction. Librarian investment in the program is also obviously crucial to accomplishing these and other organizational tasks to forward a program-wide collaboration.

Challenges

The most substantial hurdle to this program-integrated approach to information literacy instruction is surely the sizeable amount of librarian teaching time needed to fulfill the commitment to the partnering department; fortunately, CMC can often be applied to ease the workload while maintaining a high standard of student learning. At Kresge Library, for example, the instruction that is provided to WRT 160 classes makes up nearly 50% of all the sessions taught by the library faculty (Kraemer, Lombardo, & Lepkowski, 2007, p. 2). This ratio is what finally prompted the library, in 2001, to first develop online tutorials to be used to supplement the face-to-face instruction for WRT 160; the introduction of these tutorials allowed the library to pare face-to-face time from three hours down to approximately two hours with each section of WRT 160. With more than 100 sections of the course being offered at the university each school year, this revision saves a considerable chunk of time that can be directed toward other areas of the library's instructional program. It is worth noting that while there is an all-online version of the WRT 160 library instruction content that is available to serve distance education sections of the course, librarians at OU have found benefit in having students physically interact with the library facilities and staff whenever possible, and therefore do not offer the all-online version of library instruction for on-campus sections of WRT 160.

Credit-Based Online Courses

Much is still being debated in the literature about the value in academic libraries offering credit-bearing courses. As early as 1956, the argument was made that library instruction, as a vital part of college education, should be offered only as an integrated part of the general curriculum rather than as an isolated area of study (Knapp, 1956). In response to this "teaching in a vacuum" argument, it has been pointed out that in higher education, the "many segments of knowledge, taught mostly separately, expectedly come together to form the learner's unified knowledge base" (Owusu-Ansah, 2007, p. 423), anticipating that students themselves will see the value of information literacy and integrate it into their learning.

Further, it can also be argued that the vast increase in research resources and investigative avenues in the recent decades has amplified the very need for courses focused on research methods, taught by experts in the field. With the shift that has been experienced in information literacy instruction, emphasis has moved away from tool-based teaching sessions to a focus on wider concepts and transferable skills (Owusu-Ansah, 2004).

Kresge Library entered the credit-course arena in 2008 with the 2-credit online offering, LIB 200. Entitled "Research in the Information Age," the semester-long course followed the research process from topic development through to a final paper, challenging students to apply each step of the research process as it was covered in the course. To this end, students selected topics for research in the first few weeks of the semester (they could use one of the instructor's suggested topics or get approval to research a topic of their choosing) and then used those topics throughout the semester to accomplish such assignments as brainstorming keywords, finding scholarly articles in online databases, searching for related subject headings, and finally compiling an objective research paper. By linking lessons to practice throughout the semester, the students were able to directly and immediately apply concepts and see the importance of each stage.

Advantages

Offering an online credit-bearing course provides exceptional benefits to students who complete the course, to the library offering the course, and to the institution as a whole. Owusu-Ansah, in a discussion on the importance of broadening the reach of library instruction, remarked that "[college] credit is the currency of recognition, denoting the significance or value of an educational activity or experience in higher education" (2007, p. 419), thereby noting a significant advantage to libraries: a credit course will help to further legitimize to the university community the concept of and need for information literacy skills. This cause can be promoted even more emphatically should the information literacy course be integrated into the General Education curriculum at the institution.

More importantly, a credit-bearing information literacy course provides more time to demonstrate research as a holistic process and obliges students to "engage and master broader concepts" (Owusu-Ansah, 2004, p. 10) and presents students with situations intended to improve critical thinking and higher-level reasoning—opportunities that are often lacking in assignment-based sessions focusing on specific, immediate needs. Moreover, in a study comparing research papers of students who had completed a credit-bearing information literacy course and students who had not, analysis showed that regardless of discipline, students who had been exposed to the credit course tended to use a greater number of scholarly sources, have more complete citations, and earn higher grades on the assignment; these findings prompted the study's author to suggest that having learned appropriate information literacy skills in the library credit course, these students were "more adaptive, resourceful, and confident when dealing with research challenges" (Wang, 2006, p. 88).

Challenges

While there are clearly significant advantages to offering a credit-bearing library course, organizations must also consider the hurdles and considerable demands involved in such an undertaking. Among the most significant impacts of this model of information literacy instruction is the time-consuming nature of developing and teaching a full-length course; even allowing for the possibility of a partial-semester credit course, it is likely that the librarian (or librarians) responsible for teaching the class will not be free for other instructional responsibilities during the semester in which the course is held, and perhaps not while the course is being developed. Therefore, any academic library exploring this option must consider its staffing model and build flexibility into professional assignments to accommodate the course instructor. In addition, it is vital for the instructor to note that teaching an online course, regardless of class size, is generally more time-consuming than teaching the equivalent course in a face-to-face environment (Cavanaugh, 2005).

Technology tools available at the college or university must also be taken into account when initiating a credit-course. Most institutions of higher education rely on learning management software (LMS), thereby managing student access to the class and easing some aesthetic issues of designing an online course (such as the website's consistency in appearance and continuity from week-to-week). However, the deeper structure of the course must still be considered; for example, will all the class material and supplemental online resources be uploaded and stored in the LMS or on the library's server? This issue is especially relevant to an information literacy course, where the instructor may rely on existing material used by the library such as an online tutorial or digitized handouts: if a handout or tutorial is updated during the semester, merely linking to it from within the LMS--rather than storing a separate version of it in the course space--will save the professor the time of having to either edit the file multiple times or re-upload the newer version of the tutorial. Additionally, the accessibility of an e-learning or instructional support department is central to the successful preparation and maintenance of the credit-course, particularly when one is teaching online for the first time. Should unexpected technical hurdles arise (and they will), the e-learning staff will help even the most seasoned online instructor navigate the problem and find a workable solution. Further, such a department often provides instructional design assistance for library faculty new to the process of building a course.

Other significant considerations for the library must include marketing and assessment of the course. Marketing is key—particularly in preparation for the first semester when students are not yet familiar with the library's course—and an excellent resource to assist in promoting a library credit course is the institution's academic advisors, who are in contact with the very individuals who would benefit from a research course; in addition, flyers or posters located in high traffic spots around campus continue to be excellent advertising venues. In terms of assessment, information literacy instruction provides an ideal environment for authentic assessment, which employs tools beyond quizzes and tests to measure student learning; for instance, a research and writing portfolio that is compiled throughout the semester by course participants provides ample opportunities for the instructor to evaluate the understanding and application of key research concepts.

Finally, perhaps the most important ingredient in the mix is the presence of administrative support for the concept of the online library credit-course. Initially, administrative involvement certainly will be necessary to lay the groundwork for the course, including creating a rubric under which library courses will be offered, navigating the course listing process with the university's registration office, and approving release time for the course instructor. Library administrators may also need to have a say in such decisions as the number of sections of the course being offered each semester, the cap on class size for each of those sections, and also how frequently the course is offered. Ultimately, without administrative buy-in, long-term sustainability of such an initiative is unlikely, given the impact that a credit-course has on the library as a whole.

Online Learning Modules

Online learning modules constitute a key component of an overall information literacy program that aims at adapting itself to the needs of online courses. Just as seen in program-wide partnerships, these modules can serve as virtual, asynchronous equivalents of information literacy instruction when implemented in a way that matches the instructional needs of various assignments, courses, or programs. Learning modules online range from stand-alone tutorials to interactive modules integrated in subject-specific courses. There is a proliferation of tutorials on library websites that can be used by students on an as-needed basis. They have also emerged as the main way for librarians to "face the challenge" of creating instructional content which can be incorporated in online courses (Yang, 2009, p. 685). Most often tutorials focus on specific research databases and deliver the instructional content in formats, such as web pages and video screencasts.

Web-based tutorials in their most traditional form offer a largely informational approach when they convert print-based handouts for delivery online. While web-based tutorials constitute a quarter of the academic library tutorials, their effectiveness depends on the availability of some interactive features in the form of dynamic content or scripting that allows for an engaged, active learning experience (Yang, 2009, p. 688-689). The interactivity of these tutorials can make the learning process more student-centered by offering flexible navigation based on the outcomes of embedded activities instead of sequential presentation of content, as well as by creating the potential for real-time practice when students use the tutorials to guide them through the interface of a database, for instance. In the latter case, the tutorial content would occupy only part of the screen while a database is navigated simultaneously in an accompanying frame.

The more interactive the web-based tutorials are, the higher the need for technical know-how on the part of the librarians creating them. Both scripting interactive exercises and designing a guided tutorial for a research database benefits from the development of skills beyond familiarity with basic web design. The availability of many free scripts online at websites, such as HotScripts.com, alleviates the need to recreate commonly used functionality from scratch, but librarians still face the requirement to install and customize scripts and to develop additional interactivity suited to each tutorial's content. The technical infrastructure to support interactive web-based tutorials need only include support for the technology of choice and the maintenance of these resources at the server level by technical support staff within the library or by campus-wide Information Technology (IT) departments.

The presence of technical know-how within the library constitutes a primary concern in the development of online tutorials. In its absence, however, a few prepackaged information literacy tutorials exist under a Creative Commons License and can be utilized as stand-alone guides to basic information literacy skills. The Texas Information Literacy Tutorial (TILT), at http://tilt.lib.utsystem.edu, is just one popular example. Prepackaged tutorials, however, can be hard to integrate into courses other than freshman-level writing classes as they offer most instructional value to students who are just starting their higher education experience and are in need of a comprehensive introduction to information literacy.

On the other end of the technical spectrum, video screencasts as tutorials constitute about a third of academic library tutorials (Yang, 2009, p. 688). They are best suited for demonstrations of interfaces or search concepts and are commonly created using software--such as Adobe Captivate or Camtasia Studio--that outputs Flash video as a result of screen capture. Just as web-based tutorials can be of limited instructional value when created in a static and strictly informative manner, videos can be passive and have a limiting pace to them unless interactivity is incorporated to transform an informational resource into a learning experience. When informed by constructivist pedagogy, screencast tutorials enhance self-paced student-centered learning by embedding activities and allowing for a flexible, game-like engagement. Additionally, screencasts accommodate various learning styles by combining visual and narrative presentation of content, especially when they effectively implement the contiguity principle of making both modes spatially and temporally adjacent (Tempelman-Kluit, 2006, p. 366).

Creating and maintaining video tutorials that remain effective and current over time necessitate a labor-intensive process. The initial creation phase includes story-boarding the learning experience to ensure that the intended outcomes can be achieved, designing the tutorial in screen capture software, and potentially testing it with students. The literature on tutorial creation has identified multiple key factors in achieving effective learning experiences online; some of the basic principles include transparency of the terminology, consistency across sections, selectivity in scope, and accessibility (Wales & Robertson, 2008, p. 369). Over time, the maintenance of video tutorials becomes most intensive when database vendors re-design online interfaces or change their functionality. As this happens on a regular basis, video tutorials require close monitoring and frequent revisions to keep the demo sections in line with the real-time look and features of various research databases. The ongoing demand on librarians' time, however, becomes worthwhile when we consider the potential of video tutorials to turn into the building blocks of more elaborate and course-integrated information literacy modules online.

Advantages

Integrating tutorials into courses transforms them into complex learning modules and increases their instructional value by positioning only relevant tutorials within the context of course assignments: "when students are able to see a practical application of the knowledge, they are more motivated and they can more easily merge the new information with their previous experiences" (Dewald, 1999, p. 29). As more courses take place in online learning management systems (such as Blackboard, Moodle, Sakai, to name a few) integration of learning modules can be seen as the online version of course-specific information literacy instruction. Depending on the learning outcomes specific for each course, the librarian can design a learning module that pulls together video and web-based tutorials and enhances them further through the use of the active learning capabilities available within the LMS.

At Oakland University, a successful example of creating a learning module and incorporating it in a required business course involved six distinct content sections focusing on concept-based as well as tool-based topics such as keyword selection, specialized business publications, locating company financials, and similar information. Each section consisted of a brief video that offered a demo of the topic and several realistic practice activities created within the LMS. The close alternation between presenting video content and following it up with opportunities for self-guided practice allowed students to apply the new information immediately and receive feedback on the success of their application. Each activity resulted in selecting a multiple-choice answer after practicing and included the explanation of why the selected answer was correct or incorrect, reinforcing the material previously encountered in the video demo.

Challenges

Key considerations in implementing the online learning module described above include the learning outcomes of the subject course identified in collaboration with the course instructor, the pedagogy of the course so that the information literacy module fits seamlessly in the course content, and the need for assessment of student performance on the information literacy module. Each of those considerations presupposes a partnership between the librarian and the subject instructor that takes place before the academic semester. In contrast to a traditional information literacy session in a classroom, the design of an online learning module requires more advance planning as well as implementation that takes place before the content is made available online.

Online information literacy modules need not be restricted to the use of video tutorials or integration at the level of subject courses. Web-based and other tutorial formats, for instance, have been combined successfully to meet the needs of the School of Nursing at OU. In that case, the ability of our LMS to support the manual creation of a well-rounded course that is not part of the course catalog allowed for the integration of an extensive information literacy module with the whole Doctor of Nursing Practice program. Implementing module integration with departments on the program level, however, depends on the support efforts of staff who manage the campus LMS, so that all the appropriate students are "enrolled" in the learning module and have clear guidance and context regarding the program's expectations of how and when to go through the online learning experience.

Creating online information literacy modules meets the needs of the online curriculum and can provide engaging experiences through well-conceived and relevant interaction. While the time and skill demands on librarians ought not to be underestimated, the development over time of multiple tutorials of specific and narrow scope contributes to a large pool of building blocks to be utilized in learning modules for various courses. The existence of collaborative partnerships between librarians and subject faculty remains a prerequisite for effective content integration online as it has always been in the classroom. Online, however, the ability to add the librarian as an instructor within the course or to create an independent learning environment linked to the course facilitates the creation of modules that utilize fully the interactivity supported by the LMS and integrate the learning module within the assessment structure of the course.

Learning Management Systems

Learning management systems provide the main environment for online instruction and CMC between students and librarians. With the overall trend in higher education towards more course offerings that are fully online, information literacy instruction faces the need to reach a growing number of students who might have never experienced the campus library as a physical space. Pamela Jackson found in a recent study that while librarians do often collaborate with course instructors in order to integrate information literacy into face-to-face courses, they seldom do this for online courses due to barriers such as lack of faculty buy-in, high demand on librarians' time, and a steep technology learning curve (2007, p. 456-7). While collaborations might increase over time as libraries develop more online tutorials and learning modules and start using them as building blocks for integration into online courses, we cannot expect that process to scale up to the system-wide level at which every class would have a course-specific information literacy component available for students' perusal within the LMS.

At the same time, some general library resource inclusion in LMSs has already begun on many campuses as part of academic libraries' efforts to achieve greater visibility to online students; in fact, "experts encourage librarians to integrate library resources in the systems students use most" (Jackson, 2007, p. 455). Learning management systems do not offer an off-the-shelf functionality to incorporate a library presence within them, though it is not uncommon for campus-wide initiatives to lead to the inclusion of library-related content, such as a link to the library website within the main LMS interface. Such integration, however, represents only an informational approach, simply bringing the library website closer to online students and making the availability of various library services and resources highly visible. This trend of including the library, that is, links to its web site and other resources, within the LMS has not been focused on instructional content, so it can be rightfully considered *library* integration, but not necessarily *information literacy* integration. Yet the growing popularity and flexibility of discipline-specific library resource pages, referred to as subject guides, hold potential for achieving the learning objectives of information literacy instruction for more online courses than library personnel could ever accommodate, thereby bringing into grasp the concept of true IL integration into LMSs. Academic librarians see these guides as "a critical resource in helping students get started on their research in a particular subject area" (Staley, 2007, p. 119). Moreover, subject librarians are already expected to create them: 100% of respondents in a survey of academic libraries confirmed this expectation (Jackson & Pellack, 2004, p. 325).

Recent approaches to subject guides have made them into more than the lists of resources they used to be. Online content management systems like LibGuides (http://www.springshare.com/libguides/) or equivalent open source software, such as LibData (http://libdata.sourceforge.net/) or SubjectsPlus (http://www.subjectsplus.com/), are allowing subject guides to transform into subject-specific portals to both resources and common instructional materials online. A subject guide can include embedded tutorials for various aspects of the research process, widgets that allow for immediate online contact with a librarian to answer research-related questions, as well as links to subject databases and other resources. The new online systems for creating and maintaining subject guides have made this broadening of their function and definition possible without creating a steep learning curve. The systems are relatively easy to use and offer the added flexibility of compiling content dynamically, so that records for commonly used resources can be managed in a centralized way.

Advantages

Even though learning management systems do not yet provide for a direct integration with the systems for subject guide creation, they support various add-ons and plug-ins, which can enable subject-specific integration. For instance, students at Brigham Young University can view subject research guides for all the courses they are enrolled in through a custom building block in Blackboard (Washburn, 2008, p. 303). Ninety-three percent of students have reported finding those guides useful in a recent survey, while only six percent have found them simply confusing, most likely due to lack of skills in searching databases or lack of familiarity with the presented resources (Washburn, 2008, p. 308). Brigham Young University's approach illustrates a focus on developing a customized library services area within the LMS and offering subject-guides within it. An alternative approach can be implemented (depending on LMS affordances, of course) that focuses on bringing subject guides fully within each course's learning space. Kresge Library has a system in place for developing resource-specific and subject-specific guides, referred to as course pages, for each traditional information literacy session held; thus, we already have a pool of subject faculty recommending these online guides to their students. An add-on in the campus LMS can enhance our current practices by integrating a more general subject guide within each course that we have never visited or otherwise participated in, but delivering the course-specific guides when available due to collaboration between the subject faculty and the library faculty.

Challenges

The main hurdles that libraries can expect in carrying out initiatives to integrate information literacy at the LMS system-level include the need for large-scale collaborations, the crucial impact of faculty buy-in, and the necessity to still conduct active face to face information literacy sessions. Whenever any add-on is being developed for the campus learning management system, LMS administrators and programmers are called upon to work closely with the library. In addition to offering the technical implementation of the add-ons, these stakeholders bring an institutional perspective that should be taken into serious consideration; for example, since they support the online teaching on campus, LMS staffers are familiar with faculty practices and attitudes, and can suggest reasonable ways to launch an integration project. Campus-wide faculty buy-in is also crucial because any instructional or informational content integrated into their course space online could be embraced, ignored, or even seen as an encroachment on their control over the online course; therefore effective communication and marketing of IL integration into the LMS is key in gaining faculty support. Even with these collaborative elements in place, asynchronous learning modules cannot be expected to achieve the same learning objectives that a synchronous librarian-led session can, because student initiative and intellectual curiosity will be the main driving forces for the use these materials. Thus, active librarian participation, especially in research intensive courses, should still be relied on to build upon the materials presented in the learning management system.

While the main limitations of automatically-integrated subject or course guides relate to depth of instruction, their strengths justify the time, institutional partnerships, and commitment needed for their implementation. System-level integration can provide students with a continuity of access to, and familiarity with, research materials relating to their majors. Student taking courses within their majors will at the very least be exposed to the same subject-specific portal to research from within all their courses online, and to a much more customized course guide for classes with active librarian participation. In addition, some basic information literacy skills — such as developing search strategies or selecting appropriate research databases — can be successfully taught through tutorials embedded in the subject-specific portals, which would be presented even without librarian involvement. Embedded tutorials can effectively support the development of some information literacy skills if they model relevant resources and offer interactivity built upon realistic, subject-specific research examples.

COMPARISON OF IMPLEMENTATION CONCERNS

Each of the four integration models discussed above offers different advantages and limitations in terms of pedagogical flexibility, personnel needs for professional development, organizational and technical demands.

Pedagogically, the stand alone credit-based information literacy courses offer the most intense IL learning experience for students, because they take place over the course of a whole semester rather than a class session or two. At the same time, partnerships with whole programs can achieve learning outcomes of a similar magnitude by providing consistency of instruction across a large number of students, especially when the partnerships are developed strategically with programs that offer freshman-level required courses. Ideally, a comprehensive information literacy program would employ both approaches. Realistically, credit-based information literacy courses might be effectively integrated into the General Education curriculum in institutions with strong library involvement at the curriculum level. That is especially true for academic libraries whose librarians have faculty status, and, thus, participate in campus-wide service committees with their departmental colleagues.

When it comes to learning management systems, integration of instructional subject guides across the whole system offers students broad and consistent exposure to guidance and instruction, though sometimes this content can be viewed as confusing due to lack of context. Institutions with strong traditions of faculty buy-in as well as active e-learning or technology departments that administer the LMS can provide this service to the whole campus, which is especially valuable for student populations with high percentages

Table 1. Pedagogical concerns

	Advantages	Challenges
Program-Wide Partnerships	• Consistency in instruction • Ability to reach many students	• Generalized content doesn't allow for much customization from section-to-section • Can only introduce basic information literacy concepts
Credit-Based Online Courses	• In-depth research instruction • Focus on critical understanding	• Limited integration in subject areas • Reaches only the students who enroll in course
Online Learning Modules	• Student-centered instruction • Seamless integration into courses	• Traditionally focused on tools rather than concepts • Automated approach to instruction; no librarian contact
Learning Management Systems	• Focus on informational content • Ability to reach many students • Automatic placement of library resources into courses	• Cannot substitute other models

of distance learners. System-wide integration, however, cannot be viewed as a substitute for the design of synchronous or asynchronous learning modules that aim at providing highly customized information literacy instruction that pertains to very specific courses and assignments. Collaboration between librarians and subject faculty have traditionally guided and enabled this approach to IL integration.

Whatever the combination of approaches to integration, assessment is always a key concern with information literacy instruction. While assessment in a credit-based course is very well established, assessment components need to be designed and incorporated in both program-wide partnerships and in course-specific learning modules. The grade that the course instructor assigns to the IL module in either case can substantively affect both student motivation and the perceived relevance of the instructional content. Assessment is most difficult with system-wide integration of subject guides in the LMS due to the self-directed nature of their use. Their effects can most reliably, although indirectly, be measured through an examination of the quality of research assignments students submit for various courses (Table 1).

For academic libraries facing instructional overload of their personnel, online instruction might only shift the workload without necessarily alleviating the overload. As asynchronous learning modules or subject guides are developed, librarians will find themselves spending more time designing and maintaining instructional content and will therefore be in need of ongoing professional development to keep up with advances in instructional technologies. As information literacy becomes more tightly coupled with the campus curriculum through credit-based courses or program-wide partnerships, the amount of time spent on instruction-related activities will likely remain stable: teaching online in these instances merely re-invents the balance between content design and synchronous engagement with students.

Finally, each of the four approaches described above presupposes a strong administrative leadership within the academic library. As integration initiatives begin to move beyond the impromptu sessions agreed upon between subject faculty and librarians who know each other, keep in touch each semester, and have a personal working relationship, the instructional efforts of individual librarians need to contribute to the library's overall goals and complement one another. Information literacy programs that prioritize the time and focus relevant for each approach can achieve an agile structure that accommodates the various learning experiences that students are exposed to (Table 2).

Table 2. Organizational concerns

	Advantages	Challenges
Program-Wide Partnerships	• Systematic approach that lends itself to ongoing assessment	• Time-consuming coordination at faculty and administrative levels • Requires librarians to agree on standardized instructional content
Credit-Based Online Courses	• Expands library role on campus • Can enhances general education	• Flexible staffing structure needed • Depends on technical support staff
Online Learning Modules	• Content creation can be distributed among librarians • 24/7 student access to modules	• Production software may present a steep learning curve • Ongoing technical training
Learning Management Systems	• Increases awareness of IL • 24/7 student access to subject-specific guides	• Relies on collaboration between library and LMS staff

Table 3. Roles in integrating information literacy instruction

	Librarians	Subject Faculty	Administrators
Program-Wide Partnerships	• Collaborating on the development of teaching goals • Carrying out instruction consistently across sections	• Introducing the significance of IL skills to students so they understand the value of library sessions	• Cultivating and maintaining a collaborative relationship between academic department and library
Credit-Based Online Courses	• Developing, maintaining, and teaching the course	• Advising students on the availability and value of an information literacy course	• Offering support and administrative approval for course • Planning for release time for the teaching librarian
Online Learning Modules	• Connecting with subject faculty to market resources	• Including librarian in online course for seamless IL integration	• Providing needed software and technical training
Learning Management Systems	• Developing and maintaining current, relevant content	• Integrating the resources available from course page in course assignments	• Collaborating with technology units on campus

As academic libraries continue to expand their information literacy programs and adjust them to accommodate the growing number of courses taught online, and integrate new technologies, they will be implementing several, if not all, of the integration models presented here. The roles of librarians, faculty, and administrators on each campus will be influenced by the institutional structures in place, as well as by the choice of integration models. Below is a summary of the demands each model for integrating information literacy instruction places on the relevant participants (Table 3).

CONCLUSION

Computer-mediated instruction offers a new and exciting challenge to academic librarians. As librarians redefine their roles as teachers to incorporate not just instructional delivery but instructional design as well, they are building the foundation for active, meaningful participation in the online teaching arena. Even though libraries provide the locus for information literacy instruction, collaboration among subject faculty, librarians, and campus administrators is crucial to the seamless integration of information literacy instruction within the learning process.

Many additional models for integrating information literacy instruction online will likely emerge over time, but the current trends focus on the development of deliberate and purposeful IL programs within libraries. We expect those trends to play out differently in each institutional context; for example, academic libraries with separate instructional and technical personnel might ensure strong collaborations within the libraries towards the common goals of their IL program, while other institutions might invest in providing each librarian with the professional development to carry out both design and delivery of information literacy instruction. The campus-wide approaches to online learning and the need to reach distance education students will also undoubtedly advance the mission of IL programs at colleges and universities.

REFERENCES

Ali, A. (2003). Instructional design and online instruction. *TechTrends*, *47*(5), 42–45. doi:10.1007/BF02763205

American Library Association. (1989). *Presidential committee on information literacy* (final report). Chicago, il: American Library Association. Retrieved from http://www.ala.org/ ala/ mgrps/ divs/ acrl/ publications/ whitepapers/ presidential.cfm

Association of College and Research Libraries. (2010). *Information literacy competency standards for higher education*. Association of College and Research Libraries [Website]. Retrieved from http://www.ala.org/ ala/ mgrps/ divs/ acrl/ standards/ informationliteracycompetency.cfm

Bi, X. (2000). *Instructional design attributes of web-based courses*. (No. ED448746). San Antonio, TX. Retrieved from http://www.eric.ed.gov/ ERICWebPortal/ contentdelivery/ servlet/ ERICServlet? accno=ED448746

Cavanaugh, J. (2005). Teaching online - A time comparison. *Online Journal of Distance Learning Administration*, *8*(1). Retrieved from http:// www.westga.edu/ ~distance/ ojdla/ spring81/ cavanaugh81.htm

Dewald, N. H. (1999). Web-based library instruction: What is good pedagogy? *Information Technology and Libraries*, *18*(1), 26–31.

Grafstein, A. (2002). A discipline-based approach to information literacy. *Journal of Academic Librarianship*, *28*(4), 197–204. doi:10.1016/S0099-1333(02)00283-5

Hunt, F., & Birks, J. (2004). Best practices in information literacy. *Libraries and the Academy*, *4*(1), 27–39. doi:10.1353/pla.2004.0010

Jackson, P. A. (2007). Integrating information literacy into Blackboard: Building campus partnerships for successful student learning. *Journal of Academic Librarianship*, *33*(4), 454–461. doi:10.1016/j.acalib.2007.03.010

Jackson, R., & Pellack, L. J. (2004). Internet subject guides in academic libraries: An analysis of contents, practices, and opinions. *Reference and User Services Quarterly*, *43*(4), 319–327.

Knapp, P. B. (1956). A suggested program of college instruction in the use of the library. *The Library Quarterly*, *26*, 224–231. doi:10.1086/618322

Kraemer, E. W. (2003). Developing the online learning environment: The pros and cons of using WebCT for library instruction. *Information Technology and Libraries*, *22*(2), 87–92.

Kraemer, E. W., Lombardo, S. V., & Lepkowski, F. (2007). The librarian, the machine, or a little of both: A comparative study of three information literacy pedagogies at Oakland University. *College & Research Libraries*, *68*(4), 330–342.

Owusu-Ansah, E. K. (2004). Information literacy and higher education: Placing the academic library in the center of a comprehensive solution. *Journal of Academic Librarianship*, *30*(1), 3–16. doi:10.1016/j.jal.2003.11.002

Owusu-Ansah, E. K. (2007). Beyond collaboration: Seeking greater scope and centrality for library instruction. *Libraries and the Academy*, *7*(4), 415–429. doi:10.1353/pla.2007.0043

Staley, S. M. (2007). Academic subject guides: A case study of use at San Jose State University. *College & Research Libraries*, *68*(2), 119–139.

Tempelman-Kluit, N. (2006). Multimedia learning theories and online instruction. *College & Research Libraries*, *67*(4), 364–369.

Wales, T., & Robertson, P. (2008). Captivating open university students with online literature search tutorials created using screen capture software. *Program: Electronic Library and Information Systems*, *42*(4), 365–381. doi:10.1108/00330330810912052

Wang, R. (2006). The lasting impact of a library credit course. *Portal: Libraries and the Academy*, *6*(1), 79–92. doi:10.1353/pla.2006.0013

Washburn, A. (2008). Finding the library in Blackboard: An assessment of library integration. *MERLOT Journal of Online Learning and Teaching*, *4*(3), 301–316.

Yang, S. (2009). Information literacy online tutorials: An introduction to rationale and technological tools in tutorial creation. *The Electronic Library*, *27*(4), 684–693. doi:10.1108/02640470910979624

KEY TERMS AND DEFINITIONS

Academic Library: A library in a college or university. Academic libraries provide a collection of materials in support of teaching and research, as well as instructional services for learners and reference services for the whole campus community.

Course-Integrated Library Instruction: Instruction delivered by a librarian that focuses on how to conduct research, and that is integrated into the material of a semester-long course in any subject.

Information Literacy: Refers to the set of skills that enable a person to identify a need for information, to locate and evaluate relevant sources of information, to use information ethically, and to incorporate the knowledge gained in the process within their worldview.

Information Literacy Program: A coherent program within an academic library that aims at providing the campus community with the information literacy skills to conduct research successfully.

Integrating Library Instruction: Refers to efforts and initiatives by librarians that make information literacy instruction part of the learning process on campus by merging it with courses, programs, or the overall curriculum.

Learning Management Systems (LMS): Web-based systems that allow instructors to create course content and activities online, and that allow students to carry out their learning in this virtual environment. Also known as courseware or course management systems (CMS).

Learning Modules: Sets of instructional materials and activities delivered online that have a specific learning outcome and often contain an assessment component built into the learning process.

Online Library Instruction: Refers to the online delivery of library instruction, i.e. teaching students information literacy skills to carry out their research projects. Such online instruction can be either synchronous through the use conference software or asynchronous through the use of online tutorials and interactive exercises.

Online Tutorials: Instructional content online that is often delivered in the form of a video or interactive set of web pages and aims to teach a specific skill or concept.

Chapter 15
Going Viral in the Classroom: Using Emerging CMC Technologies for Social Change

Jessica J. Eckstein
Western Connecticut State University, USA

ABSTRACT

This chapter demonstrates the potential for social change in computer-mediated communication (CMC) education. A foundational discussion of emerging online technologies in terms of potentials for students and responsibilities of educators is followed with a description of a "best practices" classroom project and incorporation of case-study observations in terms of CMC benefits and challenges. The goal of this chapter is to highlight potential societal outcomes inherent in CMC education for students and educators as agents of social change.

INTRODUCTION

Classrooms can be an arena in which to endeavor for transformation of self and society. Recognizing this possibility, many teachers strive for such an ideal. Education as a practice that maximizes freedom – for students, educators, and society at large (e.g., hooks, 1994) – is exciting, particularly in a world in which new technologies are constantly emerging. The objective of this chapter is to demonstrate the feasibility of maximizing that educational freedom (and responsibility) through knowledge acquisition and distribution, using computer-mediated communication (CMC) technology in the classroom. To accomplish this goal, a foundation is derived from research on the potentials of CMC in education. Next, experiences with online teaching and a particular project case study are presented to emphasize accountability for students and teachers trying to effect social

DOI: 10.4018/978-1-61350-077-4.ch015

change. The chapter concludes with a call for inspiration and challenge.

THE POTENTIALS OF ONLINE EDUCATION

As noted by Willson (2010), the potential for social change via CMC is not without controversy. On the one hand, initial speculations of Internet networks' massive societal impacts were perhaps overstated, at least in their time- and influence-expectations. On the other hand, limiting one's focus to social communities and their online communication illustrates that CMC technologies have affected society (macro) at interpersonal (micro) levels. In this sense, speculating as to the possibilities for using the Internet or CMC for individual-level social change is not idealistic, but, as this chapter demonstrates, is evidenced in day-to-day work with CMC and students.

With each new technology that surfaces, teachers and researchers are able to harness and assess the potentials of these tools for educational purposes. As Larreamendy-Joerns and Leinhardt (2006) discuss, CMC allows universities, philanthropic organizations, and individuals to further their respective social goals. Contrary to, and perhaps not mutually exclusive with, the perception that online education is solely a tool sought by universities seeking profit (Feenberg, 1999), this chapter takes the position that the use of CMC in education is instead a method with the radical potential to challenge and re-envision, if not to uproot, dominant systems and cultural misunderstandings or stereotypes. In essence, activists and advocates of social justice can utilize technology (and have done so[1]) to advance diverse politics and encourage participation in social movements (Castells, 2007; Denning, 2002).

Thus, the potential of educating students to use CMC effectively is significant, documented, and twofold, involving (a) students' ability to operate locally and globally for social change and (b) teachers' social responsibility to conscientiously educate their student-agents-of-change.

Potential for Students and Society: Social Outcomes

Courses that implement a variety of CMC instructional materials are believed to be more effective for student learning outcomes than those using only one form of CMC or none at all (Miller & Redman, 2010). Videos, in particular, can actively involve viewers by inviting them to personally interpret the meanings of the message content, and how it applies in their own lives (Friedman, 2006). Indeed, Internet videos are a method of course content delivery preferred by students (Dey, Burn, & Gerdes, 2009; Young, 2008). Miller and Redman (2010) found that students viewing course content via online videos performed worse than face-to-face students *only* on the exam content that was not covered by the web videos; when tested on content that was covered in Internet videos, the online-video-viewing students performed significantly better than their face-to-face-instruction peers. But the benefits of CMC instructional technology (and web-based videos, in particular) are not limited to students' mastering of course content.

Students also can participate as active, influential societal members. Ever on the cutting edge of emerging technologies, students are particularly well-suited to advance alternative practices and social perspectives, learned in the classroom, into mainstream society. Eighty-five percent of eight- to 18-year olds in U.S. households own and utilize personal computers (Roberts, Foehr, & Rideout, 2005) and form and maintain social relationships online (Lenhart, Madden, Macgill, & Smith, 2007). Ninety to 97% of college students report using the Internet at least once a day (Fortson, Scotti, Chen, Malone, & Del Ben, 2007; Cotten & Jelenewicz, 2006), and these students use online media as much or more than face-to-face communication to maintain personal relationships (Heiberger &

Harper, 2008)Students, often adept at CMC, can make use of online tools in the classroom to connect educationally; use of CMC predicts increased social relationships and connectedness in school settings (Lee, 2009).

The process of thoughtful deliberation, essential for forming opinions about social practices, can be stimulated by students' CMC education projects, with the potential for reaching large audiences. The CMC education project highlighted in this chapter utilizes a "viral" approach to spur social change. This best-practices activity is especially inspiring in light of research showing online opinion-sharing can increase people's knowledge of, efficacy toward, and active involvement with socio-political issues (Min, 2007). Many educators have provided data demonstrating the success of CMC classroom methods for increasing students' social capital (Gully, 2010), collaboration with national and community non-profit groups (Calka & Black, 2010), and involvement in political campaigns (Roberson, 2010) and government social service agencies (Dimock, Kennedy, & Dimock, 2010), to name just a few published outcomes. Thus, the potential for classroom CMC to elicit social change is not overstated.

Internet sites to which anyone with a phone can contribute content allow successful communication tools to be widely disseminated and considered by users from diverse backgrounds. Essentially, filmmakers can now reach people who were inaccessible before the development of CMC, especially Web 2.0. Ultimately, making change on an individual level, with the force of CMC media, allows students to play a direct role in how potentially millions of people view and respond to issues. As Nakamura (1995) observed more than 15 years ago, the actual members of these sharing communities and social networking sites are "stubbornly optimistic" (p. 192) in their idealistic views of equalized power and potential for positive social changes, whereas scholars and theorists tend to be more pessimistic. More recently, and in keeping up-to-date with the youth culture (our students) using the Internet, one can observe the swift rise of musicians like Lady Gaga (Casserly, 2010; GaGa, 2011 with over 410 million views on her first video) or Justin Bieber (Bieber, 2011 to date the most-viewed video on YouTube with over 615 million views) and for-profit companies like Old Spice (e.g., Old Spice, 2011 with more than 35.3 million "hits" on the first video and more than 963 million hits on only the top 10 videos they have released to date) – each case suggesting that Internet social networking, as opposed to studio or traditional corporate organizations, plays a role in recognition success (Newman, 2010; Smillie, 2009).

YouTube has reported that 51% of its users view the site at least once a week and 52% of users aged 18 to 34 years report "often" sharing videos with colleagues and friends (YouTube, 2010). As a result, although even viral video producers can succeed – for example, Gaga's and Bieber's financial and [sometimes] critical success – or fail – note the number of "dislikes" on Bieber's video, it is clear that viral messages, and perhaps even those that do not number in the hundreds of millions of views, always succeed in the sense of getting heard, viewed, and discussed.

Potential for Educators: Social Responsibility

Educators who recognize the value of shaping young minds to help society can take advantage of using CMC to accomplish their goals (Sherblom, 2010). Whereas instructors may certainly have students who retain information from face to face lectures, discussions, and practical applications, this chapter contends that true educators encourage students to transmit knowledge to contexts outside the classroom; this is the nature of educational activism (Book, 1990). To undertake a social learning model of instruction, in which students are taught the skills to *do* and *act* on their own, is not merely to teach, but instead is to serve as an expert-advisor (Darling, 1990).

CMC is well-suited to a social or relational (e.g., Boyd, MacNeill, & Sullivan, 2006) model of learning, because the method allows students to assume ownership of their education in a manner conducive to equalizing power (e.g., hooks, 1994) in the classroom (Vrasidas & McIsaac, 2000). A situation conducive to learning is one where students are free to learn and to feel safe, encouraged, and potentially equal. Building this environment should be the first priority of an educator, and CMC can help facilitate this process of freedom (Gully, 2010).

To provide a quality learning experience for students using CMC instruction, educators must not only incorporate worldly, cutting-edge, and emerging media, but they also must be clear in their expectations for student learning and the potential outcomes for social change (Sherblom, 2010). Students may prefer and benefit from lectures presented via Internet videos (Dey et al., 2009), but the onus for immediacy and engagement then becomes instructors'. Clarity is a responsibility of teachers implementing CMC education. This is because, as hooks (1997) notes, there is no opportunity for challenging dialogue if students do not first understand the concepts being discussed.

Further, educators should be experts on the subject matter they convey through CMC. In the past, misinformed or uneducated, biased information was less likely to leave the classroom with the student to whom it was directly conveyed. Using CMC when teaching, however, means that course content is potentially widespread, making the educator's responsibility for accuracy greater than ever.

Academics – whether they conduct research, teach, or do both – have an added social responsibility: careful contemplation of how their CMC-distributed information is interpreted by and affects the general population. Using CMC, scholars spread ideas, opinions, and research findings that society embraces as true because they come from legitimate, academic sources (Young, 2008). Educational institutions, and the instructors who work for them, are socially imbued with a power to supply and validate knowledge and thus, in many ways, create social reality. In addition to lucidity and accuracy, a social conscience must be cultivated by the teacher/researcher who circulates education via CMC. As Larreamendy-Joerns and Leinhardt (2006) note, "regardless of the educator's explicit intent, democratizing educational experiences is an act not without societal consequence" (p. 569). For example, something as innocuous as posting online lecture slides can result in a perpetuation of stereotypes, misconceptions, and ultimately social harm derived from misquoted, misinterpreted, or biased (without explanation of that bias) statistics, quotes, or findings. Unfortunately, Internet users often draw information from the sources most easily available (e.g., Wikipedia or Google searches), without double-checking their accuracy. Because educational institutions imbue faculty with expert-level credibility, educators are doubly culpable if their errors or prejudices, however subtle, affect others. Thus, educators employing CMC have a social responsibility: to ensure clarity, in their expectations for student-posted material and through in-class discussion/analysis; to maintain current, informed understandings of the knowledge that they impart; and to endeavor to anticipate possible repercussions of their own, and their students' CMC-distributed knowledge.

BEST PRACTICES CASE STUDY

For this project, the term *viral* connotes an online word-of-mouth event in which varied methods of CMC, including email, web-sharing, and social networking sites, may be used to distribute information that is "self-replicating, exponentially increasing diffusion...and impact of the message" (Welker, 2002, p. 4). When something is viral, it is distributed online among peers because of its stimulating content (Porter & Golan, 2006). As CMC has grown in use, viral distribution, particularly of personal videos, has been channeled by everyday users to advance their own agendas. Because of their potential to go worldwide, viral films are used as an exemplar of CMC education in the activity described in this chapter.

The Viral Project

The viral film project was originally designed for use in my Communication in Abusive Relationships and Communication of Men and Masculinities courses. Illustrative examples are derived from these courses as case studies and from the author's in-the-field online teaching experiences in general. In its initial iteration, the viral project was implemented in traditional and hybrid courses. However, similar approaches to a CMC-education project design have been used in wholly online courses involving group work (i.e., students' instructor-led education and peer-collaboration both occurred via the Internet). Therefore, the viral project is intended to be adapted to a variety of disciplinary courses in an array of classroom formats, including traditional, hybrid, and online environments.

The goal of the student project is to emphasize how emerging CMC technology provides the potential for public service announcement (PSA)-style videos to go viral. Not just a class video project, this approach to CMC classroom incorporation instead allows students to apply persuasive, highly creative, and possibly provocative, edgy, and controversial approaches to mass distribution of educationally-informed content matter. Indeed, the point of this project is to showcase emerging digital and Internet technology, such as shareware, social networking sites, and digital communication tools. The specifics of professional video design and production are not within the realm of this project. Finally, students understand that true viral occurrences are naturally-spread. This project has the potential to involve truly viral education, but the initial method of accomplishing this phenomenon is artificially created in this project. Therefore, whereas it is unreasonable to expect student-posted projects to become actually viral, this feature does not detract from the activity's educational effectiveness. The ultimate goal is for students to realize their *potential* for distributing knowledge and to teach them to do so not only persuasively, but also in *educated* and *ethically responsible* ways (see Brewer, 2010; German, 2010).

The entire project involves (a) instructor preparation of students through lecture; (b) student analysis of existing online videos pertaining to a chosen topic; (c) Internet background research using credible, scholarly sources; (d) digital film design, creation, and online posting; and (e) class evaluation and project debriefing. Benefits and challenges of CMC education, as exemplified by this project, are incorporated throughout the discussion of the activity which follows.

Step One: Preparing Students

One or two class periods should be initially devoted to the topic of public health media campaigns and/or Internet distribution systems. In these lecture periods, students are presented with and discuss strategies that succeed and fail in persuasion campaigns, according to communication research. This preliminary instruction includes interpersonal and social psychological persuasion basics, media influence, and public campaign or advertising strategy effectiveness. Teachers without a background in these topics can access a number of online (e.g., Bator, 2010; Mueller, 2011) and print (e.g., Maibach & Parrott, 1995; Rice & Atkin, 2000) sources summarizing effective, research-established persuasive campaign tools. Although topically diverse, the content of these sources can be applied to any scenario or academic discipline and will aid in teaching students the best audience-specific strategies. Even for faculty without a background in persuasive communication, these preparatory lecture periods can be edifying and are necessary for a successful conclusion to the project. As noted by Rajaram (2007), whereas there may be no formula for viral files, "*video* content is king" (in Bogatin, 2007, emphasis added). Students are already cognizant of viral videos. As a result, by merely discussing viral examples in class, instructors and students can, at the very least, begin to deconstruct the trends that make videos viral and persuasively successful.

From the beginning of the semester, it is important to consistently reiterate the project's timeline and goals. Students can choose to elicit emotion, inform, educate, raise awareness, persuade, or some combination of these goals. Whatever goal(s) they choose, students should strive to make their films effectively significant. As Golan and Zaidner (2008) noted, viral messages target people "through the gut rather than the brain" (p. 970). The richness of a message, tied to visual and audio cues, ultimately determines the effectiveness of CMC in general (Vrasidas & McIsaac, 2000) and YouTube videos in particular (Bonk, 2008; Rajaram, 2007 in Bogatin, 2007). Previous films in this project have incorporated: music ranging from hip hop and rap to classical music to emotional ballads; factoids and statistics flashed across the screen in ways varying from gripping to humorous; rehearsed scenes with actors spanning from absurd and hilarious to serious and disturbing; social support resources including helpline phone-numbers and websites and how-to tips for personally assisting others; design schemes incorporating black-and-white, sepia tone, and total color spectrum views; video footage including traditional action-film shots, still images, and PowerPoint slides; content showcasing informal participant research ranging from man-on-the-street reporting to expert interviews; imagery arrayed via cartoons, photographs, and clip art; tones or moods created to be everything from contemplative to riotous; and presentation formats as diverse as background narration, first-person camera discussions, and taped live action footage. The only tactic forbidden is harming others; otherwise, everything else is fair game for projects meant to stir the public.

A common pitfall for first-time instructors and students taking courses containing CMC is the failure to fully incorporate online content with in-class material (Kaleta, Garnham, & Aycock, 2003). This project directly addresses this snare by reinforcing to students the applicability of their in-class learning to the CMC method of knowledge perpetuation. A breakdown of steps, suggestions for evaluation, and an approximate project timeline are located in Table 1.

Steps Two and Three: Building Student Expertise

The next steps in the project allow students, working in small groups, to build expertise regarding a course-related topic of their choice. Student groups research *existing online content* (Step Two) and *established, credible research* (Step Three) on their chosen topics. CMC is ideal for this activity in that both popular content and scholarly journal databases are available online. Educators concerned about a possible digital divide across socio-economic or age groups will find this challenge largely addressed in institutions of higher education, which provide technological accessibility to all students. However, the entire process should be observed closely, because some students may be unfamiliar with particular technologies or unaccustomed to using the Internet for educational purposes (Jones, Johnson-Yale, Millermaier, & Pérez, 2009; Yildiz, 2009). During this learning process, clearly supportive communication by instructors can be followed by students' supportive communication in CMC education settings (Deutschmann & Panichi, 2009). With an instructor's guidance, a secondary accomplishment of this project may be the introduction to and training in new technologies for students previously unexposed to these resources; in itself, this is an important contribution, particularly in

Table 1. Proposed timeline and assignments to implement viral project in a semester-long[a] course

Steps and Instructor Tasks	Optional Tools to Grade or Evaluate	Suggested Semester Placement
One: Preparing Students week	*Incorporate lecture material	*Lecture around 5th
Lecture on persuastion, viral media	into course quizzes/exams	*3-4 weeks before midterm*
Two: Building Expertise week	*Group summary paper of	*Assign around 7th
Assign research: Videos currently online	existing online videos	*Can finish over midterm*
Three: Building Expertise	*Group "applied bibliography"	*Due around 9th week
Assign research: Scholarly resources		*Assign immediately after Step Two papers submitted*
Four: Making the Video	*Link to completed, posted video	*Due around 13th week, gives time to distribute & get hits
Assign video production		
Verify usability of video links		*Students may finish sooner*
Five: Assessing Viral Success	*Instructor graded evaluation of each video	*Grade before viewing day
Grade videos on content & production value		
Conduct class vote to select category winners	*Instructor- & class-choice "bests"	*Hold viewing day last week
Assess most "viral" videos	*Count of hits	*Tally on last day/end of term

Notes.[a]Based on a 16-week term.

today's society, toward equalizing opportunities for restricted groups (Yildiz, 2009).

Students begin their video projects by searching for existing online films relating to their topics. Due to their prevalence in the online world, Google and YouTube are ideal places to search for what has already been done on a particular subject area. Knowing what is already online regarding their topic will aid students in designing original videos. To maximize the breadth of their searches, it is important they start thinking of topics relatively broadly. For example, if a group picks 'sibling psychological abuse' as its topic, they should begin by finding every available video on 'sibling abuse' in general, in order to retrieve a broader set of initial results.

To demonstrate skills of content-analytic research, students are expected to assess their topically relevant online videos for: (a) *Prevalence* – how many videos on this topic exist in various outlets (e.g., YouTube) or are found via online search engines, (b) *Content* – which persuasive strategies are used in the videos related to the topic, (c) *"Viralness"* – how many hits or views these videos have, and (d) *Success* – feedback/commentary these topical videos have elicited from viewers. To conclude this stage (Step Two), each group submits a summary paper of overall online content findings (a-d, listed in this paragraph).

After the groups have conducted research on existing online content, they finalize their topics. In the course of the initial online research, students often find that a chosen topic is not truly original, or is already covered exhaustively online; in such cases, chosen topics can be adjusted. Extending the previous example, if 'sibling abuse' or 'sibling psychological abuse' are not unique or specific enough to justify new online videos, groups can consider specific types of psychological (e.g., financial, degradation, shaming, hurting pets, threats) abuse directed at siblings. In other words, specificity should be emphasized continually; it is more important to cover small, specialized areas very thoroughly and accurately than to cover a broad area shallowly. Superficiality or incompleteness can lead to misinterpretation by online audiences (i.e., educators' social responsibility of clarity, discussed previously).

It is also important for students to choose a topic new to them, and one not already covered in the classroom. Students begin working from class concepts, but should not choose something that has already been explicated. For students, the point of this activity is not to parrot in a video their in-class learning. Rather, students are exercising an ability to create informed means of social change by first educating themselves. Obviously, in the role of topical course expert, instructors must verify that students do not incorporate inaccurate or biased information in their projects. Research steps (Two and Three) encourage students to conduct, or at least initiate, learning on their own, outside of a teacher's direction.

Once the groups finalize their topics, they begin the process of finding established, credible research (Step Three). Groups explore academic literature for research studies, review articles, and concept papers related to their topics. Ideal sources of scholarly information may include peer-reviewed journals, academic texts, and conference proceedings. In Step Three, students build their topical expertise. To demonstrate involvement at this stage, each group submits an academically-formatted reference list of scholarly sources that were useful in learning more about the chosen topics. Groups should include only the sources they plan on using to inform their video, not every source discovered. In this reference list, students should include each citation, properly documented, as well as their reasons for choosing each source. With this part of the assignment, reasons for choosing articles need not include content, as would a true annotated bibliography with abstracts. Instead, students should focus on listing their reasons for choosing the references – for example, why they found each one useful to inform their topic and how each applies to their project.

Step Four: Making the Film

Completed films are based on (a) class lecture material, discussion, and reading materials; (b) information gleaned from research of existing online content; and (c) scholarly literature. The fourth, and for students, the most exciting, step is the designing, filming, editing, and online uploading of the film. Although students should be concerned about presentation quality, because it affects persuasiveness, more important is the *manner* in which the film (i.e., content) is distributed and marketed. As a *viral* education project, the focus should not be video production processes, per se.

Because students tend to underestimate the length of production processes, they should schedule small-group meetings for each stage of production. An ideal viewing time for YouTube users is approximately one to four minutes in length, particularly with educational content (Bonk, 2008). Typically, to replicate mini-documentary motion pictures, four to five minutes are the standard lengths of videos produced in this project. However, for educators who wish to emphasize viral persuasion over educational content, a better time limit is no more than one minute (two minutes maximum) each, to replicate true public service announcement or commercial lengths.

In Step Four, students (1) take responsibility for their own learning and (2) adapt to new forms of technology with which they may be unfamiliar. Thus, this step transforms two common challenges for students into the accomplishment of two pedagogical goals (Kaleta et al., 2003). Further, because videos are based on background research (both their own and others' published findings), groups need to include citations – either as the referenced content comes on screen or in final reference credits – for any material used from others. For example, in my previous students' films, some groups flashed research or statistics across the screen and immediately cited the reference at the bottom of that screen shot, whereas others showed information on its own and included a final reference shot, at the end of the video, with all sources used. Some groups have creatively used fading and/or moving "movie credits" to document sources at the end.

Most colleges and universities have an audio/visual department that allows students to borrow recording equipment and provides facilities for editing productions. Students should be alerted to such campus resources when this project is introduced. However, students often have, and prefer to use, their own digital cameras, and in many cases, video capabilities on their smart phones to film projects. Further, many laptops today are equipped with user-friendly video editing software, which many students have used to complete this project in previous semesters. Each semester, almost every group has at least one member who owns or has access to video recording and editing software, either on a cell phone or laptop. For those without personal access to these media, basic computer labs on college campuses possess minimal editing tools and lab monitors sufficient to guide students through film production. Because the emphasis of this project is not on the *video* component of education, but rather is on the social, educational networking power of the Internet, seeking assistance from others for help with technicalities is within permissible realms for this project.

Although any film sharing website (e.g., Facebook's video posting function, Vimeo, Qik, or Justin.TV's SocialCam) can be used, YouTube has a number of advantages for use in this class project. Over three billion video views and hundreds of thousands of video uploads (amounting to 8 years of content per day) occur each day on YouTube (YouTube.com, 2011). Also linked to Google search and worldwide distribution (Helft, 2009), YouTube uses Adobe Flash Video software to broadcast its videos. Because of the free download access, this tool is especially advantageous for class projects. In light of research suggesting instructors' motivation for technology use is tied to perceived ease of use (Park, Lee, & Cheong, 2008), YouTube is ideal because of its accessibility (Bertolucci, 2009). YouTube is extremely user-friendly, with step-by-step instructions for uploading, responding to, and sharing (i.e., distributing outside the site) films. This is crucial, in light of research showing students' motivation to be affected by ease of use and perceived CMC usefulness (Liu, 2010). A primary criticism of the "ivory tower" of academia – and the distribution of knowledge (e.g., academic language in limited, refereed journals) – is its lack of accessibility for the masses. Online distribution of in-class research, in a societally-targeted (persuasive) style of delivery, incorporating humor, pop music, or vivid imagery, and using an easily acquired tool, addresses this concern by making sure there are few, if any, exclusions to knowledge (Young, 2008). In this way, the power that comes with information may become equalized over time (Nakamura, 1995; Sherblom, 2010).

According to Lange (2008), online sites such as YouTube, by their very nature, perpetuate social networking, a form of CMC, and an environment comprising unofficial self-governance. Films that are unacceptable or viewed negatively may not always be removed from the site (although "official" governance now dictates automatic removal of illegal or copyrighted material), but users will negatively respond to these presentations by posting public commentary. Therefore, YouTube and other distribution sites characterized by memberships exemplify (in Willson's, 2010 breakdown) a "community" gone "network" – having taken inward communication outward, having changed the norms and rules to protocols and codes, and having become more heterogeneous and dispersed. For some, the frequency of use and the immediacy and apparent diversity (e.g., cultural or demographic variety, but convergence of beliefs) of social Internet sites is felt as a true sense of idealized community, in the theoretical possibilities initially imagined by Anderson (1991). In either designation, it is clear that a tendency exists for members to identify with/apart from others and influence/be influenced accordingly.

Ultimately, each group shares a link to its posted film with the instructor, so that both students and educator can monitor technological errors or other issues that arise in posting. As a safety consideration, it is important for students to embrace being "privately public," or being prepared to share films on a massive scale while maintaining limitedly-revealed (i.e., almost anonymous) identities (Lange, 2008). Further, students completing Step Four and posting films earlier than due, have longer time online to garner *hits* (discussed in Step Five). For films to be public before in-class viewing, a final deadline should be set of at least one week before the course concludes. Links to example films from previous courses are provided in the Appendix.

Step Five: Assessing Viral Success

In addition to providing in-depth education on particular course topics, each step of the project highlights specific skills and approaches needed when communicating online. Step One clarifies success versus failure in reaching audiences, and can be measured through testing, but also will be observable through the persuasive strategies exhibited in the final videos. Steps Two and Three develop students' skills in using CMC technologies, and can be gauged by the research reports on their findings. In Step Four (video), students produce web-based PSA-style films and publicize informed knowledge innovatively through the use of sharing websites. Observed throughout, and particularly at this penultimate step, students benefit from group collaboration and task/personality/culture management (Yildiz, 2009). The final step involves complete assessment of the success of the viral project.

A day near the end of the semester should be reserved for in-class viewing of each video. After watching all films, it is important to encourage discussions about the persuasive communication (both form and content) tactics used by each group. The class can vote for its favorites, but the instructor should evaluate the best ones according to strategies conveyed in the preparatory (Step One) lecture. As incentive, extra credit can be given for "wins" in various categories, such as Most Emotional/Moving, Most Entertaining, Most Educational, or Best Overall. As additional enticement, the "viralness" of each video is rewarded. Success is assessed by the YouTube hit-count and feedback comments on the day of final viewing. Students, knowing at the beginning of the semester that the most *viral* group gets extra credit on the overall project, push themselves to complete and post projects earlier than the due date.

Of course, all students raise the hit-count on their own film by continually re-accessing it themselves. This factor can be discussed throughout the semester, to emphasize the importance of personally distributing knowledge among one's social network. Indeed, this is one of the primary ways to enact social change. Therefore, students can post films on social networking sites (e.g., Facebook, MySpace, Twitter), email links to everyone they know, and encourage third party dissemination. By creating competition to be truly viral and by encouraging use of all CMC technologies available to them, the self-promotion playing field becomes leveled. A truly viral video is a hit for its creativity and connection with popular culture at macro/global and micro/interpersonal levels.

To succeed with the ultimate goal of social change, it is crucial that educators be personally excited about course material and this project. Students consistently report more effective learning with CMC education if instructors are concerned and passionate not only about the topic, but also about the CMC method (An & Frick, 2006). This project is ideal for any course in which educators wish (a) to have students disseminate theoretical information in an applied public-distribution manner, and/or (b) to have students clarify/debunk population-specific statistical data (replacing it with valid research findings): theories of education, critical or literature analyses, sociological or psychological research findings, historical myths or developments to demystify, applications of mathematical, economic, or scientific ideas to daily life – the possibilities are endless.

SOCIETAL IMPLICATIONS AND CONCLUSION

Cutting edge, accessible technology and competent, challenging instruction are not reciprocal. Both must be constantly re-evaluated for quality in every specific situation so that CMC success in one venue (e.g., using video sites for class research projects) is not viewed as ideal for every situation (e.g., using video sites to distribute opinion essays).

Success [breeds] homogeneity at the expense of difference and variation…If successful implementations in computer-mediated learning and online instruction are used not as artifacts to support learning but rather as templates that the subject matter must fit…these tools lose their instrumental function and become yard-sticks for what is teachable and learnable. … Against this homogenization, we under-score the value of diversity in subject matters, technological means, learning styles, and implementation scenarios (e.g., stand-alone, blended instruction, synchronous online teaching). (Larreamendy-Joerns & Leinhardt, 2006, p. 595)

As educators, we are (and should be) increasingly called upon not only to impart knowledge, but also to demonstrate and encourage enactment of discipline-specific skills. Social change is a big goal. By merely providing the tools and educated guidance to students in our classrooms, societal change can be effected – if not at the global level, then certainly interpersonally in nearby social networks. Through this project and other similar activities targeting knowledge distribution on a grand scale, I and my application-focused colleagues in the communication field have encountered definitive changes in our immediate universities (e.g., Brule, 2008; Eckstein & Pinto, 2010), local communities (e.g., Calka & Black, 2010; Eckstein, 2009; German, 2010), and at state (e.g., Dimock et al., 2010; Eckstein, 2010; Roberson, 2010) and national (e.g., Eckstein et al., 2005) levels (Frey & Carragee, 2007a, 2007b; Harter, Dutta, & Cole, 2009). Social change, through knowledge gained in the classroom and university research, may be difficult, but it is not impossible.

This chapter sought to demonstrate, through incorporation of established CMC research and a personal best-practices project, the feasibility of social change through knowledge acquisition and distribution using technological resources in the classroom. A discussion of the potentials of CMC education for students and educators was followed by a presentation of a successfully implemented project case study to emphasize student and teacher roles in learning and in effecting social change. Strengths and challenges of classroom technologies were discussed throughout this chapter to illustrate the social outcomes inherent in CMC education.

In many cases, a desire to enact social change through one's students may require challenging (or supporting!) extremist, radical, or subversive ideologies. At other times, it may be ignorance that we combat. Both alternatives involve serious consideration of our social responsibilities as educators in a technologically advancing society. Incorporating CMC methods and approaches into education, as demonstrated in this chapter, allows students to feel empowered, to make their voices heard. Once educators and students overcome a sense of apathy and helplessness regarding CMC in the classroom, they are on the road to changing the world, one viral-video-potentially-reaching-millions at a time.

REFERENCES

An, Y. J., & Frick, T. (2006). Student perceptions of asynchronous computer-mediated communication in face-to-face courses. *Journal of Computer-Mediated Communication, 11*, 485–499. doi:10.1111/j.1083-6101.2006.00023.x

Anderson, B. (1991). *Imagined communities: Reflections on the origin and spread of nationalism* (Rev. ed.). London, UK: Verso.

Bator, R. J. (200). The application of persuasion theory to the development of effective proenvironmental public service announcements. *Journal of Social Issues, 56*(3). Retrieved from http:// findarticles.com/ p/ articles/ mi_ m0341/ is_ 3_ 56/ai_ 69391502/

Bertolucci, J. (2009, February 4). YouTube leads growth in U.S. online video viewing. *PC World*. Retrieved from http://www.pcworld.com/ article/ 158949/ youtube leads growth in us online video viewing.html

Bieber, J. (2011, September 2). *Justin Bieber: Baby ft. Ludacris*. Retrieved from http://www. youtube.com/ watch?v=qrO4YZeyl0I

Bogatin, D. (2007, May 7). *Google YouTube: Viral video success formula: Digital markets article*. [Web log post]. Retrieved from http://www.zdnet. com/ blog/ micro-markets/ google-youtube- viral-video- success- formula/1315

Bonk, C. J. (2008, March). *YouTube anchors and enders: The use of shared online video content as a macrocontext for learning*. Paper presented at the meeting of American Educational Research Association, New York, NY. Retrieved from http:// trainingshare com/ SFX7EED.pdf

Book, C. L. (1990). Extended discourse. In Daily, J., Friedrich, G., & Vangelisti, A. (Eds.), *Teaching communication: Theory, research, and methods* (pp. 279–291). Hillsdale, NJ: Erlbaum.

Boyd, R., MacNeill, N., & Sullivan, G. (2006). Relational pedagogy: Putting balance back into students' learning. *Curriculum Leadership, 4*(13). Retrieved from http://www.curriculum.edu.au/ leader/ relational_ pedagogy:_ putting_ balance_ back_ into_ stu,13944.html? issueID=10277

Brewer, B. (2010, April). *The end justifying the means: Ethical implications of public service announcements*. Paper presented at the meeting of Central States Communication Association, Cincinnati, OH.

Brule, N. J. (2008, April). *Abused men & shaken babies: Teaching undergrads to apply research & promote change*. Symposium education exemplar panel at the meeting of Central States Communication Association, Madison, WI.

Calka, M., & Black, L. W. (2010, April). *Doing community based partnership research in online communities*. Paper presented at the meeting of Central States Communication Association, Cincinnati, OH.

Casserly, M. (2010, October 6). The new Forbes power women. *Forbes Magazine*. Retrieved from http://www.forbes.com

Castells, M. (2007). Communication, power and counter-power in the network society. *International Journal of Communication, 1*, 238–266.

Cotten, S. R., & Jelenewicz, S. M. (2006). A disappearing digital divide among college students? *Social Science Computer Review, 24*(4), 497–506. doi:10.1177/0894439306286852

Darling, A. L. (1990). Instructional models. In Daily, J., Friedrich, G., & Vangelisti, A. (Eds.), *Teaching communication: Theory, research, and methods* (pp. 267–278). Hillsdale, NJ: Erlbaum.

Denning, D. E. (2002). Activism, hacktivism, and cyberterrorism: The internet as a tool for influencing foreign policy. In Arquilla, J., & Ronfeldt, D. (Eds.), *Networks and netwars: The future of terror, crime, and militancy* (pp. 239–288). Santa Monica, CA: RAND.

Deutschmann, M., & Panichi, L. (2009). Talking into empty space? Signaling involvement in a virtual language classroom in Second Life. *Language Awareness*, *18*, 310–328. doi:10.1080/09658410903197306

Dey, E. L., Burn, H. E., & Gerdes, D. (2009). Bringing the classroom to the Web: Effects of using new technologies to capture and deliver lectures. *Research in Higher Education*, *50*, 377–393. doi:10.1007/s11162-009-9124-0

Dimock, P., Kennedy, E. J., & Dimock, J. P. (2010, April). From just talking to taking a stand: Developing partnerships between the speech classroom and the Minnesota Genocide Intervention Network. In States Advisory Council (Sponsor), *State of Minnesota Advisory Council Report*. Symposium conducted at the meeting of the Central States Communication Association, Cincinnati, OH.

Eckstein, A. D. (2009). *Critical agents of social change: Exploiting media to jump-start radical political ideologies*. Unpublished manuscript on classroom project, University of Missouri-Rolla.

Eckstein, J. J. (2010). *Evaluation of sexual violence primary prevention training programs*. Unpublished raw data from collaboration with Connecticut Sexual Assault Crisis Services' Rape Prevention and Education Program Curricula, Connecticut Department of Public Health.

Eckstein, J. J., Fried, G., Grosser, D., Miller, B., Robinson, S., & Thakkar, U. … Wetzler, M. (2005, February). *Bugscope: Current and future directions*. Project presented at the Imaging Technology Group Forum, Beckman Institute for Advanced Science and Technology, University of Illinois, Champaign, IL.

Eckstein, J. J., & Pinto, K. (2010, October). *One man up: Young men resisting society to re-envision selves*. Paper presented at the Women & Society Conference, Poughkeepsie, NY.

Feenberg, A. (1999). Distance learning: Promise or threat? *Crosstalk*, *7*(1), 12–14.

Fortson, B. L., Scotti, J. R., Chen, Y., Malone, J., & Del Ben, K. S. (2007). Internet use, abuse, and dependence among students at a Southeastern regional university. *Journal of American College Health*, *56*(2), 137–144. doi:10.3200/JACH.56.2.137-146

(2007a). InFrey, L. R., & Carragee, K. M. (Eds.). Communication activism: *Vol. 1. Communication for social change*. Cresskill, NJ: Hampton Press.

(2007b). InFrey, L. R., & Carragee, K. M. (Eds.). Communication activism: *Vol. 2. Media and performance activism*. Cresskill, NJ: Hampton Press.

Friedman, S. L. (2006). Watching twin bracelets in China: The role of spectatorship and identification in an ethnographic analysis of film reception. *Cultural Anthropology*, *21*(4), 603–632. doi:10.1525/can.2006.21.4.603

Gaga, L. (2011, September 2). *Lady Gaga: Bad romance* [Video file]. Retrieved from http://www.youtube.com/ watch?v=qrO4YZeyl0I

German, K. (2010, April). *Citizen journalists, viral media, and the call for a new ethics of civic engagement*. Paper presented at the meeting of Central States Communication Association, Cincinnati, OH.

Golan, G. J., & Zaidner, L. (2008). Creative strategies in viral advertising: An application of Taylor's six-segment message strategy wheel. *Journal of Computer-Mediated Communication, 13*, 959–972. doi:10.1111/j.1083-6101.2008.00426.x

Gully, D. J. (2010, April). *To tweet and blog is to deliberate and engage? Exploring online and offline forms of communication to facilitate increased social capital among young citizens.* Paper presented at the meeting of Central States Communication Association, Cincinnati, OH.

Harter, L. M., Dutta, M. J., & Cole, C. E. (Eds.). (2009). *Communication for social impact: Engaging communication theory, research and pedagogy.* Cresskill, NJ: Hampton Press.

Heiberger, G., & Harper, R. (2008). Have you Facebooked Astin lately? Using technology to increase student involvement. *New Directions for Student Services, 124*, 19–35. doi:10.1002/ss.293

Helft, M. (2009, October 9). YouTube: We're bigger than you thought. *The New York Times*. Retrieved from http://bits.blogs.nytimes.com/ 2009/ 10/ 09/ youtube- were- bigger- than- you- thought/?hp

hooks, b. (1994). *Teaching to transgress: Education as the practice of freedom.* New York, NY: Routledge.

hooks, b. (1997). Keeping close to home: Class and education. In A. A. Lunsford & J. J.

Jones, S., Johnson-Yale, C., Millermaier, S., & Pérez, F. S. (2009). U.S. college students' Internet use: Race, gender and digital divides. *Journal of Computer-Mediated Communication, 14*, 244–264. doi:10.1111/j.1083-6101.2009.01439.x

Kaleta, R., Garnham, C., & Aycock, A. (2003). *Hybrid courses: Obstacles and solutions for faculty and students.* Paper presented at the 19th Annual Conference on Distance Teaching and Learning, Madison, WI. Retrieved from http:// www.uwex.edu/ disted/ conference/ Resource_ library/ proceedings/03_ 72.pdf

Lange, P. G. (2008). Publicly private and privately public: Social networking on YouTube. *Journal of Computer-Mediated Communication, 13*, 361–380. doi:10.1111/j.1083-6101.2007.00400.x

Larreamendy-Joerns, J., & Leinhardt, G. (2006). Going the distance with online education. *Review of Educational Research, 76*, 567–605. doi:10.3102/00346543076004567

Lee, S. J. (2009). Online communication and adolescent social ties: Who benefits more from Internet use? *Journal of Computer-Mediated Communication, 14*, 509–531. doi:10.1111/j.1083-6101.2009.01451.x

Lenhart, A., Madden, M., Macgill, A. R., & Smith, A. (2007). *Teens and social media: The use of social media gains a greater foothold in teen life as they embrace the conversational nature of interactive online media.* Washington, DC: Pew Internet & American Life Project.

Liu, X. (2010). Empirical testing of a theoretical extension of the technology acceptance model: An exploratory study of educational wikis. *Communication Education, 59*, 52–69. doi:10.1080/03634520903431745

Maibach, E. W., & Parrott, R. L. (Eds.). (1995). *Designing health messages: Approaches from communication theory and public health practice.* Thousand Oaks, CA: Sage.

Miller, S., & Redman, S. (2010). Improving web-based student learning through online video demonstrations. *Bulletin of the American Astronomical Society, 42*, 555–559.

Min, S. J. (2007). Online vs. face-to-face deliberation: Effects on civic engagement. *Journal of Computer-Mediated Communication*, *12*, 1369–1387. doi:10.1111/j.1083-6101.2007.00377.x

Mueller, J. (2011). *Resources for teaching social psychology: Topic: Persuasion.* Course Resources on the Web. Retrieved from http://jfmueller.faculty.noctrl.edu/ crow/ topicpersuasion.htm

Nakamura, L. (1995). Race in/for cyberspace: Identity tourism and relational passing on the Internet. *Works and Days*, *13*, 181–193.

Newman, A. A. (2010, July 16). Old Spice argues that real men smell good. *The New York Times*, B3. Retrieved from http://www.nytimes.com/ 2010/ 07/ 16/ business/ media/ 16adco.html?_r=2

Old Spice. (2011, September 2). *Old Spice: The man your man could smell like* [Video file]. Retrieved from http://www.youtube.com/ watch?v= owGykVbfgUE

Park, N., Lee, K. M., & Cheong, P. H. (2008). University instructors' acceptance of electronic courseware: An application of the technology acceptance model. *Journal of Computer-Mediated Communication*, *13*, 163–186. doi:10.1111/j.1083-6101.2007.00391.x

Porter, L., & Golan, G. J. (2006). From subservient chickens to brawny men: A comparison of viral advertising to television advertising. *Journal of Interactive Advertising*, *6*(2). Retrieved from http://www.jiad.org/ vol6/ no2/ porter/ index.htm.

Rajaram, G. (2007, May). Viral video success advice from a marketing perspective. In R. LeFurgy (Chair), *Content, consumers, community: The balance of control*. Symposium conducted at the meeting of the Interactive Advertising Bureau Digital Video Forum, New York, NY.

Rice, R. E., & Atkin, C. K. (Eds.). (2000). *Public communication campaigns* (3rd ed.). Thousand Oaks, CA: Sage.

Roberson, K. M. (2010, April). *New media technology and political campaigns: Campaigning through texts, tweets, and Facebook*. Paper presented at the meeting of Central States Communication Association, Cincinnati, OH.

Roberts, D. F., Foehr, U. G., & Rideout, V. (2005). *Generation M: Media in the lives of 8–18 year-olds*. Washington, DC: The Henry J. Kaiser Family Foundation.

Ruszkiewicz (Eds.), *The presence of others: Readings for critical thinking and writing* (pp. 85-96). New York, NY: St. Martin's Press.

Sherblom, J. C. (2010). The computer-mediated communication (CMC) classroom: A challenge of medium, presence, interaction, identity, and relationship. *Communication Education*, *59*, 497–523. doi:10.1080/03634523.2010.486440

Smillie, D. (2009, November 25). The business of Lady GaGa. *Forbes Magazine*. Retrieved from http://www.forbes.com

Vrasidas, C., & McIsaac, M. S. (2000). Principles of pedagogy and evaluation for web-based learning. *Educational Media International*, *37*(2), 105–111. doi:10.1080/095239800410405

Welker, C. B. (2002). The paradigm of viral communication. *Information Services & Use*, *22*, 3–8.

Willson, M. (2010). Technology, networks and communities. *Information Communication and Society*, *13*, 747–764. doi:10.1080/13691180903271572

Yildiz, S. (2009). Social presence in the web-based classroom: Implications for intercultural communication. *Journal of Studies in International Education*, *13*, 46–65. doi:10.1177/1028315308317654

Young, J. R. (2008, January 25). YouTube professor scholars as online video stars. *The Chronicle of Higher Education*, *54*, A19.

YouTube.com. (2010). *Fact sheet*. Retrieved from http://www.youtube.com/ t/ fact-sheet

YouTube.com. (2011). *Press room: Statistics.* Retrieved from http://www.youtube.com/ t/ press_statistics

KEY TERMS AND DEFINITIONS

Educational Activism: Using formal or informal means of instruction to influence the world and effect social change through edification of one's students.

Hits: The number of times an Internet user views a site or clicks on a hyperlink; may be used to assess the popularity of a link/site.

Information Accuracy: The social responsibility of (a) educators to certify that the knowledge they present is clear, unbiased, and free of potential societal harm or of (b) students to ensure the projects they pass on to others are accurate and credibly researched; of essential importance in viral CMC education projects.

Social Networking: In the context of CMC, building social relationships via technological means with other people, known and unknown to the online user.

Societal Change: The revolution or transformation of people's attitudes, beliefs, and/or behaviors; can be enacted by individuals or institutions globally, locally, or personally.

Viral: An online occurrence in which distribution of a file snowballs in an accelerating fashion among CMC users because of its stimulating or unique content, ultimately reaching millions of people and garnering myriad hits. True viral dissemination spreads naturally.

ENDNOTE

1 In addition to documented, published examples of this occurrence, my first-hand experiences with courses directly connecting (via various CMC tools) to (and thus, influencing) social changes include projects at the community, state, national, and international levels (e.g., Brule, 2008; A. Eckstein, 2009; J. Eckstein, 2010; Eckstein et al., 2005; Eckstein & Pinto, 2010).

APPENDIX

Example Viral Projects from My Previous Courses

Links Current as of September 2, 2011
Email jessica.eckstein@gmail.com for additional information

http://www.youtube.com/ watch?v=SCjxMvIKw-c
http://www.youtube.com/ watch?v=siAVf9_DPHo
http://www.youtube.com/ watch?v=UTCp3DIbUww
http://www.youtube.com/ watch?v=C4kxi9DnF-Q

Section 4
Online Collaboration / Online Communities

Chapter 16
The Students' Participation in WebCT:
An Activity Theory Perspective on Online Collaboration of Knowledge Construction

Urai Salam
Tanjungpura University, Indonesia

ABSTRACT

This chapter reports on the students' interaction taking place within the virtual learning environment, WebCT. It is particularly critical of computer-mediated communication (CMC) analysis regarding the knowledge construction, evidenced by the transcripts of their online discussions. The guiding theory, Activity Theory, provides perspectives that can help instructors and teachers understand emerging learning participation mediated by computers. From this perspective, computers do not simply facilitate learning that could have been done without their presence. Rather, computer mediated communication could alter the entire flow of knowledge construction processes.

INTRODUCTION

Universities are now increasingly encouraged to utilize virtual or electronic learning environments (VLEs) to expand the process of education. In general, VLEs are defined as web-based online environments that are relatively open systems, allowing for interactions and knowledge sharing, as well as providing access to a wide range of information resources (Caladine, 2008; Holmes & Gardner, 2006; Littlejohn & Higgison, 2003; Fisher & Coleman, 2001-2002). This definition covers a broad range of a VLEs' features, ranging from static, informative individual web pages to

DOI: 10.4018/978-1-61350-077-4.ch016

interactive *bulletin boards*; from asynchronous email messages and WebCT discussion forums to synchronous chat rooms and e-conferences.

Evidences from previous studies show that Virtual Learning Environments have enriched the context for learning processes (Bonk, 2009; Salam, 2009; Dutton, Cheong, & Park, 2004; Lange, Suwardy, & Mavondo, 2003; Bunt-Kokhuis, 2001; Bonk & King, 1998). These studies have found that global networks of computers have contributed to the ease of accessing vast resources for learning, as well as to promoting interpersonal interaction. Classrooms are no longer limited by physical buildings; learners may reach virtual resources located in other universities and are able to easily communicate with people around the world.

Situated in such environments, learning is not simply defined as knowledge acquisition; rather, it includes the skills to manage information, as put forth by Rüschoff and Ritter's (2001) argument that "learning should be regarded as a process of information gathering and knowledge processing" (p.224).

This approach requires learners to be active and interactive toward their environments. The notion of *active* learning is highlighted in contrast to a traditional approach, which treats learners as passive information recipients (McLuckie & Topping, 2004; Hughes & Daykin, 2002). The notion of *interactive* learning, on the other hand, refers to the students' interaction with both physical resources and other people (Sims, 2003; Tam, 2000). The VLEs provide learners with opportunities to construct their understanding by conjoining in online communication (Sims, 2003; Yakimovicz & Murphy, 1995). Here, as noted by Bonk and King (1998), the construction of knowledge has been represented by learners' contribution and interaction through electronic collaboration.

Activity Theory emphasizes the interconnected relationship between individuals, tools, and sociocultural contexts. Learning does not exist in isolation; rather it involves participants within a cultural context in which individuals engage with each other. The VLEs are in fact electronic environments that enable users to communicate with broader communities at any time and from any place.

The current study aims to explore the Education Faculty students' participation in such environments. Particularly, this chapter will analyze online collaboration processes of knowledge construction.

THEORETICAL BACKGROUND

To explore the collaborative process of knowledge construction, we need to examine the sociocultural view of activities. This perspective views learning as being distributed across people and tools. Hutchins (1995) argues that higher order thinking cannot be understood by simply studying individual cognition per se; rather, one should investigate the whole system in which the individual operates. This view shares much with Activity Theory in considering a system of people and technology as they engage together.

Leontiev (1978) introduces the concept of *activity* as we understand it today. He conceptualizes activities as micro systems that are complex processes driven by objects and motives (Leontiev, 2005b). In relation to the social nature of human activities, Leontiev argues that an activity never stands by itself; rather, it is connected with other activities; "the action of a single given person [is] under conditions of the activity of other people, that is, it presumes a certain joint activity" (Leontiev, 2005b, p. 62). An object is seen as something that is realized through individual actions that are goal-driven. Leontiev (2005a), furthermore, proposed that activities can be described in three different ways, or three "functionally subordinated hierarchical levels" (Kaptelinin, Nardi, & Macaulay, 1999, p. 29): the *activity* level, the *action* level, and the *operation* level.

Activities are seen by Leontiev to consist of distinct actions or series of actions, which in turn consist of operations (Leontiev, 2005a). Activities are undertaken in order to fulfill *motives*. Leontiev explains that "the sign of an activity is that the object and the motive coincide" (2005b, p. 63); when the motive is taken away, the activities collapse. In other words, motives can be seen as major objectives in activities. For example, in the case of the current study, the students' participation in the courses is meant to achieve success in their studies. Such a relationship is described as *activity* and *motive*. When the 'success in their studies' is removed from the context, constituting actions will never happen.

To accomplish one activity, an individual needs to perform several actions. Leontiev defines action as "a process that is directed at a conscious goal" (Leontiev, 2005b, p. 62). It can be seen as a basic component of an activity (Leontiev, 2005a, p. 73). The goal of an action is an object that guides the action. Meanwhile, a goal can be broken down into sub-goals in order to meet a common goal (Leontiev, 2005b). For example, in order to gain success in their studies, the students are required to participate in several actions, including participation in discussion forums, giving group presentations, and writing essays.

As one moves down the levels of actions, one crosses the border between conscious and automatic processes, between *action* and *operation*. An action contains several operations. Leontiev describes an operation as follow:

If the given content of an action emerges depending on the object (goal) of the action, then it is not an operation; if, on the other hand, the given content emerges in the action depending on the conditions in which the goal is given, then it is an operation (Leontiev, 2005a, p. 74).

In other words, *operations* can be described as functional sub-units of actions that are carried out automatically (Kaptelinin, Nardi, & Macaulay, 1999). They do not have their own goals, but adjust actions to specific situations. Operations, to continue with the example above, may include tasks such as the students logging on to their email accounts, reading email messages (not a discussion forum), and downloading files. However, as Kaptelinin et al. (1999) argue, operations can become actions when goal orientation is involved. For example, the reading of email messages can be classified as actions, rather than operations, when the students have obvious goals in doing this activity; likewise it becomes an operation when it turns into being routinized.

As previously mentioned, over time operations can, in some cases, become unconscious actions, thereby blurring the distinction between the two. What can be seen as an action one day, can be classified as an operation the next. Similarly, the distinction between action and activity can also become blurred when goals and motives tend to overlap or be used interchangeably in certain cases. *Activity must, therefore, be viewed as part of a complex but contextualized system that must be studied in its entirety.* Consequently, activity theorists are not concerned with 'doing' as a disembodied action or non-contextual one, but are referring to "doing in order to transform something" with the focus on the contextualized activity of the activity system as a whole (Engeström, 2001).

For the present study, the conceptions outlined above are paramount. In their daily lives, the students are believed to have interactions with rich environments including the Virtual Learning Environments (the VLEs) that they can access at any time and from anywhere. With the rapid development of ICT, environments are not only envisaged as tangible artifacts, but also virtual ones. Technologies can be portable and wearable, so that the students can easily access resources and information as well as their online classes. Some authors, including Barab, Schatz, and Scheckler (2004), Issroff and Scanlon (2002), and Zurita and Nussbaum (2007), argue that by having access to those environments, the students are exposed to abundant resources and are

Table 1. Emerging content analysis model (Salam, 2009, pp. 175-177)

Category	Description
Sharing Appreciation:	- Greeting - Showing interest - Inquiring
Sharing Content:	- Summarizing - Describing the content - Quoting without comments
Sharing Experience:	- Giving examples - Linking the text with real life experiences
Sharing Comprehension:	- Giving elaboration - Linking across resources - Drawing conclusion - Accommodating different views

connected to broader communities. Their constant interaction with rich environments will reshape their learning endeavors.

THE STUDY

The participants in this study were the undergraduate Education Faculty students from two different units of the Early Childhood courses. The study offered three levels of participation: completing the survey (N=68), giving permission to access the online discussion transcripts (N=49), and participating in the interviews (N=10). However, only ten students agreed to participate in all levels of participation. Therefore, only those ten participants were selected to be reported on in this study; they were named S1-S10. In addition, five other students (called Peer 1-Peer 5) who did not join the interviews were sometime reported to serve as contexts in analysis.

The survey found that the majority of the participants were in the 19-22 age group. They were relatively young and thus were expected to have been familiar with the Internet. Most of them were second year students, indicating that they would have had previous experience in using the Virtual Learning Environments (VLEs) during their first year of study.

The analysis of knowledge construction was derived from 699 messages of two classes following two levels of analysis. On the first level, the analysis was meant to identify ideas contained in those messages; it was to determine how individual messages or some parts of messages functioned in the entire flow of communication. In this method, the classification is based on the function of communication within the contexts of the discussion forums; they range from *sharing appreciation, sharing content, sharing experience, to sharing comprehension* (see Table 1 below):

1. *Sharing Appreciation* category emphasizes the practical purposes of communication.
2. *Sharing Content* indicates that the participants seem to have started showing their perspectives on the subjects being discussed.
3. In *Sharing Experiences* category, the participants made every effort to show their understanding by giving examples from real life situations.
4. The last category is *Sharing Comprehension.* At this level the participants reflected on the issues being discussed and made some elaboration on them to suggest their understanding.

On the second level, the focus of the analysis was to map the ideas in the whole process of

Figure 1. Activity theory's approach to content analysis

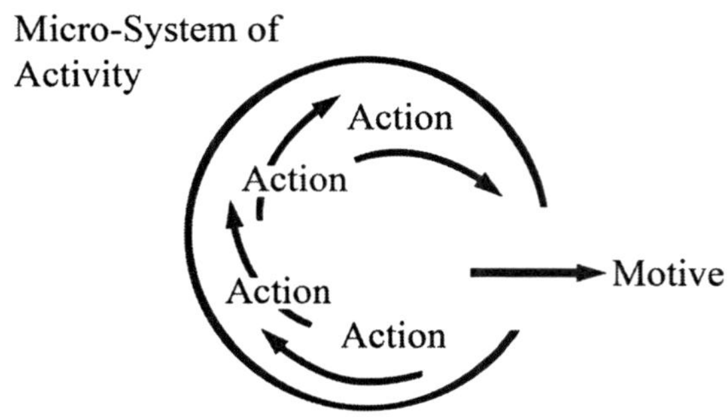

Figure 2. Activity micro-system of one episode discussion

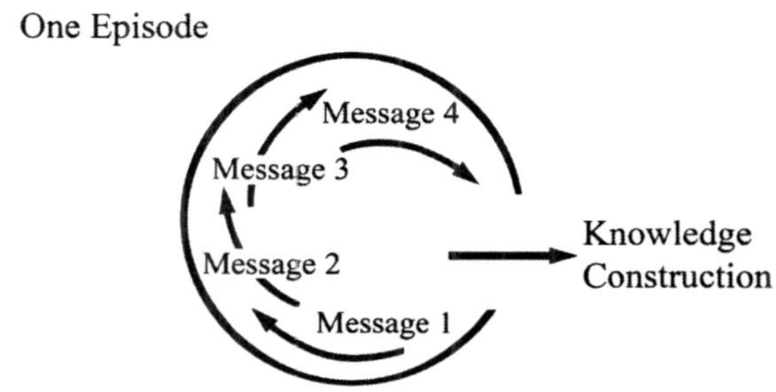

knowledge construction. An Activity Theory framework was used to perform the analysis.

From Activity Theory perspectives, the construction of knowledge starts from a micro-system of an *activity* (see Figure 1). Leontiev (2005a) describes a micro-system as being composed of motives, action, and operation. Any activity within the system is driven by a series of actions and operations to accomplish the motive. In this framework, certain actions or operations are inseparable from the others. They form *collaboration*.

In the case of this study, one episode was considered as a micro-system of a discussion forum. In that context, the students' postings were regarded as distinct actions or series of operations to engineer some sort of negotiation of meaning, understanding, and development of ideas; they form a collaborative process of knowledge construction. The analysis should not separate one posting from the others; instead, they are seen as a whole process as described in Figure 2 below. This figure shows that Message 1 is linked to Message 2 and so forth; they overlap and refine each other to accomplish one mission, knowledge construction. In this way, analyzing isolated postings may lead to an incomplete understanding of the collaborative process of learning in the context of online discussion forums.

THE RESULTS

First Level Analysis

It was interesting to note that the students were not interested in contributing short one greeting messages like "hi-how-are-you" expressions in the discussion board. When they were required to post messages, they were almost always pertaining to something about the subject. This phenomenon was observed in the discussion board related to this study. The log tracking recorded that 92% of messages were coded as *sharing* perspectives, while the remaining were *social interaction*.

The following sections will present the evidence from the data. However, the quotes are not displayed in their entirety here; rather selected quotations are used to serve as examples.

Sharing Appreciation

Almost one third of the messages fell into the *Sharing Appreciation* category. At this level, the emphasis was less on reflection and more on social interaction. Here, the students were greeting, inquiring, showing interest, and summarizing their participation in the discussion. Three messages below show these activities.

In Message 3.2, for example, S3 showed her interest in the topic being introduced. In that particular message, she had not yet indicated her perspectives on the reading. However, by saying: "*the eye opening part was to be able to see these certain traits in myself*," S3 indicated her eagerness to get involved in the discussion, and in other messages, she actually expressed her opinions about it. Meanwhile, in Message 6.1, S6 expressed his uncertainty about the ongoing assignment. He invited the other students to share with him what they thought about it. Similarly, S7, in Message 7.1, did not refer to a particular reading chapter; instead, she gave a light suggestion on how to react to cultural differences.

S3: Hi again, I found the whole concept of individualist/collectivist interesting and eye opening. I had never really thought about it before. The eye opening part was to be able to see these certain traits in myself [Message 3.2, Date: 31 July 22:26].

S6: Hey all my fellow Laurie's, Just wondering what in the heck is our part in this assignment? Just been thinking about what we r to do in regards to this assignment, maybe it will come to me sooner than later....hopefully some of us have an idea of what's doing. Anyways catch ya all later [Message 6.1, Date: 29 July 00:05].

S7: Sorry for the late reply. I found both your responses interesting. CL, why do you feel insulted when people say Australians don't have any culture? Perhaps this unit will empower you to defend your views on this statement and debate the next person that tells you this (BTW, I've had similar jokes on the cultural centre in Frankston!). [Message 7.1, Date: 27 July 17:39]

Overall, in this category, the participants rarely exhibited explicit critical arguments in the messages. They merely commented on how they felt about the subjects or the reading materials. However, the analysis did not end in this category, but continued, in order to uncover the complete picture of the students' engagement in learning processes. Despite the lower levels of cognitive engagement, such postings are nonetheless considered to be sharing perspectives because, at least through those postings, the students demonstrated their interest and involvement in the online Discussion Forum. In many cases, such postings invited additional responses that triggered a deeper discussion, as shown at the next levels.

Sharing Content

Unlike *Sharing Appreciation*, in *Sharing Content* the students attempted to summarize the reading content and repeat others postings' ideas. They did not reflect on the reading or the projects they were working on. Even though the postings in the category demonstrate an effort to comment on the readings, they were often superficial comments that did not provide evidence of the students' understanding of the subjects. In the messages below, the participants described the content of the chapter that was introduced in that particular week. Even though the students included some direct quotes from the readings, they did not show their understanding of those materials by elaborating on the quotations.

In Message 1.3, for example, S1 indicated that he had made some observations about the materials from the reading lists, and found a quote relevant to his situation. In that opportunity, he brought relevant materials that might benefit other members of the groups. The quote itself might look simple; but when others did the same, all members of the group were able to develop their understanding from the collected materials. Similarly, S2 pointed out a few key points from the chapter.

S1: There is some great information that can be found on the reading lists for this unit. I have been looking at the relationships between the parents and Phoebe, and finding that her behaviour could be in fact a result of her relationship as the parents are the main secure attachment for Phoebe, and her needs may not be fulfilled.

"Children whose attachment needs were not fully met might be inhibited, aggressive, or unable to share or take turns" [Message 1.3, Date: 5 August 14:55].

S2: The reading I found interesting was chapter one of "Gonzalas Mena".

The chapter also discussed how different cultures bring up their infants. I think these things are important to have an understanding of if working in a childcare setting. Ensuring we do not stereotype families because they may have different styles of upbringing to ourselves [Message 2.2, Date: 22 July 10:30].

Still in this category, the participants showed their agreement with their fellow students. Instead of elaborating on the issues posted by other students, they repeated each other's ideas and paraphrased the language. For example, in Message 7.4, the student expressed her agreement with other students; she repeated other postings by paraphrasing the main points without further elaboration. It was the same case with S9, who basically confirmed his agreement with S7. These students did not really show their own understanding of the materials, rather they merely responded to the postings and made messages identical with the original ones.

S7: I agree that looking at the child as separate from their culture is rather pointless, I prefer the idea of culture influencing the child … and as a result the child influencing their culture. In response to your final point, have faith in yourself! I'm sure after some experience it will become very obvious if a child is just very unique, or developmentally delayed! [Message 7.4, Date: 8 August 08:51].

S9: I found it quite interesting to read about Whiting & Whiting Psycho-cultural model. However, although I agree with the main idea presented in the model - 'that maintenance systems determine to a large extent the learning environment in which a child grows up, and the learning environment influences the child's behavior and development' (p44) [Message 9.4, Date 7 August 07:23].

At this level of participation, the students made efforts to share something about their readings. Even though they simply posted summaries of the readings and agreement with others or posted some relevant quotes, this actually indicated that they had read the materials or showed some evidence that they had started working on their projects. By communicating their efforts, they at least had contributed to the progression of constructing understanding.

Sharing Experience

At this level, the participants made every effort to show their understanding by giving examples from real life situations. Here, the participants tried to link what they understood to what they had experienced. They showed some degree of understanding, as one would not be able to give appropriate examples without sufficient understanding. For instance, all of the messages below include discussions about individualist versus collectivist cultures. Each student presented his or her understanding about the issues and linked them with their personal experiences. In one episode, S10 claimed that he improved his understanding about individualist-collectivist cultures after reading the assigned chapter. He also mentioned one episode in his experiences that related to the contradiction between the individualist and collectivist. Similarly, S3 and S4 each showed some level of understanding and shared their particular perspectives about the collectivist and individualist by presenting their own contexts.

S3: .. for me I believe I was very much a collectivist when I was younger, now I would say I have a balance of both. As teachers I feel it is important for us to have been both collectivists and individualists. This will help make all children and their families to feel comfortable with the program [at school]. Also, it helps the children feel comfortable with the wider group setting. I agree with the individualist approach as I feel it is important to have the accompanying skills but I also believe it is very much important to have collectivist skills as well. In everyday life both sets of skills come into play on a daily basis. The example of the student teacher and the teacher with the crayons shows how both styles are just as valid but need to suit the children. So the challenge for us as teachers will be to identify which style will be appropriate and most effective within our program... [Message 3.2, Date: 31 July 22:26].

S10:... As I was reading the chapter, I was thinking of which one I would put myself in as a teacher in training.... I think childcare workers and preschool teachers would be more towards the individualists. ... On my last kindergarten placement, I was encouraging a child to put on his shoe. But the teacher said to me "he won't put it on because his mum does everything for him and she will always, so don't bother." I was a bit shocked at the time. But after reading this chapter, I have a greater understanding of that situation. The mum's view was different to mine. In relation to that situation, it's called individualist vs. collectivist [Message 10.2, Date: 5 July 11:48].

S4:... I come from what can be generally termed as collectivist culture. As Gonzalez-Mena says, no culture is entirely collectivist or individualist. From reading the Fleer & Williams-Kennedy text, I found myself laughing at some of the examples they gave because it looked so much like something I would see in my own culture. For example, this line: *'Sometimes we don't know the kid's name, but we all know the family – that's so and so; you don't need the name but you need the connection. But as the teacher you need the name for the roll!'* [italic is in original].

A lot of the time, especially in gatherings like Chinese New Year banquets, I am always 'so-and-so's daughter' and it just seems like everyone else is 'X's cousin on Y's father's side'.... Perhaps with the Aboriginal people, whom I see as members of a collective culture, their families and friend networks are so large, that to keep track of them you need connections instead of names [Message 4.6, Date: 31 July 01:12].

At this level, the participants shared their perspectives by giving authentic experiences that were derived from real life situations. In this way, not only did the students evidence their understanding, they also enriched the explanation about the subject matter.

Sharing Comprehension

At the *Sharing Comprehension* level, the participants showed their understanding not only by presenting their summary of the resources, but also by linking across resources to draw some sort of conclusion, and analyzing arguments. More than 30% of the postings fell into this category.

At this highest level, the participants reflected on the discussion by accommodating different opinions while presenting their own views. In these discussions, they read the messages thoroughly and then responded to them. In many cases, they summarized the ideas presented by other participants and then presented their own elaboration of the issues. In that way, they shared their views and ultimately improved their understanding. S10 and his classmates discussed the topic of knowing children's culture. This topic was introduced by Peer 3 in relation to the first week reading list. Peer 3 made a few points regarding her understanding of the chapter, and Peer 1 added the elaboration. Message 10.1 was a response. Here, S10 attempted to accommodate the ideas from fellow students, in addition to linking them to his own view from the chapter.

Similarly, Message 4.20 below was also a response to Peer 3, who introduced a new topic in week 11. She pointed out some issues from Rogoff's Chapter 8, and tied them into their practicum experiences. S4 corroborated Peer 3's points in her message below. S4 quoted Peer 3's message (in italic) to expand on her explanation. In her message, S4 presented more examples from her own experience when she had the practicum, and also conveyed that she had found a meaningful case about the issue being discussed. In that way, the elaboration was not only based on the text from the chapter but also from the peers' perspectives.

S10: Hey Peer 3 to your point that "teachers learn about each culture from parents in their program" I totally agree and I have had an experience where a teacher was [not] aware of the language a child spoke at home, which really surprised me! How could a teacher not know what language one of the children in their class speak at home!!! I believe a strong part of a teacher's job is to work with parents to help the children they are teaching. Not just finding out the children's interests and so on but also finding out about the parents, their beliefs and values on schooling, there experience about school and so on!! Also with this point I think it is important as a teacher to provide the parents with information about our beliefs and values and why we believe that. This could be done in a form of an introductory letter at the start of the year, interviews with parents etc…Hey Peer 1 I totally agree with your point about how your own values affect the way you view and accept another culture. It is very easy to say that we will always be open minded and of course we will always try and stay open minded but as you said our values and beliefs that have been developed from our culture will always affect what we believe. From the readings I have learnt how difficult/complicated culture is and as a teacher it is just an important aspect that we have to be aware of and respect. However as Peer 1 mentioned how many values should you set aside, I believe that we shouldn't set any aside, however is that possible?? [Message 10.1, Date: 24 July 16:03].

S4: Hey Peer 3! Definitely agree with you about the practicum. For me the most obvious improvement I saw in everyone was during the science unit last year. … Also, I found your point about social referencing interesting, and I have also been able to relate it to Rogoff's discussion in Chapter 9 about how that cultural strategy does not necessarily apply to all children as individuals have different preferences in different circumstances. [You said] "*I learnt that I needed to use a low tone of voice (firm voice) and facial expression to send the message to the child that I really mean it. … I was amazed how effective those verbal/nonverbal expressions were to send children messages, and I could build my confidence in dealing with behavior management situations*". I personally found that this worked for most children. But there was always one or two students who took that as something they needed to challenge. So the more firm/strict I was with him, the more he rebelled. But once I took a deep breath and sort of smiled and spoke nicely, and explained everything in a different way, he was more than happy to cooperate. I find it VERY challenging to sort of stop and rewind the situation, especially as I'm getting increasingly pissed off. But yes, it's interesting to see how children's home experiences and cultural practices, negative or positive, affect how they interpret the verbal/non-verbal messages you are sending out [Message 4.20, Date: 8 October 02:48].

In other messages at this level, the participants showed their coverage of the issues across resources. Here the students did not simply present a summary of what they had read in a particular chapter of the week, but they linked the reading they had done with other resources, to show their observation across resources that they had had access to. For instance, in Message 7.8, S7 indicated that she had read a number of chapters from Fleer, Siraj-Blatchford, Rogoff, and Gonzales-Mena. She managed to draw a general theme from these resources and share them with other peers. Likewise, S9 introduced a new topic for the week five reading. Chiefly, he shared his reading of the chapters. He highlighted the main ideas from the texts and posted them in Message 9.4 below.

S7: One link I found between all the texts was the idea of culture being non stagnant – in Fleer (p58) "All cultural groups change over time - cultures are always changing" (and the example of how asking questions was once forbidden and is now encouraged further supports this idea), Siraj-Blatchford (P113) "Cultures are dynamic and changing and are rightly defined by their members, "each culture builds on a moving historical base". We discussed this in depth last wk so I won't go further into that but it was nice to see a common theme in such a diverse range of texts! [Message 7.8, Date: 11 August 15:57].

S9: A common link in all the readings was the importance of recognizing cultural diversity in Early Childhood Settings. … The huge diversity of experiences was highlighted, making it clear that different cultures do not just involve language differences but varying interests and activities, family and community relationships and attachments. The readings, specifically Gonzalez-Mena and Siraj-Baltchford & Clarke, displayed how cultural policies in Early Childhood Settings, are highly necessary so this diversity is acknowledged and taken into account when planning programs. Fleer & Williams-Kennedy demonstrate the importance of being aware of these cultural differences by using the example of questioning in Indigenous cultures. Although asking questions may be highly valued in Western cultures, Fleer & Williams-Kennedy describe how in Indigenous cultures, questioning can be seen as impolite [Message 9.4, Date: 15 August 00:48].

What is more, in the discussion board the participants, indeed, advanced their arguments and demonstrated the use of critical thinking. In the messages below, the participants managed to examine the opinions put forth by the other participants and suggested different perspectives on the issues. For example, in Message 7.3, S7 corrected the notion of "failure" as introduced by Peer 5 in the previous message. S7 expanded the meaning of "failure" to refer to encouragement, not a negative meaning. Likewise, S4 produced analytical arguments about the testimony in the literature that claims that individuals place aside their values or culture when making judgments on other cultures. S4 argued that there should be a compromise between two parties to accommodate the differences in culture. For her, individuals may always have choices of which culture to promote and which culture is appropriate for educating the children.

S7: Peer 4, I really liked your analogy to the ANZACs - its true they held a collectivist ideal, and I think much of the honor they received stems from that. I also agree that in recent time, there's been a real move in school towards developing interpersonal education, whilst still encouraging intra-personal skills. Peer 5, I really related to your views on "failure" as an opportunity to encourage the child. I don't feel failure has to be a negative experience-it is a platform to work from and improve on. You've probably read the debates in the media about changing the term 'fail' on school reports - I think this is a bit over the top - no matter what you call it, children are not stupid (even if they have failed!) and we'll still interpret the result as a fail. I agree with Peer 5 that we should use the failure as an opportunity to encourage children, and also to look at our own teaching practices and how they can be improved to suit every child, instead of taking the easy road and not using the term "fail"! [Message 7.3, Date: 1 August 09:54, bold was added].

S4: I agree with you. I don't think its professional or possible to ignore your own values, I don't think that will work. It's really a matter of compromise and give and take, in my opinion. Imagine if you were working in childcare, and a mother comes in and tells you, my child is used to having everything done for him by his nanny or maid, so I expect you to do the same. Of course you would think, "no way!" In that case (I assume!) you would then try to explain to the mother, that this is not how its done here, and the value system of not just yourself, but the childcare, is for the children to learn independence. You wouldn't mind helping the child out and design a program for his needs and help him along with his developmental skills, that's your job, but ultimately you're not about to drop all your values and say, 'Yes, sure, I'll do it because I respect your culture and how in your family's culture children are waited on hand and foot.' [Message 4.1, Date: 21 July 20:53].

From this analysis, the online discussion forums provided the students a venue to extend peer collaboration in knowledge construction. In some way, this supported it in a better way than face to face communication. As commented by one participant, the system (WebCT) preserved their discussions, allowing them to return more easily to their ideas and to study them from a variety of perspectives. He said, "*using this discussion board, we can know all other postings... sometimes other friends have brilliant ideas that I never thought of before.*"

The next level of analysis will look into the process of meaning negotiation, as the students proceed with collaboration in knowledge construction.

Second Level Analysis

While the first level of analysis classifies the postings into four groups, these classifications were seen as the dynamic and the progression of knowledge building, not simply the methods to measure the students' cognitive engagement as in Henri's (1992), Ng and Murphy's (2005), Zhu's (2006), and Fahy, Crawford, and Ally (2001) studies. From an Activity Theory perspective, one action is not isolated from other actions. Therefore, in this study, each category contributed to certain phases of the understanding, from an Activity Theory perspective. For example, within the context of meaning making '(a) sharing appreciation' was not considered to be less important or lower in cognitive content than '(b) sharing content', nor was '(b) sharing content' considered lower than '(d) sharing comprehension,' and so forth. In the whole process, each level of contribution was significant in creating context for collaboration. In many cases, '(d) sharing comprehension' would not emerge without being initiated by simple questions (sharing appreciation). In fact,

Figure 3. One episode of online discussion

Topics	Contributors	Time
Week 2	Peer 1	Fri 29 July 23:19
Re: Week 2	Teacher	Sat 30 July 10:42
Re: Week 2	S4	Mon 1 Aug 00:30
Re: Week 2	Peer 2	Tue 2 Aug 17:15
Re: Week 2	Peer 1	Tue 2 Aug 19:52
Re: Week 2	S4	Tue 2 Aug 23:16
Re: Week 2	Peer 1	Wed 3 Aug 22:55
Re: Week 2	S4	Thu 4 Aug 21:00
Keep up the good work!	Teacher	Sat 30 July 10:36
Week 2 Response	S10	Sat 30 July 11:48

the data showed that higher order thinking usually resulted from interactive discussion initiated by simple statements.

To elaborate the online collaboration, one episode of the discussion, as shown in Figure 3 below, is referenced. The interaction between the participants that occurred in that episode was a particularly good example of collaborative construction of knowledge through social negotiation. The students constructed their understanding through collaborative efforts brought about by the differences in their perspectives on certain issues. They shared their concerns, their readings, their perspectives, and their experiences with each other. On the whole, the interaction in that episode showed how their understanding developed from a series of postings and the contribution of all the individuals.

In that episode, first, Peer 1 introduced the topic of the Week-2 reading by expressing her frustration with not understanding the chapter that she was reading. She brought up several key issues that she thought were difficult to understand. On the one hand, this comment could mean that she invited others to share their understanding of their reading on those issues. On the other hand, she actually presented her own perspectives and opinions about the issues. In the other posting, she expressed her disagreement with S4 about the issue of "stereotyping." For instance, she gave examples from what she had experienced in her daily life. Finally, she indicated that she had gained some understanding after the discussion, and she corrected herself about the issue she presented in the previous posting. She said, *"I do agree with what you said, and looking back on my posting, I should replace the word, 'stereotype,' with, 'having a general view.'"*

Next, Peer 2 made a point by herself without referring to any parts of the chapter. However, she clearly presented her point of view about the topic by taking herself as an example, i.e. how she perceived the way to understand someone's culture.

Unlike Peer 2, S4 responded to the discussion by tying in the current topic to the reading chapter that the students had been previously assigned. She summarized the content of the chapter in the form of a discussion, in addition to writing her personal opinion on the issue. After others had commented on her previous posting, in the second posting, she tried to re-elaborate her ideas by giving more examples on key issues, in order to correct the wrong impression made by other members about her previous posting. In other words, the students negotiated the meaning. She also gave examples with real life observations about how the topic (culture) should be understood. In the end, she showed her appreciation to the other students who had helped her sharpen her understanding and had corrected her overgeneralization.

Like Peer 1, S10 also expressed his difficulty in understanding the chapter content. He brought up some of the issues that he thought he would need help with. Nevertheless, he managed to link what he read with what had happened in his real life situations. Particularly, he tried to use the information from the chapter in order to understand what he had experienced in his teaching placement. In this way, he elevated the issues so that they were relevant to everybody in the group.

In this episode, the teacher built upon the discussion. She posted two messages. First, she advised that the students should not worry about the chapter's difficulty. She tried to maintain the optimism of the students by saying that it was too early to judge that the topic "Culture" was frustrating. The reading list that was prepared in the course was arranged in such a way that could help them understand the issue more easily. Second, at the end of that episode, she congratulated the students in managing to keep the discussion productive.

In the case of this study, the collaboration that took place in the Discussion Forums provided venues for the students' cognitive engagement. The discussions facilitated the cognitive processes through social interaction where the distributed participants shared their resources, views, and responsibilities. The focus was not knowledge as an entity but as a process of communicating and coordinating actions. The episode shown in Figure 3 provided an illuminating example of how the knowledge was distributed across the participants. For example, one participant introduced the topic by expressing her confusion about the topic; the other participants shared their views and perspectives on the subjects; meanwhile others linked it to their real life experiences. In this way, the participants cultivated knowledge and experiences amongst themselves; they managed to construct shared understanding out of diverse, conflicting, and sometimes confusing subject matters.

During the collaboration, the thinking of each individual was inevitably influenced by the thinking of the other members taking part in the discussion. Even though some of the participants only expressed their frustration or difficulty in understanding a certain chapter such as the example of Peer 1 and S10 in the above episode, they triggered further discussion. *In terms of the Activity Theory, they created tension that elevated the condition to move forward and recreated the context. The tension that existed within the group provided both the energy and the operations that drove the collaboration and intensified the discussion.* The tensions in that episode were shown through different perspectives, different experi-

Figure 4. Knowledge construction cycle in the discussion forum

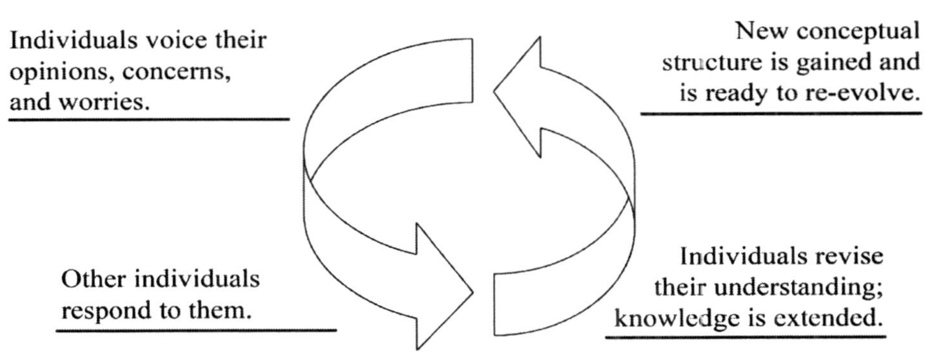

ences, and disagreement as well as worries and concerns expressed by the students. The tension in the episode infused energy to *re*-cycle the process of knowledge construction.

The cycle of knowledge construction is presented in Figure 4. This model represents the findings emerging from the data that engineers a knowledge cycle. In this study, the knowledge construction cycle was represented in four phases. It began when the participants voiced that their groups applied this new understanding or to some extent reshaped their understanding in Phase 3. In this way, the knowledge is extended. Also, when the idea is accepted by the group, it becomes the core conceptual structure. According to Engeström (1999; 1987) when this happens, the cycle is complete and is ready to re-evolve as shown in Figure 4.

The data indicated that all phases might contain any types of sharing from sharing appreciation, contents, experiences, to sharing comprehension. When Peer 1 elicited the issue of "knowing a child's culture," she received a variety of responses from the other discussion participants. One of the participants linked the topic with the concept of "stereotype," while the others tried to elaborate on it from the previous course readings, other literature, and their daily life experiences. The data suggested that through the Discussion Forums, the students gained some sort of understanding and corrected their previous assumptions about "knowing a child's culture." Overall, the knowledge was expanded through the whole of the contributions.

CONCLUSION

The study has shown that the students co-elaborated on the topics by posting their ideas, perspectives, and experiences. As the system (WebCT environment) stored their discussion, it provided the students with a broader range of ideas and explanation of the topics. During their discussion, the students did not simply passively receive information, but they shared what they had in their minds (prior knowledge) and coordinated it with others' ideas. In this way, knowledge was created and shared through interpersonal interactions. In other words, the act of knowledge construction was mediated by the collaborative discussion within technology-rich environments. The students contributed to building knowledge by sharing their reading, interpretation, views, and experiences that were relevant to the topics being discussed. In this way, they had more conscious control over memory-related processes, which could be described as higher mental function.

REFERENCES

Barab, S., Schatz, S., & Scheckler, R. (2004). Using activity theory to conceptualize online community and using online community to conceptualize activity theory. *Mind, Culture, and Activity*, *11*(1), 25–47. doi:10.1207/s15327884mca1101_3

Bonk, C. J. (2009). *The world is open: How web technology is revolutionizing education.* San Francisco, CA: Jossey-Bass.

Bonk, C. J., & King, K. S. (Eds.). (1998). *Electronic collaborators: Learner-centered technologies for literacy, apprenticeship, and discourse.* New Jersey: Lawrence Erlbaum.

Bunt-Kokhuis, S. d. (2001). On-line learning at universities in developing countries: From leap-frogging to antelope-jumping—Specific needs and solutions. *Higher Education in Europe*, *26*(1), 241–246. doi:10.1080/03797720120082633

Caladine, R. (2008). *Enhancing e-learning with media-rich content and interactions*. Hershey, PA: Information Science Publishing. doi:10.4018/978-1-59904-732-4

Dutton, W. H., Cheong, P. H., & Park, A. (2004). An ecology of constraints on e-learning in higher education: The case of a virtual learning environment. *Prometheus*, *22*(2), 131–149. doi:10.1080/0810902042000218337

Engeström, Y. (1987). *Learning by expanding: An activity-theoretical approach to developmental research*. Helsinki, Finland: Orienta-Konsultit Oy.

Engeström, Y. (1999). Activity theory and individual and social transformation. In Engeström, Y., Miettinen, R., & Punamaki, R. L. (Eds.), *Perspectives on activity theory* (pp. 19–38). Cambridge, UK: Cambridge University Press.

Engeström, Y. (2001). Expansive learning at work: Toward an activity theoretical reconceptualization. *Journal of Education and Work*, *14*(1), 133–156.

Fahy, P., Crawford, G., & Ally, M. (2001). Patterns of interaction in a computer conference transcript. *International Review of Research in Open and Distance Learning*, *2*(1).

Fisher, M., & Coleman, B. (2001-2002). Collaborative online learning in virtual discussions. *Journal of Educational Technology Systems*, *30*(1), 3–17. doi:10.2190/CRDK-VNR8-43DX-V4DB

Henri, F. (1992). Computer conferencing and content analysis. In Kaye, A. R. (Ed.), *Collaborative learning through computer conferencing* (pp. 117–136). Heidelberg, Germany: Springer.

Holmes, B., & Gardner, J. (2006). *e-Learning: Concepts and practice*. London, UK: SAGE Publications.

Hughes, M., & Daykin, N. (2002). Towards constructivism: Investigating students' perceptions and learning as a result of using an online environment. *Innovations in Education and Teaching International*, *39*(3), 217–224. doi:10.1080/13558000210150036

Hutchins, E. (1995). *Cognition in the wild*. Cambridge, MA: MIT Press.

Issroff, K., & Scanlon, E. (2002). Using technology in higher education: An activity theory perspective. *Journal of Computer Assisted Learning*, *18*(1), 77–83. doi:10.1046/j.0266-4909.2001.00213.x

Kaptelinin, V., Nardi, B. A., & Macaulay, C. (1999). Methods and tools: The activity checklist: A tool for representing the "space" of context. *Interaction*, *6*(4), 27–39. doi:10.1145/306412.306431

Kuutti, K. (1996). Activity theory as a potential framework for human-computer interaction. In Nardi, B. A. (Ed.), *Context and consciousness: Activity theory and human-computer interaction* (pp. 17–44). Cambridge, MA: MIT Press.

Lange, P. d., Suwardy, T., & Mavondo, F. (2003). Integrating a virtual learning environment into an introductory accounting course: Determinants of student motivation. *Accounting Education*, *12*(1), 1–14. doi:10.1080/0963928032000064567

Leontiev, A. N. (1978). *Activity, consciousness, and personality*. Englewood Cliffs, NJ: Prentice-Hall.

Leontiev, A. N. (2005a). The fundamental processes of mental life. *Journal of Russian & East European Psychology*, *43*(4), 72–75.

Leontiev, A. N. (2005b). The genesis of activity. *Journal of Russian & East European Psychology*, *43*(4), 58–71.

Littlejohn, A., & Higgison, C. (2003). *A guide for teachers: e-learning series No.3, learning and teaching support network (LTSN)*. York, England: Generic Centre.

McLuckie, J., & Topping, K. J. (2004). Transferable skills for online peer learning. *Assessment & Evaluation in Higher Education*, *29*(5), 563–584. doi:10.1080/02602930410001689144

Ng, K. C., & Murphy, D. (2005). Evaluating interactivity and learning in computer conferencing using content analysis techniques. *Distance Education*, *26*(1), 89–109. doi:10.1080/01587910500081327

Rüschoff, B., & Ritter, M. (2001). Technology-enhanced language learning: Construction of knowledge and template-based learning in the foreign language classroom. *Computer Assisted Language Learning, 14*(3-4), 219–232. doi:10.1076/call.14.3.219.5789

Salam, U. (2009). *Students' participation in virtual learning environments. Unpublished Ph.D thesis*. Melbourne, Australia: Monash University.

Sims, R. (2003). Promises of interactivity: Aligning learner perceptions and expectations with strategies for flexible and online learning. *Distance Education, 24*(1). doi:10.1080/01587910303050

Tam, M. (2000). Constructivism, instructional design and technology: Implication for transforming distant learning. *Journal of Educational Technology & Society, 3*(2).

Yakimovicz, A. D., & Murphy, K. L. (1995). Constructivism and collaboration on the Internet: Case study of a graduate class experience. *Computers & Education, 24*(3), 203–209. doi:10.1016/0360-1315(95)00015-E

Zhu, E. (2006). Interaction and cognitive engagement: An analysis of four asynchronous online discussions. *Instructional Science, 34*, 451–480. doi:10.1007/s11251-006-0004-0

Zurita, G., & Nussbaum, M. (2007). A conceptual framework based on activity theory for mobile CSCL. *British Journal of Educational Technology, 38*(2), 211–235. doi:10.1111/j.1467-8535.2006.00580.x

KEY TERMS AND DEFINITINONS

Episode: A series of postings in the Discussion Forums that addresses a particular theme.

Knowledge Construction Cycle: The progression of meaning negotiation that takes place in the Discussion Forum.

Micro-System: The unit analysis of the knowledge construction cycle.

Sharing Appreciation: The posting category that emphasizes practical purposes of communication like greeting and enquiring.

Sharing Comprehension: The students' elaboration on reading materials that they have read posted in their discussion forums.

Sharing Content: The students' summaries and/or quotations from their reading lists, posted in their discussion forums.

Sharing Experience: The students' postings containing their efforts to suggest their understanding by linking the reading with real life experiences.

VLEs: Web-based system environments, not exclusively course management system, that integrate communication space and resources databases.

WebCT (Web Course Tools): A course management system providing faculty and students an online presence for their discussion forums.

Chapter 17
Building Virtual Communities: Can We Talk?

René Tanner
Arizona State University, USA

Tricia Amato
Independent Scholar, USA

ABSTRACT

Communication deepens learning and builds community. Online classes are built around text-based discussions, and while studies show that students are learning, students do not feel connected in this type of environment. In order to bridge this gap, instructors must use technology and new approaches to build a sense of community and connection. This chapter examines the benefits of and challenges to online discussion and provides solutions to improve interactivity.

INTRODUCTION

Studies show that when students interact with each other (Zakaria, Chin, & Daud, 2010) and with the instructor, learning improves (Gokhale, 1995; Zhao, Lui, Lai, & Tan, 2005). In particular, in-class discussions create opportunities for students to examine material and form new connections with it, making discussions an integral part of the education process (Bruffee, 1984; Dixson, 2010; Larson & Keiper, 2002; Blumenfeld, Kempler, & Krajcik, 2006). Nagel, Blignaut and Cronjé (2009), found a strong positive relationship between discussion participation and grades. When interaction occurs regularly, a learning community is created that connects students to content through collaborative communication. In traditional, on-campus classrooms, the formation of learning communities occurs naturally, as students chat with each other and the instructor, engage in topical discussions, and form study groups. This is not necessarily the case in online classes, however.

Both students and faculty voice concerns about online communication between the two groups, wondering if students can be socialized to the college experience in an online classroom,

DOI: 10.4018/978-1-61350-077-4.ch017

and students expressing similar concerns about the lack of face-to-face interaction with faculty (Ouzts, 2006). Student burnout and feelings of isolation can both be attributed to a low sense of classroom community and may also be a factor in retention in online classes (Bambara, Harbour, Davies, & Athey, 2009; Rovai & Jordan, 2004), where dropout rates are significantly higher than for traditional classes (Carpenter, Brown, & Hickman 2004; Carr, 2000; Lynch, 2001). What is more, some students perceive that they learn more in traditional classes (Rovai & Barnum, 2003); while others perceive that the medium has little effect on their learning (Tesone & Ricca, 2008). Numerous studies indicate that student performance is similar in fully online and face-to-face classes (Bernard et al., 2004; Cavus & Ibrahim, 2007; Sitzmann, Kraiger, Stewart, & Wisher, 2006) and courses that blend online features into traditional courses create the most successful situation for students (Boyle, Bradley, Chalk, Jones, & Pickard, 2003, Rovai & Jordan, 2004; Sitzmann, Kraiger, Stewart, & Wisher, 2006; U.S. Department of Education [USDE], Office of Planning, Evaluation, and Policy Development, 2009; Verduin & Clark, 1991).

A recent meta-analysis of postsecondary online education, sponsored by the United States Department of Education (USDE) in 2009, suggested that online courses may actually be superior to traditional face-to-face courses. However, a careful scrutiny of the USDE report found that while the meta-analysis included 51 studies, not all of them were adequate for the comparison. In fact, only 28 of the studies compared face-to-face classes with those taught solely online. The remaining 23 studies were from blended courses, which are structured similarly to face-to-face courses, with the addition of an online element. In addition, over half of the studies used to compare the face-to-face instruction to online instruction were brief educational lessons, some as short as 15 minutes, which are not directly comparable to semester-long courses (Jaggars and Bailey, 2010). These findings have put the recent USDE study, which has been often cited as evidence that online education can be superior to traditional education, in a new light (Wojciechowska, 2010).

Another factor to consider when examining online courses are concerns that low-income and lower-achieving students may struggle with online classes due to a lack of access to a reliable high-speed Internet connection (Rainie, Estabrook, & Witt, 2007) as well as a lack of needed social and psychological skills, including the ability to ask for help and to be self-motivated (Liu et al., 2007), both of which contribute to success in online classes. Additionally, Figlio, Yin and Rush (2010) found that Hispanic, male and low-achieving students were more likely to do well in a class presented via live lectures than one in which the same lectures were presented over the Internet.

Online courses can provide many different opportunities for participation, including synchronous and asynchronous instructor-lead discussions, informal chat sessions, discussion forums, group work, and visits to supplemental websites, including social media sites. Often the instructor sets criteria for what constitutes successful participation (i.e.: a minimum word count for posts, content expectations, number of posts per thread, etc.), and participation does increase when students receive credit for doing so.

Interestingly, students who sense classroom community in face-to-face courses (McKinney, McKinney, Franiuk & Schweitzer, 2006) and online classes perceive that they have learned more (Rovai, 2002), and are more satisfied with the instructor (Richardson & Swan, 2003). While the formation of community through discussion is a key part of the online experience, community does not form naturally or without intervention. Furthermore, if classroom community is not developed, it can influence student and faculty perceptions of learning success and create feelings of isolation among students (Orvis & Lassiter, 2006; Rovai & Wighting, 2005).

Most online courses include a discussion forum, which attempts to mimic the type of discussion that happens in traditional classrooms. Unfortunately, participation in these discussions is not always productive. In many cases, students post only what is required of them or graded by the instructor (Vonderwell, 2003) and do not build on previous statements made by classmates, which can result in a string of unrelated comments, rather than a conversation (Thomas, 2002). Even more detrimental to the online community dynamic is when there are a large number of "read-only" participants or "lurkers," who visit the class site and read posts, but who do not respond with their own thoughts and ideas (Nagel, Blignaut & Cronje, 2009). While these participants are learning by observing (Beaudoin, 2002), their lack of participation can make other participants uncomfortable and disrupt the formation of virtual communities (Nagel, Blignaut & Cronje, 2009). Factors such as these, which inhibit the formation of productive learning communities, are of increasing concern now that online education has entered the mainstream.

Online courses present the most viable option for many students, such as those who want to return to school to complete a degree but still need to continue working, have families to care for, or cannot relocate for education. This type of education works best for self-directed learners who typically value dialog over structure, but it does not work as well for dependent learners who tend to value the opposite (Rovai, 2000). Older adults (or nontraditional students) who make up the majority of online learners possess specific learning characteristics: they are highly motivated, focused, and have a strong concept of what they expect from their education (Cole & Watts, 2009; Knowles, 1984). In order for online education to be truly successful and improve low retention rates, instructors need to engage students with the material and with each other (Smith, 1994). Therefore, it is important to understand the dynamics of a brick-and-mortar classroom community and work to build this sense of community into distance courses, especially as technology and changes in pedagogy (Bonk & Cunningham, 1998, p. 27) move us away from teacher-centered instruction and toward student-centered instruction.

In this chapter, the role of group discussion in online courses is explored and its unique opportunities and challenges are evaluated. The chapter examines how student involvement in online courses can mirror that in traditional courses, resulting in the type of connections that students value in their education. The end of this chapter provides techniques and suggestions for building a more connected online classroom. The chapter concludes that a sense of community is critical to the perceived value of online education and educators and course developers must continue to innovate in order to engage students.

THE PURPOSE OF DISCUSSION IN THE CLASSROOM

While the spontaneity of an animated in-person dialog is missing, it appears that online discussions have certain advantages over face-to-face dialog. One advantage may be that in the online environment, students are able to reflect on the discussion and craft their messages before posting, allowing them to collect their thoughts and formulate meaningful responses, which deepens learning. Another reason may be that online discussions create a class "study hall" environment where students learn from one another and spend more time on the material. In fact, when strictly online education is compared to blended education (where the traditional classroom is supplemented with an online component), the blended environment produces the best conditions for academic success (Sitzmann, Kraiger, Stewart, & Wisher, 2006). Blended learning can also deepen learning by providing opportunities to deliver content online and allowing more class time for discussion (Salter, Pang, & Sharma, 2009).

A case study of learning processes in online and face-to-face courses by Heckman and Annabi (2005) demonstrated that asynchronous discussions generate high levels of cognitive activity, which are equal or superior to face-to-face discussions. Students also adopted aspects of the instructor's role in courses with asynchronous discussions (Heckman & Annabi, 2005). Research shows that students in lecture courses perform better when their courses are supplemented with an online discussion (Meyer 2003). However, it is difficult for an instructor to facilitate large online discussions and provide individual feedback, which students enjoy. Ertmer et al. (2010) experimented with peer feedback to student-discussion posts via an automated rating system. Interestingly, students who received feedback from peers felt more confident about posting to the discussion. Bongey, Cizadlo, & Kalnbach (2005) used a course management system to supplement a large lecture class with an online grade book and student "self-tests," all of which improved students' test scores.

Reflection Deepens Learning

When a concept or idea can be carefully considered without the pressure of a quick response, creative thinking can occur and learning deepens (Caine & Caine, 1994). Likewise, it is also believed that students engage in higher-order information processing when they reflect on their peers' comments, and as a result they form responses that are the product of the interaction. Text-based, asynchronous discussions provide more opportunity for students to reflect on what is being discussed (Rovai & Jordan, 2004). Responding to discussions also encourages students to take a position. This may result in the topic going beyond the standard or "safe" realm of ideas. The reflective nature of online discussions can also bring students (and instructors) into conflict with their own preconceptions as the discussion unfolds and varying points of view are expressed (Thomas, 2002). Additionally, students often reveal their thought processes in their posts. In a blended course, a student who rarely speaks in class may write thoughtful posts and add insights to the discussion. The online environment gives such a student an opportunity to contribute, where the traditional environment does not. Reflection is a powerful component of learning, helping students to craft a point of view. This is clearly a strength of the online/reflective discussion forum.

Online Forums Minimize the Fear and Stress

Stress is detrimental to learning. Often in the traditional classroom there is both pressure to participate and a perceived judgment on the sharing. Students are generally concerned about the way they are perceived by fellow classmates in brick and mortar classrooms, and are less likely to expose themselves by asking questions or expressing ideas (Vonderwell, 2003). Fear in the classroom is a very real phenomenon that can significantly limit learning (Caine & Caine, 1994). For some students, this discomfort and fear of exposure can affect their ability to use critical thinking skills, thus reducing the chance that they will connect new information to existing knowledge, thereby inhibiting learning (Caine & Caine, 1994).

In a qualitative case study that examined the asynchronous communications of online undergraduate students, Vonderwell (2003) found that students felt freer to ask questions of their professor as a result of the inherent anonymity of the online classroom, and that in the online classroom they were less likely to be concerned about how other students perceived them. The relative anonymity of the online environment can be an asset to students who experience stress in a traditional classroom (Simonson, Smaldino, Albright & Zvacek, 2000).

Unlike face-to face dialog, text-based communication (as in chats or discussion forums) focuses attention on what is said, rather than who the speaker is. Nonverbal cues are eliminated and more impartial (unbiased) communication results

(Thomas, 2002). Women, for example, can feel left out of traditional discussions where male students often dominate the discourse (Herring, 1999). Likewise, cultural differences can impede a student's participation in classroom discussion. However, this tendency can be lessened in the online classroom, where text-based conversation is more prevalent and there are fewer cues, physical and auditory, to use for stereotyping (Cheung & Hew, 2004; Vonderwell, 2003).

In addition, the online environment may be more appealing to introverted or shy students. It has been observed in classrooms with both an online and in-class component that students can have very different behavior in each environment. For example, students who are quiet in the classroom and reluctant to ask questions or talk about themselves are more willing to do these things in the online environment. One student in René Tanner's class posted fun and interesting photographs in the "sandbox" or play area, which generated quite a bit of "hallway" discussion. Another seemingly shy student in face-to-face interactions posted an intriguing video to the class forum. This video got a lot of attention from the class as the class commented on it. Participation in online class discussions is generally higher than in traditional classrooms by as much as 70% (King, 1994). Possibly it is the sharing by all participants that makes an online class more advantageous to all learners.

ONLINE DISCUSSION CHALLENGES

Several problems have been identified with online discussions, including ineffective communication (Brooks & Jeong, 2006), inadequate evaluation of peers' ideas (Rourke & Anderson, 2002), and the tendency of students to read and respond to only the latest post even when doing so does not advance an existing, more meaningful discussion (Hewitt, 2005). Moreover, as previously stated, studies show that students are most likely to participate in discussion boards if participation is required or graded (Vonderwell, 2003). This results in significant numbers of responses but, as students and instructors can confirm, the result can be a series of disjointed posts that may not move the conversation into deeper, more nuanced territory (Vonderwell, 2003). Additionally, since students often only comment on each other's posts when required to do so (Thomas, 2002; Coole & Watts, 2009), they are not fully engaged in reflection and discussion, which results in more superficial learning.

Student make up, experience, and expectations can inhibit the effectiveness of discussions as well. Some students may view their relative lack of expertise in the subject being discussed as a roadblock to participation and may refrain from posting (Johnson & Pitcock, 2008). And since many online learners are considered "non-traditional" in that they are older, already have a significant amount of life experience, and often have considerable demands on their time, they may not see the immediate value of online discourse (Johnson & Pitcock, 2008). Additionally, students who are unfamiliar with the class structure may not readily participate until they feel more comfortable with the interface and process (Thomas, 2002). Therefore, it is important to articulate what is expected of students in online discussions and to explain that each student plays a significant role in the learning outcomes for all students (Johnson & Pitcock, 2008).

Students are generally more satisfied with an online class when they sense the presence of the instructor (Russo & Benson, 2005). This includes providing feedback on assignments, timely responses to student questions, and meaningful participation in discussions (Ouzts, 2006). Lack of instructor involvement can lead to overall dissatisfaction with the course and low participation levels (Ouzts, 2006). Students in one study expressed frustration over having to wait for feedback and answers to questions in the online environment. They also observed that the instructor was present and responsive at the beginning of the semester, but as the term progressed, response times increased (Vonderwell, 2003).

As important as an engaged instructor presence is, instructors must be careful not to dominate the discussion, which may suppress students' active participation (Rovai, 2007). The instructor has three main roles in the online environment: regulator, monitor and guide (Olivares, 2008). As regulator, the instructor provides opportunities for interaction and sets guidelines for the nature of that interaction. This can be in the form of setting up discussions, assignments and the like, as well as encouraging social interaction among students. As monitor, the instructor poses questions, reads posts and responds to them in a way that moves the conversation forward (Olivares, 2008). Responses to students' posts are most effective when they pose a new question or reinforce something already stated. Correcting students in discussions should be done with care; reinforcing thoughtful and accurate posts is a more productive strategy. The final role identified by Olivares is that of guide. As a guide, the instructor provides feedback in the form of graded assignments, letting students know what is expected of them and what they can expect from the instructor, and informing them (in private) when they do not meet those expectations (Olivares, 2008). Ultimately, instructors should be aware of their behavior and stay actively involved in the dialog without becoming the focus.

TIPS FOR A BETTER ONLINE EXPERIENCE

Online programs provide considerable benefits to students. The flexibility they offer enables many so-called "nontraditional" students to pursue degrees that would otherwise be out of reach. There will be a growing need and pressure to provide online programs as more jobs will require advanced degrees or certificates and schools will continue to experience budgetary issues and diminishing classroom space. It is evident from the literature that students appreciate the flexibility of online courses, but they also find the courses lacking the connectedness of the traditional classroom. This leads to the realization that the online classroom needs some nuanced changes in order to improve learning and protect what is best about the online experience—the freedom for students to attend class on their schedule and be the primary driver of their educations.

For some instructors, building community may seem natural, while for others, and especially those making the transition into online teaching, it may be a challenge. Online class facilitation can require new pedagogical approaches to teaching (Baran & Correia, 2009) and new technical skills (Berge, 2007, p. 73). Therefore, instructors should be open to trying new approaches and technologies and understand that they will need to do more, especially in the beginning, to set the stage for a more collaborative environment. In this section various ways in which instructors and course developers can create an atmosphere that encourages connection and participation among students, are explored. Many of these ideas were gathered from instructors who have used them successfully in their classes and were shared during a roundtable discussion led by the authors at a conference for online educators in 2010.

Getting off on the Right Foot

Because the face-to-face element is missing in the online environment, it generally takes longer for students to feel connected to each other and the course (Johnson & Pitcock, 2008, p. 96). It is therefore necessary for instructors to expend more effort in the beginning of a course introducing students to the class and to each other. Some participants in the roundtable have done the following to get dialog going and encourage participation: 1) lead an "introduce-yourself" discussion where students greet one another and share information about themselves such as hobbies and career goals, and 2) have students send photos of themselves to the instructor who then makes a movie or slide show of all the students in the class. One instructor reported that in his classes there was more interaction between students who had introduced themselves to each other.

Students want to get to know the instructor as well, and this can be done by posting a photo and short biography to the classroom web page. Short videos can also be used to illustrate key points and to introduce new topics, so students can see the instructor in action. One teacher at the roundtable related the importance of expressing a personality online. She accomplished this by telling stories in her posts and posting cartoons to the class.

Clearly Outline Expectations

In a 2010 panel of online students enrolled in the Montana University system, students noted a disparity between the traditional and online class requirements. Students who had never taken an online class were surprised to learn how much "self-teaching" and personal initiative were required to succeed. Many students noted that the workload for online classes was generally greater than for the same course taught on campus, a fact they learned by comparing syllabi with their peers who were taking the same class face-to-face. These students suggested that the instructors of online classes inform their students of the greater expectations and responsibilities at the very beginning of the course.

Introduce Students to the Classroom Technology

Since the very nature of an online classroom is different from a traditional classroom, teachers can help prepare students to make the most of online classes by walking them through the virtual classroom and the technology necessary for successful class participation (Johnson & Pitcock, 2008, p. 93–94). At the roundtable discussion, some participants shared that they create quizzes for their students based on the structure and content of the classroom. This requires students to familiarize themselves with the syllabus and various areas (such as drop boxes, document sharing areas, and chat rooms) in order to successfully complete the quiz. This practice not only provides an opportunity for the student to get acquainted with how the class works but enables the instructor to recognize early on any problems students may have with the technology.

Set Rules for Participation

In online programs, students can be located anywhere in the world and possibly come from a variety of cultural backgrounds with differing social norms for participation. It is important, therefore, for instructors to state the course rules for discussion at the beginning of class. Expectations such as being civil, polite, avoiding stereotypes, and not using foul language may seem obvious, especially to experienced students, but need to be stated nonetheless. By clearly stating the rules of engagement, students are put at ease and can freely participate. Rules can go further than the basics, and can include such topics such as responding to classmates by name, setting time frames when discussions are open for posting, requiring that posts be of a certain length, and setting a minimum number of posts for each discussion. If discussions are graded, a rubric will help students understand how to write an appropriate post.

Be Available to Students

A fact of traditional and online classes alike is that students rarely come to office hours. However, in the online environment, it may be even more important for students to feel that they can easily connect with the instructor. Tricia Amato tells her students that they can chat with her any time she is available online, which is pretty much all the time. She finds that students take advantage of this and ask questions about projects or just say hello. The majority of these chat requests come late at night when the students are online and doing their work. It may also be a good idea, as one of the roundtable instructors suggested, for instructors to tell students that they are interested in their thoughts and concerns about the course. This lets students know they are being heard.

Engage in Active Participation

A key role of the instructor in the online classroom is to act as a facilitator of student interactions (Orvis & Lassiter, 2006). This means that instructors should be actively engaged in the student discussions by posing questions, drawing connections, and building on discussion threads. An instructor sees the "big picture" and can encourage the collaborative environment by drawing together students' posts and building on them. The instructor can also redirect discussions that move off topic or become redundant. Instructor feedback should be positive and challenge the students to think beyond the obvious, while drawing upon their own experiences and education. It is important, however, for the instructor to remain an active participant without dominating the conversation. Instructors should avoid negative feedback or providing the answer to students in their posts.

Peer-to-Peer Interaction

As mentioned above, students feel connected when they receive feedback from the instructor or their peers. Therefore, it is important for the instructor to be present in the course but all the feedback cannot, and should not, come from only the instructor. To this end, instructors often require peer review of projects and papers, enabling students to engage with each other on another level and to practice giving feedback, an important life skill. Reviewing classmates' papers and projects also provides students with a sense of how they are performing relative to the other students, something that is difficult for students to assess in online courses.

When it comes to discussion posts, instructors may wish to consider requiring students to respond to a minimum number of classmates and specify that they cannot respond to the same people all the time. When students respond to a classmate's post, they should be instructed to build on that post to create a conversation, and not simply agreeing with each other. By requiring thoughtful responses to posts, the topic being discussed will take on a greater significance and deeper connections will be made.

Design the Course to Facilitate Connections

One general criticism the authors have heard from students is that the various course sections and activities (readings, discussions, and assignments) are not coordinated in any meaningful way. The students have related that they feel like they are doing "busy work," especially in the discussion forums, where discussion topics often do not appear to relate to what they are learning (Montana Student Panel). If students feel that they are only contributing to indicate they are participating in the course, they may be less likely to post meaningful responses. Instructors should keep in mind that students report that online courses are generally more work than traditional courses, and the discussions and assignments should be designed so that they build on the readings and on previous class topics. That way, important concepts are reinforced and the students see the interrelationships of various aspects of the course. In the online format, where spontaneous dialog is not an option, this can be more difficult than it seems.

Build on Posts with Synchronous Seminars or Chats

Even the best-designed courses, where the students engage in meaningful dialog in the discussion forums, can benefit from real-time interaction. This may not always be possible, depending on the course structure, but if technology allows, an instructor should try to find a time when everyone can get together and "chat." This can be in the form of live (synchronous) seminars, or even a chat session built around a specific topic or problem that everyone can work on together. Whether an instructor chooses to schedule such a session weekly, or just once or twice per term, it can go a long way toward creating a sense of common purpose among the students.

Bring in Experts to Enhance the Course Experience

Instructors can capitalize on the talent present at their respective institutions and enhance the experience for the students by inviting experts on specific topics to record short video "lectures," and post the lecture online to the course. This enables the students to get an additional perspective on the topic, and allows the course to take on a more expansive character. Experts in virtually any subject can easily be found on the web as well. For example, TED videos (www.ted.com) offer innovative and thought-provoking perspectives covering a wide range of topics. These videos can be embedded into courses. Guest experts can enhance a course in a variety of other ways. For example, the instructor can: ask a guest expert to moderate a discussion, bring in a librarian to help students prepare for a research project, or invite an author to read from his or her book.

Make the Best Use of Available Technology

Technology changes daily, and in the past several years a plethora of new applications have been developed, while some older ones have gained popularity. These applications, when used appropriately by instructors, can increase the opportunities for student interaction and participation in online courses. Examples include, but certainly are not limited to: Skype, Oovoo, Ning, Audacity, Camtasia, TeacherTube, Adobe Connect and Jing, not to mention blogging, podcasting, Twitter, and RSS feeds. Instructors who stay current in their knowledge and use of new developments, and who learn new technologies, have more tools at their disposal to engage their students. It is important to note, however, that technology should not be used simply because it is available. It needs to make sense in the overall scheme of the class and reinforce the existing course platform. Too many applications can confuse even the most tech-savvy students.

Protect Students' Privacy

Encouraging students to share opinions and ideas is a great way to build community. However, it is important for students not to share personal information, such as their physical addresses, with the other students. Also, if they have the opportunity to meet with their classmates, they should do so in a public space. Common sense rules and safety precautions, such as those that apply to online dating, are also advised in the online classroom.

CONCLUSION

The National Center for Education Statistics projects that between 2006 and 2017 there will be a shift in admission patterns with older students (25 years and older) increasing their enrollment in higher education by 19%, while that of younger students (under age 25) will increase by only 10% (USDE, 2009). Between 2000 and 2009, the number of persons age 25 to 29 obtaining a master's, professional or doctoral degree increased 59% from 994,000 to 1,579,000 (US Census Bureau, 2009). In addition, more jobs are expected to require advanced degrees—about 2.5 million jobs are predicted to require a master's or doctoral degree between 2008 and 2018 (Wendler et al.,

2010). These trends will undoubtedly increase the demand for distance education from nontraditional students who have numerous obligations and may already be on a career path. Retention will continue to be a challenge for online courses, partly due to the time-strapped nature of the students who enroll, but also due to the perception of a lower benefit of online education, a perception held by both students and faculty. However, there is some evidence that lecture capture systems that facilitate the creation of audio-visual recordings of presentations that can be posted to a course management system can improve the online experience and improve retention (Brittan-Powell, 2008). Using tools that create a strong sense of community will continue to improve retention in online classes.

Students are more connected to social networking spaces than ever before; however, it is difficult to replicate this natural tendency in the online classroom. Educators should explore ways in which to reinvent their discussion spaces to be more like social networks, with rules for participation so that everyone can feel empowered and safer. Safer spaces require rules for etiquette and conversation (no tolerance for flaming, etc.). Taking advantage of the features that work well in social networking environments could include adding options that are similar to those found in Twitter or Facebook. For example, enabling students to add class comments to their Twitter feeds, and see other students' posts in real time, and giving them the ability to interact in visually interesting virtual spaces, structured like Facebook. Ultimately, instructors and course designers must think outside the "brick-and-mortar box" and use technology to create new learning spaces for online classrooms.

REFERENCES

Bambara, C., Harbour, C., Davies, T., & Athey, S. (2009). The lived experience of community college students enrolled in high-risk online courses. *Community College Review*, *36*(3), 219–238. doi:10.1177/0091552108327187

Baran, E., & Correia, A. (2009). Student-led facilitation strategies in online discussions. *Distance Education*, *30*(3), 339–361. doi:10.1080/01587910903236510

Beaudoin, M. F. (2002). Learning or lurking? Tracking the "invisible" online student. *The Internet and Higher Education*, *5*(2), 147–155.. doi:10.1016/S1096-7516(02)00086-6

Berge, Z. L. (2007). Motivate and manage: Key activities of online instructors. In Spector, J. M. (Ed.), *Finding your online voice: Stories told by experienced online educators* (pp. 73–82). Mahwah, NJ: Lawrence Erlbaum Associates.

Bernard, R. M., Abrami, P. C., Lou, Y., Borokhovski, E., Wade, A., & Wozney, L. (2004). How does distance education compare with classroom instruction? A meta-analysis of the empirical literature. *Review of Educational Research*, *74*(3), 379–439. doi:10.3102/00346543074003379

Blumenfeld, P. C., Kempler, T. M., & Krajcik, J. S. (2006). Motivation and cognitive engagement in learning environments. In R. K. Sawyer (Ed.), *The Cambridge handbook of the learning sciences* (pp. 475–488). New York, NY: Cambridge University Press. Retrieved from http://131.193.130.213/media/ Blumenfeld.et.al_ 2006_ Cambridge_ handbook_ LS.pdf

Bongey, S. B., Cizadlo, G., & Kalnback, L. (2005). Using a course management system (CMS) to meet the challenges of large lecture classes. *Campus-Wide Information Systems*, *22*(5), 252–262.. doi:10.1108/10650740510632172

Bonk, C. J., & Cunningham, D. J. (1998). Searching for learner-centered, constructivist, and sociocultural components of collaborative educational learning tools. In Bonk, C. J., & King, K. S. (Eds.), *Electronic collaborators: Learner-centered technologies for literacy, apprenticeship, and discourse* (pp. 25–50). Mahwah, NJ: Lawrence Erlbaum Associates.

Boyle, T., Bradley, C., Chalk, P., Jones, R., & Pickard, P. (2003). Using blended learning to improve student success rates in learning to program. *Learning, Media and Technology*, *28*(2), 165–178.

Brittan-Powell, C. (2008). Preparing and retaining students in online courses. *International Journal of Instructional Technology and Distance Learning*, *5*(9). Retrieved from http://www.itdl.org/ Journal/ Sep_08/ article03.htm.

Brooks, C. D., & Jeong, A. (2006). Effects of pre-structuring discussion threads on group interaction and group performance in computer-supported collaborative argumentation. *Distance Education*, *27*(3), 371–390. doi:10.1080/01587910600940448

Bruffee, K. A. (1984). Collaborative learning and the "conversation of mankind.". *College English*, *46*(7), 635–652. doi:10.2307/376924

Caine, R. N., & Caine, G. (1994). Threat and the brain. In *Making connections: Teaching and the human brain* (pp. 69–86). Menlo Park, CA: Addison-Wesley Pub. Co.

Carpenter, T. G., Brown, W. L., & Hickman, R. C. (2004). Influences of online delivery on developmental writing outcomes. *Journal of Developmental Education*, *28*(1), 14–16, 18, 35.

Carr, S. (2000). As distance education comes of age, the challenge is keeping the students. *The Chronicle of Higher Education*, *46*(23), A39–A41.

Cavus, N., & Ibrahim, D. (2007). Assessing the success rate of students using a learning management system together with a collaborative tool in web-based teaching of programming languages. *Journal of Educational Computing Research*, *36*(3), 301–321. doi:10.2190/T728-G676-4N18-6871

Cheung, W. S., & Hew, K. F. (2004). Evaluating the extent of ill-structured problem solving process among pre-service teachers in an asynchronous online discussion and reflection log learning environment. *Journal of Educational Computing Research*, *30*(3), 197–227. doi:10.2190/9JTN-10T3-WTXH-P6HN

Coole, H., & Watts, M. (2009). Communal e-learning styles in the online classroom. *Research in Education*, *82*(1), 13–27.

Dixson, M. D. (2010). Creating effective student engagement in online courses: What do students find engaging? *Journal of the Scholarship of Teaching and Learning*, *10*(2), 1–13.

Ertmer, P., Richardson, J., Lehman, J., Newby, T., Cheng, X., Mong, C., & Sadaf, A. (2010). Peer feedback in a large undergraduate blended course: Perceptions of value and learning. *Journal of Educational Computing Research*, *43*(1), 67–88.. doi:10.2190/EC.43.1.e

Figlio, D. N., Rush, M., & Yin, L. (2010). *Is it live or is it internet? Experimental estimates of the effects of online instruction on student learning.* NBER Working Paper No. 16089. Cambridge, MA: National Bureau of Economic Research.

Gokhale, A. A. (1995). Collaborative learning enhances critical thinking. *Journal of Technology Education*, *7*, 22–30.

Heckman, R., & Annabi, H. (2005). A content analytic comparison of learning processes in online and face-to-face case study discussions. *Journal of Computer-Mediated Communication*, *10*(2). Retrieved from http://jcmc.indiana.edu/ vol10/ issue2/ heckman.html.

Herring, S. C. (1999). The rhetorical dynamics of gender harassment on-line. *The Information Society*, *15*(3), 151–167..doi:10.1080/019722499128466

Hewitt, J. (2005). Toward an understanding of how threads die in asynchronous computer conferences. *Journal of the Learning Sciences*, *14*(4), 567. doi:10.1207/s15327809jls1404_4

Jaggars, S. S., & Bailey, T. (2010). *Effectiveness of fully online courses for college students: Response to a department of education meta-analysis*. New York, NY: Community College Research Center, Teachers College, Columbia University. Retrieved from http://ccrc.tc.columbia.edu/ Publication.asp? UID=796

Johnson, E., & Pitcock, J. (2008). Preparing online instructors: Beyond using the technology. In Orvis, K., & Lassiter, A. (Eds.), *Computer-supported collaborative learning: Best practices and principles for instructors* (pp. 89–113). Hershey, PA: Information Science Publishing. doi:10.4018/978-1-59904-753-9.ch005

King, K. M. (1994). Leading classroom discussions: Using computers for a new approach. *Teaching Sociology*, *22*(2), 174–182. doi:10.2307/1318563

Knowles, M. S. (1984). *The adult learner: A neglected species*. Houston, TX: Gulf Pub. Co., Book Division.

Larson, B. E., & Keiper, T. A. (2002). Classroom discussion and threaded electronic discussion: Learning in two arenas. *Contemporary Issues in Technology & Teacher Education*, *2*(1). Retrieved from http://www.citejournal.org/ vol2/ iss1/ socialstudies/ article1.cfm.

Liu, S., Gomez, J., Khan, B., & Yen, C. (2007). Toward a learner-oriented community college online course dropout framework. *International Journal on E-Learning*, *6*(4), 519–542.

Lynch, M. M. (2001). Effective student preparation for online learning. *The Technology Source Archives*. Retrieved from http://technologysource.org/ article/ effective_ student_ preparation_ for_ online_ learning/

McKinney, J. P., McKinney, K., Franiuk, R., & Schweitzer, J. (2006). The college classroom as a community: Impact on student attitudes and learning. *College Teaching*, *54*(3), 281–284.. doi:10.3200/CTCH.54.3.281-284

Meyer, K. A. (2003). Face-to-face versus threaded discussions: The role of time and higher-order thinking. *Journal of Asynchronous Learning Networks*, *7*(3), 55–65.

Nagel, L., Blignaut, A. S., & Cronjé, J. C. (2009). Read-only participants: A case for student communication in online classes. *Interactive Learning Environments*, *17*(1), 37–51. doi:10.1080/10494820701501028

Olivares, O. J. (2008). Collaborative vs. cooperative learning: The instructor's role in computer supported collaborative learning. In Orvis, K. L., & Lassiter, A. L. R. (Eds.), *Computer-supported collaborative learning: Best practices and principles for instructors* (pp. 20–39). Hershey, PA: Information Science Publishing. doi:10.4018/978-1-59904-753-9.ch002

Orvis, K. L., & Lassiter, A. L. (2006). Computer-supported collaborative learning: The role of the instructor. In Ferris, S. P., & Godar, S. (Eds.), *Teaching and learning with virtual teams* (pp. 158–179). Hershey, PA: Information Science Publishing. doi:10.4018/978-1-59140-708-9.ch007

Ouzts, K. (2006). Sense of community in online courses. *The Quarterly Review of Distance Education*, *7*(3), 285–296.

Rainie, L., Estabrook, L., & Witt, E. (2007). *Information searches that solve problems*. Washington, DC: Pew Internet & American Life Project.

Richardson, J., & Swan, K. (2003). Examining social presence in online courses in relation to students' perceived learning and satisfaction. *Journal of Asynchronous Learning Networks*, *7*(1), 68–88.

Rourke, L., & Anderson, T. (2002). Using peer teams to lead online discussions. *Journal of Interactive Media in Education*, *1*(1), 1–21. Retrieved from http://www-jime.open.ac.uk/ 2002/ 1/ rourke- anderson- 02-1.pdf.

Rovai, A. P. (2000). Building and sustaining community in asynchronous learning networks. *The Internet and Higher Education*, *3*(4), 285–297. doi:10.1016/S1096-7516(01)00037-9

Rovai, A. P. (2002). Sense of community, perceived cognitive learning, and persistence in asynchronous learning networks. *The Internet and Higher Education*, *5*(4), 319–332..doi:10.1016/S1096-7516(02)00130-6

Rovai, A. P. (2007). Facilitating online discussions effectively. *The Internet and Higher Education*, *10*(1), 77–88. doi:10.1016/j.iheduc.2006.10.001

Rovai, A. P., & Barnum, K. T. (2003). On-line course effectiveness: An analysis of student interactions and perceptions of learning. *Journal of Distance Education*, *18*(1), 57–73.

Rovai, A. P., & Jordan, H. (2004). Blended learning and sense of community: A comparative analysis with traditional and fully online graduate courses. *International Review of Research in Open and Distance Learning*, *5*(2). Retrieved from http://www.irrodl.org/ index.php/ irrodl/ article/ view/192.

Rovai, A. P., & Wighting, M. J. (2005). Feelings of alienation and community among higher education students in a virtual classroom. *The Internet and Higher Education*, *8*(2), 97–110. doi:10.1016/j.iheduc.2005.03.001

Russo, T., & Benson, S. (2005). Learning with invisible others: Perceptions of online presence and their relationship to cognitive and affective learning. *Journal of Educational Technology & Society*, *8*(1), 54–62.

Salter, D., Pang, M. Y. C., & Sharma, P. (2009). Active tasks to change the use of class time within an outcomes based approach to curriculum design. *The Journal of University Teaching & Learning Practice*, *6*(1), 27–38.

Simonson, M., Smaldino, S., Albright, M., & Zvacek, S. (2000). *Teaching and learning at a distance: Foundations of distance education*. Upper Saddle River, NJ: Merrill.

Sitzmann, T., Kraiger, K., Stewart, D., & Wisher, R. (2006). The comparative effectiveness of web-based and classroom instruction: A meta-analysis. *Personnel Psychology*, *59*(3), 623–664.. doi:10.1111/j.1744-6570.2006.00049.x

Smith, K. A. (1994). Cooperation in the college classroom. In *What works: building effective collaborative learning experiences* (pp. 65–70). National Center on Postsecondary Teaching, Learning and Assessment.

Tesone, D. V., & Ricci, P. (2008). Student perceptions of web-based instruction: A comparative analysis. *MERLOT Journal of Online Learning and Teaching*, *4*(3), 317–324.

Thomas, M. (2002). Learning within incoherent structures: The space of online discussion forums. *Journal of Computer Assisted Learning*, *18*, 351–366. doi:10.1046/j.0266-4909.2002.03800.x

US Census Bureau. (2009). Newsroom: Census bureau reports nearly 6 in 10 advanced degree holders age 25-29 are women. Retrieved October 14, 2010, from http://www.census.gov/ newsroom/ releases/ archives/ education/ cb10-55.html

U.S Department of Education, National Center for Education Statistics. (2009). *Digest of Education Statistics, 2008* (NCES 2009-020), Table 188 Retrieved from http://nces.ed.gov/ fastfacts/ display.asp?id=98

U.S Department of Education, Office of Planning, Evaluation, and Policy Development. (2009). *Evaluation of evidence-based practices in online learning: A meta-analysis and review of online learning studies*. Washington, D.C. Retrieved from http://www2.ed.gov/ rschstat/ eval/ tech/ evidence- based- practices/ finalreport.pdf

Verduin, J. R., & Clark, T. A. (1991). *Distance education: The foundations of effective practice* (1st ed.). San Francisco, CA: Jossey-Bass.

Vonderwell, S. (2003). An examination of asynchronous communication experiences and perspectives of students in an online course: A case study. *The Internet and Higher Education*, *6*(1), 77–90. doi:10.1016/S1096-7516(02)00164-1

Wendler, C., Bridgeman, B., Cline, F., Millett, C., Rock, J., Bell, N., & McAllister, P. (2010). *The path forward: The future of graduate education in the United States*. Princeton, NJ: Educational Testing Service.

Wojciechowska, I. (2010, July 16). News: Continuing debate over online education. *Inside Higher Ed*. Retrieved from http://www.insidehighered.com/ news/ 2010/ 07/ 16/ online

Zakaria, E., Chin, L. C., & Daud, M. Y. (2010). The effects of cooperative learning on students' mathematics achievement and attitude towards mathematics. *Journal of the Social Sciences*, *6*(2), 272–275. doi:10.3844/jssp.2010.272.275

Zhao, Y., Lei, J., Yan, B., Lai, C., & Tan, H. S. (2005). What makes the difference? A practical analysis of research on the effectiveness of distance education. *Teachers College Record*, *107*(8), 1836–1884. doi:10.1111/j.1467-9620.2005.00544.x

KEY TERMS AND DEFINITIONS

Asynchronous: in online education, asynchronous discussions are text-based discussions, which can occur over a period of days, with students adding to the discussion according to their schedules.

Blended Education or Courses: courses that meet regularly in a physical classroom and also have an online presence. The online presence is usually facilitated through the use of asynchronous discussion boards or other social media resources.

Chat: synchronous, text-based communication.

Facilitator: an instructor who encourages a collaborative class environment and promotes discussion between students.

Nontraditional Student: generally older students who many have a significant amount of life experience and less opportunity to attend face-to-face courses because they currently have a career and/or family obligations.

Self-Directed Learner: an internally motivated student who takes initiative and responsibility for their learning. This type of student is particularly well suited for an online class.

Synchronous: in online education, synchronous discussions are those in which all students participate in real time.

Chapter 18
Using Traditional Classroom Facilitation Methods in Online Discussion

Maura Valentino
University of Oklahoma, USA

ABSTRACT

This chapter studies the impact of the facilitator on the effectiveness of an online discussion forum. The study examined, categorized, and statistically analyzed 224 discussions from eight online courses. The results demonstrated that facilitators are presented with many opportunities to increase the probability of a successful learning experience and that facilitators should be encouraged to develop new approaches to improve the quality of online discussions by leveraging the unique qualities of an online discussion environment.

INTRODUCTION

"Discussion is a valuable and inspiring means for revealing the diversity of opinion that lies just below the surface of almost any complex issue... To see a topic come alive as diverse and complex views multiply is one of the most powerful experiences we can have as learners and teachers" (Brookfield, 1999, p. 3).

DOI: 10.4018/978-1-61350-077-4.ch018

Educators interested in improving their students' learning experiences have devoted considerable time and effort to the study of the ways in which classroom discussion can be used as an effective tool. It is not surprising that such efforts have focused primarily on face-to-face discussions conducted in a classroom environment. Given the growing importance of Internet delivered courseware to educational institutions, it is important that the pedagogical tools used in connection with this new educational environment be examined to determine if traditional tools can continue to function effectively.

Patterned after Internet user forums and often intended to replace the role of traditional classroom discussions as a forum for students and teachers to explore opinions and ideas and to ensure successful knowledge transfer, the discussion board has become a key component in the delivery of many online classes. Traditional face-to-face classroom discussions have been shown to promote improved learning outcomes and teachers are well trained in facilitating such discussions to help ensure their success. This chapter will study whether equally effective use is being made of discussion boards used in connection with courseware delivered via the Internet.

THESIS

"Democratic discussion works best when a large number of students participate, when they do so on many different issues, and when what they contribute adds depth and subtlety to the discussion" (Brookfield, 1999, p. 9).

This chapter proposes that, as with discussions conducted in a traditional classroom setting, the involvement of the facilitator is a key element in ensuring the effectiveness of a discussion in an online environment. To study this thesis, 224 class conversation threads from eight online courses were examined. These conversation threads were selected from a variety of courses conducted via the Internet as a part of an MSLIS program at a private University in the Northeast.

LITERATURE REVIEW

A review of the relevant literature revealed extensive scholarship on the topic of classroom participation and the facilitator's role in improving the effectiveness of such discussions. However, almost all such literature focused on face-to-face discussions conducted in a traditional classroom environment. Only a small percentage of the literature focused specifically on issues relating to discussions conducted in an online learning environment or on the differences between the classroom and online experience.

Facilitating Discussion in a Classroom Environment

Bean and Peterson (1998) explored the effects of grading classroom participation and conclude that when students are aware of the value of class participation and the percentage of the final grade that it represents, they prepare more thoroughly to participate in class discussions. Bean and Peterson also discussed various ways to structure a participatory classroom and the importance of the instructor having a clear outcome for student participation in mind. They conclude that as well as encouraging class participation, instructors must also assess classroom participation.

Reynolds and Nunn (1997) conducted a survey of both faculty and students and both agreed that "praise, humor, and a supportive classroom atmosphere" encouraged an increased level of student participation in classroom discussions. While they concede that the existing studies differ on the effectiveness of a professor's behavior in increasing classroom participation, they conclude that facilitators should engage in behaviors thought to increase participation.

Ahern, Peck and Laycock (1992) studied computer-mediated discussions and classroom discussions and how teacher interaction improves student interaction in both environments. They explored the correlation between teacher talk and student interaction. They conducted an experiment involving eighty students, to determine if the conversational style of the teacher influenced student comments. The teachers either 1) only asked the discussion group a question, 2) provided the discussion group with a statement, or 3) provided individual comments to in response to each student. The study concluded that when the facilitator provided individual comments in response to students, a greater level of student-to-student interaction occurred than in either of the other two scenarios.

Christensen (1995) studied the same three styles in traditional classrooms and agrees that when the instructor employs a conversational style there is a corresponding increase in the amount of student-to-student interaction. She also found that students were more likely to participate when the class was small and the students were graded on participation, reinforcing some of Bean and Peterson's conclusions.

When considering classroom discussion, Kristina S. Bruss (2009) notes that many classroom discussions do not prove effective. She contends that this is the result of the fact the students are not taught how to be good classroom participators. She posits strategies to teach these skills, including the creation of criteria for assessment including "mastery of material, quality of ideas, effectiveness of argumentation," and providing feedback.

Katherine C. Reynolds and Claudia E. Nunn (1998) used a survey to study whether instructor behavior influenced student participation in freshman courses. They found that the teacher has the ability to create an atmosphere that encourages open discussion.

Dallimore, Hertenstein and Platt (2004) asked students to rate what instructor behavior they determined to be important to quality classroom discussion. Students cited facilitation, asking questions and constructive feedback as the most important behaviors.

Auster and MacRone (1994) agree that asking analytical questions and asking for student opinions are important factors to increasing classroom participation. Michaels, O'Connor, and Resnick (2007) also agree that teachers can increase the quantity and quality of their classroom discussions by asking open-ended questions or prompting for further explanation from a student.

Facilitating Discussions in an Online Environment

Camilla Gant (2007) explores a rubric for assessing student participation in online discussions that combines analytic and holistic approaches to grading. She advocates specific criteria to earn points.

Susan Edelstein and Jason Edwards (2002) add that "the facilitator must design and manage the threaded discussion to direct students in achieving the intended learning outcomes." They concur that the wording of the topic question as well as and interaction during discussion threads are important factors designing a successful learning experience for students.

Betty Cox (2008) studied discussion boards from three online courses to determine the different types of student interactions. She studied whether or not they asked each other questions, if they shared personal and professional experiences, and if they addressed each other by name. She found students engaging in these behaviors with increasing frequency as the semester progressed.

Hutchins (2003) encourages more studies on the effect of instructor behavior on student learning outcomes in online courses, including instructional immediacy. She agrees that the instructor should be a facilitator as opposed to the "giver of knowledge."

Betts (2009) discusses the importance of effective communication in online courses and of using communication theory to improve instructor facilitation. She discovered that teacher traits, including providing feedback and being supportive, correlate to classroom participation. She also found that students are more likely to participate if they think it will affect their grade.

Differences between the Online Environment and the Classroom Discussion Environment

Sherry L. Markel (2001) touts the benefits of discussion for long-term learning. She also asserts

Table 1. Categories

Category	Description
Discussion	The comment acknowledges other comment and leads the discussion forward.
Positional	The comment is on topic, but does not acknowledge other comments. The comment states an opinion on the topic and may answer the original question.
Facilitating	The comment is posted by the facilitator of the discussion thread. A facilitating comment must ask a question or lead the conversation forward. Other comments by the facilitator are categorized as Discussion.
Off Topic	This comment is off topic.
Beginning	The first comment in each discussion thread.

that in the traditional classroom structure, students comfortable with verbal interaction take over the class, leaving no room for those less comfortable with such exchanges. However, in an asynchronous discussion environment students have time to organize their thoughts and there is time for everyone to participate. J. B. Arbaugh (2000) agrees and adds that students who are quiet in a classroom environment may be more comfortable participating in an online environment because there is no competition for the teacher's attention or for the attention of other students. Arbaugh (2000) also adds that an archive of the conversation is available in a way that is not possible in a classroom environment.

METHODOLOGY

To analyze the amount and manner in which instructors facilitate online discussions, 224 discussion threads were analyzed. Each comment, except for the opening comment, was analyzed and assigned a category. Five categories were chosen based on the thesis that facilitation promotes discussion in an online environment. Differentiation between comments that only responded to the opening instructions and comments that truly advanced the conversation was crucial. Student comments that acknowledged the posts previous to their own post and integrated those thoughts into their own post were assigned to the discussion category. Also, student posts that advanced the conversation by posing questions as well as stating an opinion were also considered discussion. Conversely, those posts that simply stated the opinion of the author or only responded to the opening instructions without regard to comments that preceded them were considered position comments. Any comment authored by the class professor or an assigned student facilitator were assigned the category of facilitation. Since the first comment in any thread can only respond to the opening post, these were assigned their own category, "Beginning." Comments that were completely off the topic were assigned to the "Off Topic" category.

The categories available to be assigned to each comment were as follows (see Table 1).

Once each comment was rated by the author of this chapter, a colleague was asked to review a sample of the discussion threads and categorize each comment. This reviewer agreed with the categories assigned by the author for 90% of the comments they analyzed.

The results by thread were then tabulated and analyzes, determining how many total comments were contained in each thread and how many individuals participated in each thread. In addition, the total number of comments per thread was tabulated, as well as the percentage of total comments per thread that fell into each category. The results were sorted to determine if a rise in the percentage of facilitating comments correlated to a rise in the percentage of discussion comments and a decline in the percent of positional comments.

The off topic comments appeared so rarely that they were not analyzed.

The results were then sorted by percentage of discussion comments per thread to determine which threads contained zero discussion comments and which threads contained only discussion comments. The ratio of discussion comments was reached by dividing the number of discussion comments per thread by the total number of all comments minus the beginning comment and any facilitating comments. The opening comments of the threads that contained zero discussion comments and the threads that contained only discussion comments were then analyzed.

Lastly, the percentage of discussion comments per thread was divided by course for all 224 threads and the instructions in the syllabus and in the opening comments for all threads were analyzed.

RESULTS

When the percentage of facilitating comments to the percentage of discussion comments was compared, a direct correlation could be seen. When the percentage of facilitating comments was compared to the percentage of positional comments an indirect correlation was seen.

No facilitation occurred in 128 out of 224 discussion threads. These discussion threads were still highly conversational with an average percentage of discussion comments of 74%. However, as the percentage of facilitating comments increased, so did the percent of discussion comments. Once the facilitation rate increased to at least 5%, the percentage of discussion comments averaged 88% showing a direct correlation ($r = .25$, $p = 0$).

In the128 threads with no facilitation, 13 had no discussion at all. Whereas, once the facilitation percentage had risen above zero, the remaining 96 threads only contained one thread with no discussion. The correlation did not continue as directly once the facilitation rate increased above 5%. For example, at a 6% rate of facilitation, the percentage of discussion comments was 90% whereas at an 11% rate it was 81%. The highest rate of discussion, 100%, fell at a 15% facilitation rate (see Table 2, Figure 1).

Conversely, in the 71 discussion threads that had no facilitation, the positional comments averaged 25%. Once the facilitation comments increased to 5% or more, the highest percentage of positional comments was 18%. Once the facilitation comments increased to 30%, the positional comments were zero ($p = .53$, $r = .005$). See Table 3, Figure 2 for full results.

One unexpected result was a clear correlation between the number of comments per student in each thread and the level of discussion. When each student participated more than once in a discussion thread, the level of the discussion improved ($r = .80$, $p = .04$). However, the number of comments per thread seems to have had no effect. When the average number of comments per student was 1,one the percentage of discussion comments was 58%. When the average number of comments per student rose to more than four

Table 2. Percent of facilitating comments versus percent discussion comments

Percentage Facilitating	Average % Discussion Comments as Facilitating Rises
0	74
5	90
6	90
7	86
8	91
10	87
11	81
13	89
14	94
15	100
17	83
21	88
25	69
Over 30	100

Table 3. Percent of facilitating comments versus percent positional comments

Percentage Facilitating	Average % Positional Comments as Facilitating Rises
0	25
5	10
6	10
7	14
12	17
13	11
15-20	15
21-30	18
>30	0

per student, the percentage of discussion comments rose to 89% (see Table 4, Figure 3).

The opening comments in each discussion thread were analyzed for those discussions that had zero discussion comments and consisted of a series of positional comments and those that were composed of 100% discussion comments. It was found that threads with no discussion comments had very specific instructions and that these instructions were worded like an assignment. They often asked the students to summarize or quote readings for the week. However, the opening comments for those threads containing 100% discussion comments were open-ended questions, often with long explanations of the issues at hand. In other cases they consisted of a link to an article posted with a request to read and then discuss the article (Figure 4).

Table 4. Comments per student versus percent discussion comments

Comments per Student	% of Discussion Comments As Comments Per Students Increase
1	58
1-2	85
2-3	81
3-4	90
>4	89

The last item that was analyzed was which courses had the least and greatest percentage of discussion comments. Most classes averaged between 75% and 85% discussion comments. One class averaged only 5%. It was found that in this class, participating in online discussion comprised 15% of the course grade. Instructions in the syllabus were limited to one sentence saying discussion should be substantive with an emphasis on quality of ideas. One class averaged 98%. In this

Figure 1. Average % discussion as facilitating rises

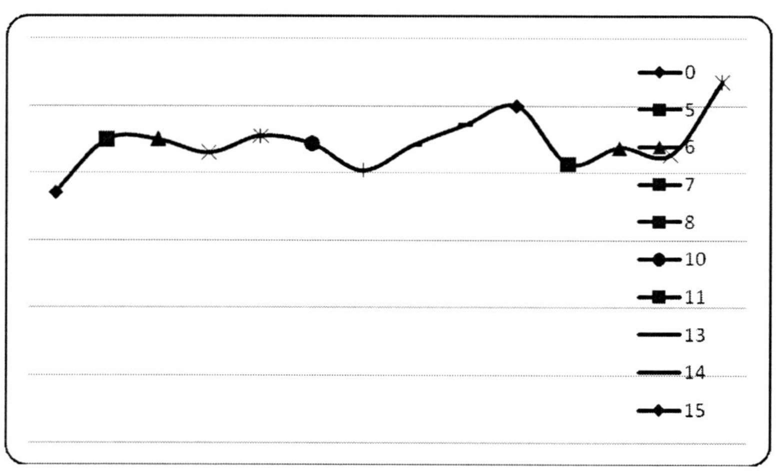

Figure 2. Percent positional

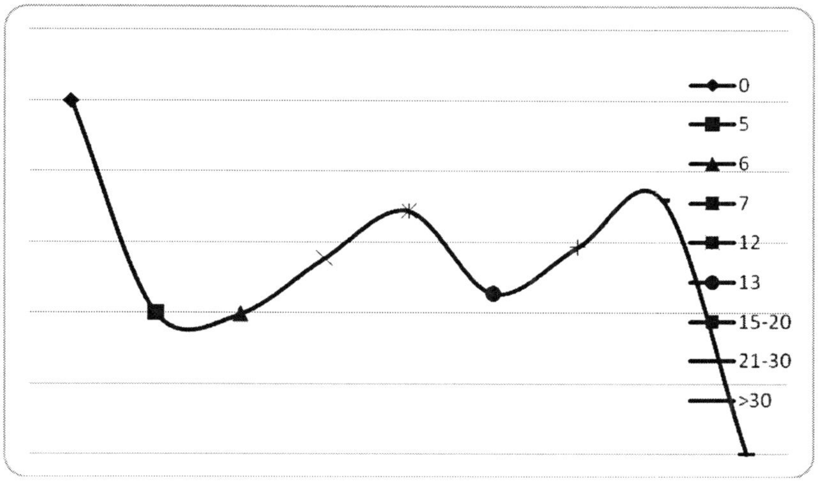

class, participating in online discussion also comprised 15% of the course grade. Instructions in the syllabus consisted of five paragraphs of detailed information including charging the students with beginning the discussion threads with challenging questions. Each student was also charged with answering and discussing questions posed by other students with quality responses that cited the lectures, other materials and personal experience. Each student was guided to synthesize ideas and add to the discussion. This syllabus also promised feedback on the quality of each individual's discussion three times during the semester.

DISCUSSION

The result of this study suggests three ways in which the facilitator can influence the effective-

Figure 3. % discussion as comments per student increase

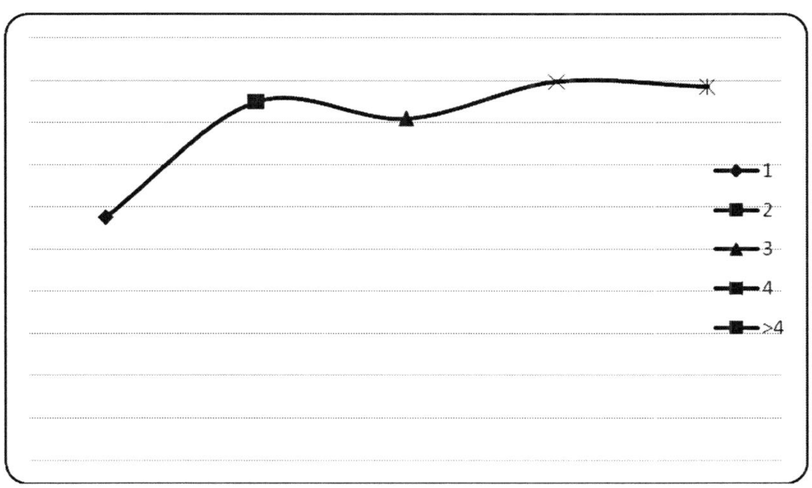

Figure 4. % discussion as number of comments per thread increases

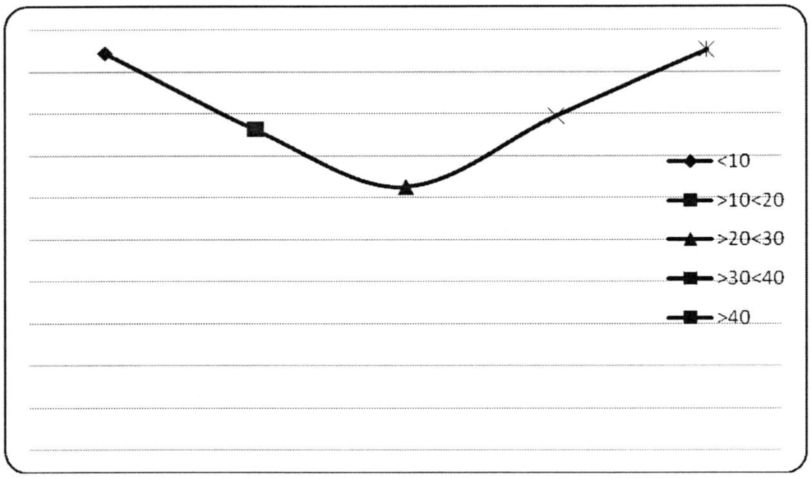

ness of an online discussion: through the class syllabus, through the opening comment, and through mid-discussion facilitation.

Class Syllabus

Instructions in the syllabi for the courses from which the analyzed discussion threads were generated varied from a partial sentence explaining that class participation will count for 15% of the grade to detailed explanations as to what was required for full credit. It is generally agreed that if a student is aware of what is expected, the student will comply with those expectations. It can be seen in the results of analyzing these discussion threads that while it is difficult to make a direct connection to what is explained in the syllabus and the percentage of discussion comments, a detailed explanation of what is expected from the student can affect outcomes positively. However, the same results can also be attained if these expectations are discussed elsewhere. For example, in one case where the syllabus simply stated that discussion was worth 15% of the final grade the resulting discussion thread contained a 96% discussion rate. However, opening posts from this class were the most open-ended and the discussions were well facilitated. The class with the lowest rate of discussion also had very brief instructions asking for an emphasis on quality of ideas. The opening comments for this class were worded like assignments, asking for quotes or summaries of the week's readings and these discussions did not contain mid-discussion facilitation. Another professor offered detailed instructions on participating in discussions, which included asking students to address comments made previous to their own in the thread, rather than creating a series of individual posts. For this class, the overall discussion percentage was 84%, which is quite high, but not the highest recorded. However, when a few outliers are removed, the percentage rises to 90%. Interestingly, the three outliers with low percentage of discussion are all very short threads with few facilitating comments. Bruss (2009) emphasizes that students must be taught how to be good participators. This observation would seem to reinforce this conclusion.

Several of the syllabi had detailed instructions on how to compose a good discussion comment. The class that had the highest percentage of discussion comments had detailed instructions for posting on class discussion boards. The instructions included asking the students to be substantive and to challenge other students. Students were also charged to cite other materials and introduce topics not covered in the lecture and reading materials. This is also the only class that required that the students create the questions that began each discussion thread. The other two classes that offered instructions on how to post in the syllabus had an 80% and an 84% discussion rate. It is worth noting that the class with an 80% discussion rate had student facilitators, and though there were detailed instructions on how to post generally, the instructions for facilitating a discussion were brief but included the idea of asking students questions in order to facilitate the conversation. It is interesting to note that students in the class with the least amount of discussion were required to post each week, whereas the students in the class with the most discussion were not required to post each week or on each topic. Quality over quantity was emphasized.

It is agreed across the literature that if a student knows that they will be graded on discussion, participation thorough preparation and increased participation is encouraged. The syllabi for all 224 analyzed discussion threads mentioned that the student would be graded on discussion participation. As the average percentage of discussion comments for all 224 threads is 78%, the overall percentage of positional comments is only 19%, and off topic comments are limited to nine examples, the overall quality of discussion was very good, demonstrating the effectiveness of grading discussion in an online environment.

An unexpected outcome of the analysis of these discussion threads was also the most direct correlation. The more individual posts each student adds to the thread, the more the thread functions as a useful discussion. It has been discussed that requiring posts for every discussion thread is not conducive to a good discussion; however requiring students to post multiple times in each discussion thread in which they participate will increase the level of discussion. Therefore, when facilitators establish the requirements for full credit for the discussion component of an online course, rather than requiring a single discussion comment in every thread, it may prove more effective to require multiple comments in the discussions in which the students choose to participate.

Opening Comment

The second way that the instructor interacts with students regarding discussion boards is in the opening comment that begins each discussion thread. This interaction was analyzed by considering the opening comments in threads that had no discussion comments and those that had only discussion comments. Again, much of the literature findings conclude that asking questions is an effective way to encourage class participation. In these results it can be seen that the same holds true with online discussions. It was found that the opening comments of the threads that comprised only discussion comments asked open-ended questions or simply posted an article with brief directions to read and then discuss the article. The opening comments in the threads with no discussion comments were worded like assignments, with very specific instructions and directions to summarize or quote readings. In these cases, each student responded directly to the proffered assignment and not to each other's comments.

Students generally do as they are asked; if they are asked to complete an assignment they do so, whether it is in a discussion thread or in another format. Therefore, it is important that the facilitator specifically and clearly instructs the students to discuss the topic and does not structure the opening comment in a way that might be perceived by the student as an assignment. As Christensen (1995) determined, a conversational style encourages discussion in a classroom. The results of this study demonstrate that the same techniques can encourage discussion in an online environment.

Facilitating Mid-Discussion

It is generally agreed in the literature that discussion of a topic aids in retention and understanding of that topic. Cox (2008) touts the benefits of "learner-to-learner" interaction. Many teachers see the value in this approach. Particularly at the graduate level, course discussion is often a significant component of the curriculum. Teachers want students to have the same advantage in an online course and often set up discussion threads so that students can discuss relevant topics. In a traditional classroom setting, teachers facilitate discussion to ensure that students remain focused and on topic and to be sure that salient points are discussed. Most teachers would not explain the topic of discussion and then leave the room allowing the students to progress with the discussion on their own. As with a classroom discussion environments, teachers who are facilitating online discussions must remain involved by posting mid-discussion comments in response to student comments.

As seen in the results section, in many cases students are capable of engaging in relevant discussions without facilitation by the instructor. However, in some instances, what results is a series of position papers expressing each student's opinions and thoughts on the topic with no relation to other students' ideas. While position papers are a valuable tool to determine whether students understand the readings and the relevant topics under consideration, they do not provide all the benefits of discussions and cannot serve to replace a well-developed discussion. Instructors in traditional classes do in many cases require position papers, but these papers are not used as a replacement for classroom discussion. The same methodology is recommended for an online learning environment.

Another negative outcome that can be avoided through successful mid-discussion facilitation occurs when one student misinterprets a reading or an idea presented in a lecture and therefore moves the discussion down a line of inquiry where the basis of the inquiry is incorrect. The resulting discussion often proves to be useless or even detrimental to promoting an understanding of the topic at hand. This was evident in one discussion thread where there was no facilitation.

While with no facilitation, there is often a good conversation, when a teacher or a facilitator also participates in the conversation and uses traditional classroom facilitating techniques, a useful conversation is almost guaranteed. Among the threads with at least one facilitating comment, there was only one thread with no discussion at all.

Some professors assigned a different student to facilitate the discussion each week. This had a mixed result. Each student fulfilled the assignment and participated in the discussion every few comments. However, the students did not receive instruction in how to facilitate the discussion. Some student facilitators were excellent and lead the discussion forward by questioning students by name or asking for clarifications of previous comments. Other student facilitators participated in the conversation frequently, but did not facilitate in any traditional sense. Interestingly, the average percent of discussion comments was the same between the students that actually facilitated and those that only participated in the discussion. These results indicate that any type of facilitator interaction increases discussion, even if some facilitators are less skilled than others. This fact may also be a result of something similar to the Hawthorne Effect. The Hawthorne Effect proposes that the act of observing someone at completing a task changes the very way in which the observed party completes the task. In this case, the students know that they are being monitored when they notice the facilitator entering the discussion every few comments and therefore may change their behavior and be encouraged to increase their participation.

While the comments in these threads were analyzed in multiple ways, there are many variables in such an analysis that cannot be measured. For example various topics are of different interest levels to each individual student. Some topics inspired students to return again and again to the discussion thread. Others only prompted the minimum required participation to fulfill the requirements for the course. Now way was found to measure this effect or predict what topic may be more or less interesting and engaging. Students are individuals and the groups of students that create these discussion threads are each unique in composition and character. In one course, a facilitator may find that all of the students are very interested in one topic and then discover the following semester that the same topic does not seem as inspiring to the new group of students.

Personal factors also may influence a student's level of participation in an online discussion. A particular student may normally participate by posting several comments in a discussion thread, but if personal factors such as employment demands or illness intervene, the same student's participation may suddenly reduce to a basic level. For example, if a course requires a minimum number of posts per week the student may just read the opening comment and post comments without reading and responding to other comments. This variable was also difficult to quantify.

Interestingly, in several discussion threads, while some students where conducting a productive discussion each day of the week, other students only posted at the end of the week or at the end of the course. These comments often did not acknowledge the discussion that had occurred prior to their own comments, but just expressed a position on the topic. It is reasonable to surmise that these students only participated to fulfill the discussion requirement for that week or for the course itself.

Additional Observations

Classroom discussion is a synchronous activity. If the thread of a conversation changes to a new topic it can be difficult for the facilitator to return the conversation to its original track. If a student does not have the time to respond before the topic changes or the class moves on to a new activity that comment is never related and considered. However, in an online asynchronous discussion environment, students can add their opinion to various conversation threads and discuss several topics at once. Even within the same thread, it is much easier to revisit previous comments as students can quote the previous comment to which they are responding. Arbaugh (2000) discusses the value of having the ability to refer to an archive in an online discussion board. Markel (2001) comments that an online environment is good for students who are too shy to speak in class, as these students are often comfortable discussing their ideas when they express themselves in writing. An online discussion environment can be an effective tool for such a student.

CONCLUSION

This study has demonstrated that facilitators of class discussions in an online environment are presented with many opportunities to increase the probability of a successful learning experience. Asking for quality discussion comments and explaining what is meant by a quality comment encourages improved results over simply requiring that a student post a certain number of comments per week. Opening comments that contain open-ended questions encourage increased and more effective discussions than opening comments structured as specific assignments. A facilitator who participates frequently in the conversation can keep the discussion on track and increase the discussion's relevance and effectiveness.

Instructors should be encouraged to develop unique approaches to improve the quality of online discussions. Instead of simply attempting to make online discussions resemble face-to-face classroom interactions, facilitators can improve the learning experience by leveraging the unique qualities of an online discussion environment.

For example, this study revealed online discussions often degenerate into a series of position papers rather than evolving into a true discussion. However, facilitators might reduce the chance of this outcome by assigning several students each week to initiate threads by quoting or summarizing the assigned readings and then stimulating discussion by inviting other students to respond to their ideas. This would take advantage of the strengths of an online discussion by allowing multiple discussions to develop on the various topics and by encouraging students to respond directly to each other's ideas.

S. Michaels et al. (2007) notes that the qualities of a classroom discussion that lead to learning include "listening, reflecting, proposing and incorporating alternative views." The same is true of discussions conducted in an online environment. Indeed, since an asynchronous online environment provides ample time to absorb the views of others, to reflect on those views, and to formulate a well-reasoned and thoughtful response, it might be said that online discussions provide an important opportunity to improve the overall quality of educational discourse, and if developed effectively, point the way to exciting new opportunities for educators and students alike.

REFERENCES

Ahern, T. C., Peck, K., & Laycock, M. (1992). The effects of teacher discourse in computer-mediated discussion. *Journal of Educational Computing Research*, *8*(3), 291–309. doi:10.2190/HFPW-JYR3-YMBE-0J7D

Arbaugh, J. B. (2000). Virtual classroom versus physical classroom: An exploratory study of class discussion patterns and student learning in an asynchronous internet-based MBA course. *Journal of Management Education*, *24*(2), 213–233. doi:10.1177/105256290002400206

Auster, C. J., & MacRone, M. (1994). The classroom as a negotiated social setting: An empirical study of the effects of faculty members' behavior on students' participation. *Teaching Sociology*, *22*, 289–300. doi:10.2307/1318921

Bean, J. C., & Peterson, D. (1998). Grading classroom participation. *New Directions for Teaching and Learning*, *74*(Summer), 33–40. doi:10.1002/tl.7403

Betts, K. (2009). Lost in translation: Importance of effective communication in online education. *Online Journal of Online Distance Learning Administration, 12*(2). Retrieved from http://www.westga.edu/ ~distance/ ojdla

Brookfield, S. D., & Preskill, S. (1999). *Discussion as a way of teaching: Tools and techniques for democratic classrooms*. San Francisco, CA: John Wiley & Sons.

Bruss, K. S. (2009). Improving classroom discussion: A rhetorical approach. *The Journal of General Education*, *58*(1), 28–46. doi:10.1353/jge.0.0034

Christensen, L. J. (1995). Classroom situations which lead to student participation. *Proceedings from Annual Meeting of the Speech Communication Association*. San Antonio, TX.

Cox, B. (2008). Developing interpersonal and group dynamics through asynchronous threaded discussions: The use of discussion board in collaborative learning. *Education*, *128*(4), 1–16.

Dallimore, E. J., Herenstein, J. H., & Platt, M. B. (2004). Classroom participation and discussion effectiveness: Student-generated strategies. *Communication Education*, *53*(1), 103–115. doi:10.1080/0363452032000135805

Edelstein, S., & Edwards, J. (2002). If you build it, they will come: Building learning communities through threaded discussions. *Online Journal of Distance Learning Administration*, *5*(1). Retrieved from http://www.westga.edu/ ~distance/ ojdla.

Gant, C. (2007). Assessing asynchronous discussion: An exploratory hybrid model. *Online Journal of Distance Learning Administration*, *10*(3). Retrieved from http://www.westga.edu/ ~distance/ ojdla.

Hutchins, H. M. (2003). Instructional immediacy and the seven principles: Strategies for facilitating online courses. *Online Journal of Distance Learning Administration*, *6*(3). Retrieved from http://www.westga.edu/ ~distance/ ojdla.

Markel, S. L. (2001). Technology and education online discussion forums: It's in the response. *Online Journal of Distance Learning Administration*, *4*(2). Retrieved from http://www.westga.edu/ ~distance/ ojdla.

Michaels, S., O'Connor, C., & Resnick, L. B. (2007). Deliberative discourse idealized and realized: Accountable talk in the classroom and in civic life. *Studies in Philosophy and Education*, *27*(4), 283–297. doi:10.1007/s11217-007-9071-1

Reynolds, K. C., & Nunn, C. E. (1997). Engaging freshmen in classroom discussion: Interaction and the instructor techniques that encourage it. *Proceedings from Annual Meeting of the Association for the Study of Higher Education*. Albuquerque, NM.

KEY TERMS AND DEFINITIONS

Asynchronous Discussion: A discussion where participants may or may not be participating at the same time. Multiple discussions and topics can occur at the same time. Online discussions are asynchronous.

Conversation Archive: The past discussion threads stored in the discussion forum. These are available at any time to review past discussions.

Discussion Forum: An online software environment where students or the professor can begin discussion threads.

Discussion Thread: A discussion on a particular topic in an online environment. One person begins a new topic and as each person replies to that topic a thread is created.

Facilitator: A person who leads a classroom or online discussion and guides that discussion to produce learning outcomes.

Online Class: A class that has no scheduled classroom time, but instead has required activities completed in each individual's geographic location, managed in an online environment.

Position Paper: A paper assigned to be sure students understand the assigned readings, prior to discussion, containing the student's opinions on the topics, rather than a summary.

Synchronous Discussion: A discussion where all the participants are present for the discussion at the same time. They generally only have one topic at a time and then move onto another topic. Classroom discussions and chat-rooms are synchronous.

Compilation of References

Adlakha, V. G., & Aggarwal, A. K. (2009). Minimal functionalities of course management systems: A faculty perspective. *International Journal of Web-Based Learning and Teaching Technologies*, *4*(2), 26–42. doi:10.4018/jwltt.2009040102

Ahern, T. C., Peck, K., & Laycock, M. (1992). The effects of teacher discourse in computer-mediated discussion. *Journal of Educational Computing Research*, *8*(3), 291–309. doi:10.2190/HFPW-JYR3-YMBE-0J7D

Aigrain, P. (2005). Positive intellectual rights and information exchanges. In Ghosh, R. A. (Ed.), *CODE: Collaborative ownership and the digital economy* (pp. 287–316). Cambridge, MA: MIT Press.

Aitken, J. E., & Shedletsky, L. J. (2002). Using electronic discussion to teach communication courses. *Communication Education*, *51*(3), 325–331. doi:10.1080/03634520216512

Akins, F., Check, E., & Riley, R. (2004). Technological lifelines: Virtual intimacies and distance learning. *Studies in Art Education*, *46*(1), 34–47.

Alemán. A. M. M. N., & Wartman, K. L. (2008). *Online social networking on campus: Understanding what matters in student culture* (1st ed.). New York, NY: Routledge.

Alexander, B. (2006). Web 2.0: A new wave of innovation for teaching and learning? *EDUCAUSE Review*, *41*(2), 33–44.

Ali, A. (2003). Instructional design and online instruction. *TechTrends*, *47*(5), 42–45. doi:10.1007/BF02763205

Allen, I. E., & Seaman, J. (2010). *Learning on demand: Online education in the United States, 2009*. Needham, MA: Sloan Consortium.

Alonso, G., Casati, F., Kuno, H., & Machiraju, V. (2004). *Web services: Concepts, architectures, and applications*. Wiesbaden, Germany: Springer.

American Library Association. (1989). *Presidential committee on information literacy* (final report). Chicago, il: American Library Association. Retrieved from http://www.ala.org/ ala/ mgrps/ divs/ acrl/ publications/ whitepapers/ presidential.cfm

An, Y.-J., & Frick, T. (2006). Student perceptions of asynchronous computer-mediated communication in face-to-face courses. *Journal of Computer-Mediated Communication*, *11*(2). Retrieved from http://jcmc.indiana.edu/ vol11/ issue2/ an.html. doi:10.1111/j.1083-6101.2006.00023.x

Anderson, R. P., & Wilson, S. P. (2009). Quantifying the effectiveness of interactive tutorials in medical library instruction. *Medical Reference Services Quarterly*, *28*(1), 10–21. doi:10.1080/02763860802615815

Anderson, R. H., Bikson, T. K., Law, S. A., & Mitchlee, B. M. (1995). *Universal access to e-mail: Feasibility and societal implications*. Santa Monica, CA: The Rand Corporation.

Anderson, B. (1991). *Imagined communities: Reflections on the origin and spread of nationalism* (Rev. ed.). London, UK: Verso.

Andrews, T., Curbera, F., Dholaki, H., Goland, Y., Klein, J., Leymann, F., et al. (2003). *BPEL4WS: Business Process Execution Language for Web Services,* version 1.1.

Araujo, A. L. (2009). Instrumental and social influences on adoption of collaborative technologies in global virtual teams. In Salmons, J., & Wilson, L. (Eds.), *Handbook of research on electronic collaboration and organizational synergy* (pp. 400–408). Hershey, PA: Information Science Reference.

Arbaugh, J. B. (2000). Virtual classroom versus physical classroom: An exploratory study of class discussion patterns and student learning in an asynchronous internet-based MBA course. *Journal of Management Education, 24*(2), 213–233. doi:10.1177/105256290002400206

Association of College and Research Libraries (ACRL). (2000). *Information literacy competency standards for higher education.* Retrieved from http://www.ala.org/ala/ mgrps/ divs/ acrl/ standards/ informationliteracy competency.cfm# Baecker, R., Birnholtz, J., Causey, R., Laughton, S., Rankin, K., Mak, C., … Wolf, P. (2007). Webcasting made interactive: Integrating real-time videoconferencing in distributed learning spaces. *Lecture Notes in Computer Science, 4558*, 269-278.

Association of College and Research Libraries. (2010). *Information literacy competency standards for higher education.* Association of College and Research Libraries [Website]. Retrieved from http://www.ala.org/ ala/ mgrps/ divs/ acrl/ standards/ informationliteracycompetency.cfm

Auster, C. J., & MacRone, M. (1994). The classroom as a negotiated social setting: An empirical study of the effects of faculty members' behavior on students' participation. *Teaching Sociology, 22*, 289–300. doi:10.2307/1318921

Baddeley, A. D. (1999). *Human memory.* Boston, MA: Allyn & Bacon.

Baecker, R., Moore, G., & Zijdemans, A. (2003). Reinventing the lecture: Webcasting made interactive. *Proceedings of HCI International 2003.*

Baecker, R., Wolf, P., & Rankin, K. (2004). The ePresence interactive webcasting and archiving system: Technology overview and current research issues. In J. Nall & R Robson (Eds.), *Proceedings of World Conference on E-Learning in Corporate, Government, Healthcare, and Higher Education 2004* (pp. 2532-2537). Chesapeake, VA: AACE.

Bakhtiari, S., & Shajar, H. (2006). Globalization and education: Challenges and opportunities. *International Business & Economics Research Journal, 5*(2). Retrieved September 19, 2010, from http://www.cluteinstitute- onlinejournals.com/ PDFs/ 2006232.pdf

Bambara, C., Harbour, C., Davies, T., & Athey, S. (2009). The lived experience of community college students enrolled in high-risk online courses. *Community College Review, 36*(3), 219–238. doi:10.1177/0091552108327187

Banks, J. A. (2004, Summer). Teaching for social justice, diversity, and citizenship in a global world. *The Educational Forum*, , 68.

Banks, J. A. (2003). Educating global citizens in a diverse world. *New Horizons for Learning.* Retrieved February 27, 2010, from http://www.newhorizons.org/ strategies/ multicultural/ banks2.htm

Barab, S., Schatz, S., & Scheckler, R. (2004). Using activity theory to conceptualize online community and using online community to conceptualize activity theory. *Mind, Culture, and Activity, 11*(1), 25–47. doi:10.1207/s15327884mca1101_3

Baraldi, P., Manginelli, A. A., Maieron, M., Liberati, D., & Porro, C. A. (2007). An ARX model-based approach to trial by trial identification of fMRI-BOLD responses. *NeuroImage, 37*(1), 189–201. doi:10.1016/j.neuroimage.2007.02.045

Baran, E., & Correia, A. (2009). Student-led facilitation strategies in online discussions. *Distance Education, 30*(3), 339–361. doi:10.1080/01587910903236510

Barjak, F. (2006). The role of the internet in informal scholarly communication. *Journal of the American Society for Information Science and Technology, 57*(10), 1350–1367. doi:10.1002/asi.20454

Barley, S. R. (1986). Technology as an occasion for structuring: Evidence from observations of CT scanners and the social order of radiology departments. *Administrative Science Quarterly, 31*(1), 78–108. doi:10.2307/2392767

Barnes, N. G., & Mattson, E. (2009). *Social media and college admissions: Higher-ed beats business in adoption of new tools for third year.* Dartmouth, MA: Center for Marketing Research - University of Massachusetts Dartmouth.

Baron, C. L. (2004). *Designing a digital portfolio*. Indianapolis, IN: New Riders.

Bartle, R. (2003). Not yet you fools! *Game+Girl=Advance*. Retrieved from http://www.gamegirladvance.com/ archives/2003/07/28/ not_yet_you_fools.html

Basile, A., & D'Aquila, J. M. (2002). An experimental analysis of computer–mediated instruction and student attitudes in a principles of financial accounting course. *Journal of Education for Business*, *77*(3), 137–143. doi:10.1080/08832320209599062

Bastos, F. (2010). A 21st-century art room: The remix of creativity and technology. *Art Education*, *63*(1), 4–5.

Bator, R. J. (2000). The application of persuasion theory to the development of effective proenvironmental public service announcements. *Journal of Social Issues, 56*(3). Retrieved from http://findarticles.com/ p/ articles/ mi_ m0341/ is_ 3_ 56/ai_ 69391502/

Batson, T. (2005). The electronic portfolio boom: What's it all about? *Campus Technology*. Retrieved February 15, 2011, from http://www.campustechnology.com/article.asp?id=6984

Baym, N. K. (2000). *Tune in, log on: Soaps, fandom, and online community*. Thousand Oaks, CA: Sage.

Baym, N. K. (2006). Interpersonal life online. In Lievrouw, L. A., & Livingstone, S. (Eds.), *The handbook of new media* (pp. 35–54). London, UK: Sage.

Bean, J. C., & Peterson, D. (1998). Grading classroom participation. *New Directions for Teaching and Learning*, *74*(Summer), 33–40. doi:10.1002/tl.7403

Beaudoin, M. F. (2002). Learning or lurking? Tracking the "invisible" online student. *The Internet and Higher Education*, *5*(2), 147–155..doi:10.1016/S1096-7516(02)00086-6

Bell, L., Peters, T., & Pope, K. (2007). Get a (second) life! *Computers in Libraries*, *27*(1), 10–15.

Bell, S. (2003). Using webcasts as a teaching tool. *TechTrends*, *47*(4), 10–14. doi:10.1007/BF02763506

Bell, F., & Whatley, J. (2004). Trans-national online activities for students—a pragmatic approach. In *ILIA—Innovative Learning in Action, issue one: Internationalising the curriculum* (pp. 13–18). UK: University of Salford.

Bell, F., & Zaitseva, E. (2005). Only connect: Complexities in international student communication. *E-learning*, *2*(4), 342–355. doi:10.2304/elea.2005.2.4.341

Benoit, P. J., Benoit, W. L., Milyo, J., & Hansen, G. J. (2006). *The effects of traditional vs. Web–assisted instruction on student learning satisfaction*. Columbia, MO: University of Missouri Press.

Benson, A. D., Johnson, S. D., Taylor, G. D., Treat, T., Shinkareva, O. N., & Duncan, J. (2005). Achievement in online and campus–based career and technical education (CTE) courses. *Community College Journal of Research and Practice*, *29*, 369–394. doi:10.1080/10668920590921589

Berge, Z. L. (2007). Motivate and manage: Key activities of online instructors. In Spector, J. M. (Ed.), *Finding your online voice: Stories told by experienced online educators* (pp. 73–82). Mahwah, NJ: Lawrence Erlbaum Associates.

Bernard, R. M., Abrami, P. C., Lou, Y., Borokhovski, E., Wade, A., & Wozney, L. (2004). How does distance education compare with classroom instruction? A meta-analysis of the empirical literature. *Review of Educational Research*, *74*(3), 379–439. doi:10.3102/00346543074003379

Bertolucci, J. (2009, February 4). YouTube leads growth in U.S. online video viewing. *PC World*. Retrieved from http://www.pcworld.com/ article/ 158949/ youtube leads growth in us online video viewing.html

Betne, P., & Castonguay, R. (2008). On the role of mathematics educators and librarians in constructivist pedagogy. *Education*, *129*(1), 56–79.

Betts, K. (2009). Lost in translation: Importance of effective communication in online education. *Online Journal of Online Distance Learning Administration, 12*(2). Retrieved from http://www.westga.edu/ ~distance/ ojdla

Bi, X. (2000). *Instructional design attributes of web-based courses*. (No. ED448746). San Antonio, TX. Retrieved from http://www.eric.ed.gov/ ERICWebPortal/ content-delivery/ servlet/ ERICServlet? accno=ED448746

Bieber, J. (2011, September 2). *Justin Bieber: Baby ft. Ludacris*. Retrieved from http://www.youtube.com/ watch?v=qrO4YZeyl0I

Bloom, B. S. (1956). *Taxonomy of educational objectives*. New York, NY: David Mckay Company.

Blumenfeld, P. C., Kempler, T. M., & Krajcik, J. S. (2006). Motivation and cognitive engagement in learning environments. In R. K. Sawyer (Ed.), *The Cambridge handbook of the learning sciences* (pp. 475–488). New York, NY: Cambridge University Press. Retrieved from http://131.193.130.213/ media/ Blumenfeld.et.al_ 2006_ Cambridge_ handbook_ LS.pdf

Boehm, D., & Aniola-Jedrzejek, L. (2006). Seven principles of good practice for virtual international collaboration. In Ferris, S. P., & Godar, S. (Eds.), *Teaching and learning with virtual teams* (pp. 1–31). Hershey, PA: Information Science Publishing. doi:10.4018/978-1-59140-708-9.ch001

Boehm, D., Kurthen, H., & Aniola-Jedrzejek, L. (2010). Do international online collaborative learning projects impact ethnocentrism? *E-Learning and Digital Media, 7*(2). Retrieved from http://www.wwwords.co.uk/ elea

Boellstorff, T. (2008). *Coming of age in Second Life: An anthropologist explores the virtually human*. Princeton, NJ: Princeton University Press.

Bogatin, D. (2007, May 7). *Google YouTube: Viral video success formula: Digital markets article*. [Web log post]. Retrieved from http://www.zdnet.com/ blog/ micro-markets/ google-youtube- viral- video- success- formula/1315

Boisot, M. (2002). *Knowledge assets: Securing competitive advantage in the information economy*. New York, NY: Oxford University Press.

Bongey, S. B., Cizadlo, G., & Kalnback, L. (2005). Using a course management system (CMS) to meet the challenges of large lecture classes. *Campus-Wide Information Systems, 22*(5), 252–262..doi:10.1108/10650740510632172

Bonk, C. J. (2009). *The world is open: How web technology is revolutionizing education*. San Francisco, CA: Jossey-Bass.

Bonk, C. J., & King, K. S. (Eds.). (1998). *Electronic collaborators: Learner-centered technologies for literacy, apprenticeship, and discourse*. New Jersey: Lawrence Erlbaum.

Bonk, C. J., & Cunningham, D. J. (1998). Searching for learner-centered, constructivist, and sociocultural components of collaborative educational learning tools. In Bonk, C. J., & King, K. S. (Eds.), *Electronic collaborators: Learner-centered technologies for literacy, apprenticeship, and discourse* (pp. 25–50). Mahwah, NJ: Lawrence Erlbaum Associates.

Bonk, C. J. (2008, March). *YouTube anchors and enders: The use of shared online video content as a macrocontext for learning*. Paper presented at the meeting of American Educational Research Association, New York, NY. Retrieved from http://trainingshare.com/ SFX7EED.pdf

Book, C. L. (1990). Extended discourse. In Daily, J., Friedrich, G., & Vangelisti, A. (Eds.), *Teaching communication: Theory, research, and methods* (pp. 279–291). Hillsdale, NJ: Erlbaum.

Booth, D., Haas, H., McCabe, F., Newcomer, E. I., Champion, M. F., & Orchard, D. (Eds.). (2004, February 11). Web services architecture. *W3C Working Group Note*. Retrieved February 15, 2011, from http://www.w3.org/ TR/ ws-arch/

Bordia, P. (1997). Face-to-face versus compute-mediated communication: A synthesis of the experimental literature. *Journal of Business Communication, 34*, 98–118. doi:10.1177/002194369703400106

Bosin, A., Dessì, N., Fugini, M. G., Liberati, D., & Pes, B. (2006a). Applying enterprise models to design cooperative scientific environments. [Wiesbaden, Germany: Springer.]. *Lecture Notes in Computer Science, 3812*, 281–292. doi:10.1007/11678564_25

Bosin, A., Dessì, N., Liberati, D., & Pes, B. (2006b). Learning Bayesian classifiers from gene-expression microarray data. [Wiesbaden, Germany: Springer.]. *Lecture Notes in Computer Science, 3849*, 297–304. doi:10.1007/11676935_37

Boughton, D. (2000). *The shaping of visual arts in education* (pp. 956–969). Routledge International Companion to Education.

Boyd, D. M., & Ellison, N. B. (2007). Social network sites: Definition, history, and scholarship. *Journal of Computer-Mediated Communication, 13*(1), 210–230. doi:10.1111/j.1083-6101.2007.00393.x

Boyd, R., MacNeill, N., & Sullivan, G. (2006). Relational pedagogy: Putting balance back into students' learning. *Curriculum Leadership, 4*(13). Retrieved from http://www.curriculum.edu.au/ leader/ relational_ pedagogy:_ putting_ balance_ back_ into_ stu,13944.html?issueID=10277

Boyle, T., Bradley, C., Chalk, P., Jones, R., & Pickard, P. (2003). Using blended learning to improve student success rates in learning to program. *Learning, Media and Technology, 28*(2), 165–178.

Bradner, E., Kellogg, W., & Erickson, T. (1999). The adoption and use of "Babble": A field study of chat in the workplace. *Proceedings of the European Conference on Computer Supported Cooperative Work (ECSCW '99), Copenhagen, Denmark*, (pp. 139-158). Retrieved from http://www.ecscw.org/ 1999/08.pdf

Brandon, D. P., & Hollingshead, A. B. (1999). Collaborative learning and computer-supported groups. *Communication Education, 48*, 109–126. doi:10.1080/03634529909379159

Bretag, T. (2006). Developing 'third space' interculturality by using computer-mediated communication. *Journal of Computer-Mediated Communication, 11*(4), 981–1011. doi:10.1111/j.1083-6101.2006.00304.x

Brewer, B. (2010, April). *The end justifying the means: Ethical implications of public service announcements*. Paper presented at the meeting of Central States Communication Association, Cincinnati, OH.

Brittan-Powell, C. (2008). Preparing and retaining students in online courses. *International Journal of Instructional Technology and Distance Learning, 5*(9). Retrieved from http://www.itdl.org/ Journal/ Sep_08/ article03.htm.

Bronack, S., Riedl, R., & Tashner, J. (2006). Learning in the zone: A social constructivist framework for distance education in a 3D virtual world. *Interactive Learning Environments, 14*(3), 219–232. doi:10.1080/10494820600909157

Brockfield, S. D., & Preskill, S. (1999). *Discussion as a way of teaching: Tools and techniques for democratic classrooms*. San Francisco, CA: John Wiley & Sons.

Brooks, C. D., & Jeong, A. (2006). Effects of pre-structuring discussion threads on group interaction and group performance in computer-supported collaborative argumentation. *Distance Education, 27*(3), 371–390. doi:10.1080/01587910600940448

Brophy, J., & Bawden, D. (2005). Is Google enough? Comparison of an internet search engine with academic library resources. *Aslib Proceedings, 57*(6), 498–512. doi:10.1108/00012530510634235

Brown, B. W., & Liedholm, C. E. (2004). Student preferences in using online learning resources. *Social Science Computer Review, 22*, 479–492. doi:10.1177/0894439304268529

Bruffee, K. A. (1984). Collaborative learning and the "conversation of mankind.". *College English, 46*(7), 635–652. doi:10.2307/376924

Brule, N. J. (2008, April). *Abused men & shaken babies: Teaching undergrads to apply research & promote change*. Symposium education exemplar panel at the meeting of Central States Communication Association, Madison, WI.

Bruns, A., & Humphreys, S. (2005). *Wikis in teaching and assessment – The M/Cyclopedia project*. Paper presented at The 2005 International Symposium on Wikis, October 16-18, San Diego, CA. Retrieved from http:// snurb.info/ files/ Wikis% 20in% 20Teaching% 20and% 20Assessment.pdf

Bruss, K. S. (2009). Improving classroom discussion: A rhetorical approach. *The Journal of General Education, 58*(1), 28–46. doi:10.1353/jge.0.0034

Bryant, C. (2010). A 21st-century art room: The remix of creativity and technology. *Art Education, 63*(2), 43–48.

Buckley, W., & Smith, A. (2007). Application of multimedia technologies to enhance distance learning. *RE:view, 39*(2), 57–65. doi:10.3200/REVU.39.2.57-65

Buffington, M. L. (2008). New media art education. *Art Education, 61*(3), 36–41.

Bunt-Kokhuis, S. d. (2001). On-line learning at universities in developing countries: From leap-frogging to antelope-jumping—Specific needs and solutions. *Higher Education in Europe, 26*(1), 241–246. doi:10.1080/03797720120082633

Button, K., Johnson, M. J., & Furgerson, P. (1996). Interactive writing in a primary classroom. *The Reading Teacher*, *49*(6), 446–454.

Caine, R. N., & Caine, G. (1994). Threat and the brain. In *Making connections: Teaching and the human brain* (pp. 69–86). Menlo Park, CA: Addison-Wesley Pub. Co.

Caladine, R. (2008). *Enhancing e-learning with media-rich content and interactions*. Hershey, PA: Information Science Publishing. doi:10.4018/978-1-59904-732-4

Calka, M., & Black, L. W. (2010, April). *Doing community based partnership research in online communities*. Paper presented at the meeting of Central States Communication Association, Cincinnati, OH.

Cambridge, D. (2010). *Eportfolios for lifelong learning and assessment*. New York, NY: Jossey-Bass.

Cambridge, D., Cambridge, B., & Yancey, K. (2009). *Electronic portfolios 2.0: Emergent research on implementation and impact*. Sterling, VA: Stylus Publishing.

Caracelli, V. J., & Greene, J. C. (1997). Crafting mixed-method evaluation designs. *New Directions for Evaluation*, (74): 19–32. doi:10.1002/ev.1069

Carbaugh, D. (2005). *Cutures in conversation*. Mahwah, NJ: Lawrence Erlbaum Associates.

Carbaugh, D. (1995). The ethnographic communication theory of Philipsen and associates. In Cushman, D. P., & Kovacic, B. (Eds.), *Watershed research traditions in human communication theory* (pp. 269–297). Albany, NY: State University of New York Press.

Carey, J. (2009). *Communication as Culture* (rev. ed.). New York, NY: Routledge.

Carpenter, T. G., Brown, W. L., & Hickman, R. C. (2004). Influences of online delivery on developmental writing outcomes. *Journal of Developmental Education*, *28*(1), 14–16, 18, 35.

Carr, N. (2010). *The shallows: What the internet is doing to our brains*. New York, NY: W. W. Norton & Company.

Carr, S. (2000). As distance education comes of age, the challenge is keeping the students. *The Chronicle of Higher Education*, *46*(23), A39–A41.

Carte, T. A., & Chidambaram, L. (2004). A capabilities-based theory of technology deployment in diverse teams: Leapfrogging the pitfalls of diversity and leveraging its potential with collaborative technology. *Journal of the Association for Information Systems*, *5*(11/12), 448–471.

Carte, T. A., Chidambaram, L., & Becker, A. (2006). Emergent leadership in self-managed virtual teams. *Group Decision and Negotiation*, *15*(4), 323–343..doi:10.1007/s10726-006-9045-7

Caruso, J. B., Smith, S. D., & Salaway, G. (2009). *The ECAR study of undergraudate students and Information Technology*. 2009: EDUCAUSE Center for Applied Research.

Cassell, J., & Tversky, D. (2005). The language of online intercultural community formation. *Journal of Computer-Mediated Communication*, *10*(2).

Casserly, M. (2010, October 6). The new Forbes power women. *Forbes Magazine*. Retrieved from http://www.forbes.com

Castek, J., Leu, D. J. Jr, Coiro, J., Gort, M., Henry, L. A., & Lima, C. O. (2007). Developing new literacies among multilingual learners in the elementary grades. In Parker, L. (Ed.), *Technology-mediated learning environments for young English learners: Connections in and out of school* (pp. 111–154). Mahwah, NJ: Lawrence Erlbaum Associates.

Castells, M. (2007). Communication, power and counter-power in the network society. *International Journal of Communication*, *1*, 238–266.

Cavanaugh, J. (2005). Teaching online - A time comparison. *Online Journal of Distance Learning Administration*, *8*(1). Retrieved from http://www.westga.edu/ ~distance/ojdla/ spring81/ cavanaugh81.htm.

Cavus, N., & Ibrahim, D. (2007). Assessing the success rate of students using a learning management system together with a collaborative tool in web-based teaching of programming languages. *Journal of Educational Computing Research*, *36*(3), 301–321. doi:10.2190/T728-G676-4N18-6871

Cheung, W. S., & Hew, K. F. (2004). Evaluating the extent of ill-structured problem solving process among pre-service teachers in an asynchronous online discussion and reflection log learning environment. *Journal of Educational Computing Research*, *30*(3), 197–227. doi:10.2190/9JTN-10T3-WTXH-P6HN

Christensen, E., Curbera, F., Meredith, G., & Weerawarana, S. (2003). Web service definition language (WSDL) 1.1. *W3C Note*. Retrieved February 15, 2011, from http://www.w3.org/ TR/ 2001/ NOTE- wsdl- 20010315

Christensen, L. J. (1995). Classroom situations which lead to student participation. *Proceedings from Annual Meeting of the Speech Communication Association*. San Antonio, TX.

Clark, R. A., & Jones, D. (2001). A comparison of traditional and online formats in a public speaking course. *Communication Education*, *50*, 109–124. doi:10.1080/03634520109379238

Common Core State Standards Initiative. (2010). *English language arts standards*. Retrieved February 14, 2011, from http://www.corestandards.org/the-standards/english-language-arts-standards/introduction/key-design-considerations

Connolly, T. M., MacArthur, E., Stansfield, M. H., & McLellan, E. (2007). A quasi-experimental study of three online learning courses in computing. *Journal of Computers and Education*, *49*(2), 345–359. doi:10.1016/j.compedu.2005.09.001

Conrad, D. L. (2002). Engagement, excitement, anxiety, and fear: Learners' experiences of starting an online course. *American Journal of Distance Education*, *16*(4), 205–226. doi:10.1207/S15389286AJDE1604_2

Contractor, N. S., Whitbred, R., Fonti, F., Hyatt, A., Jones, P., & O'Keefe, B. (1998). *Self-organizing communication networks in organizations: Validation of a computational model using exogenous and endogenous theoretical mechanisms*. Paper presented at the International Communication Association, Jerusalem, Israel. Retrieved October 18, 2009, from http://hyperion.math.upatras.gr/ commorg/nosh/comp.html

Coole, H., & Watts, M. (2009). Communal e-learning styles in the online classroom. *Research in Education*, *82*(1), 13–27.

Cotten, S. R., & Jelenewicz, S. M. (2006). A disappearing digital divide among college students? *Social Science Computer Review*, *24*(4), 497–506. doi:10.1177/0894439306286852

Coutu, L. (2000). Communication codes of rationality and spirituality in the discourse of and about Robert S. McNamara's "In retrospect". *Research on Language and Social Interaction*, *33*(2), 179–211. doi:10.1207/S15327973RLSI3302_3

Cox, B. (2008). Developing interpersonal and group dynamics through asynchronous threaded discussions: The use of discussion board in collaborative learning. *Education*, *128*(4), 1–16.

Cramton, C. D., & Hinds, P. (2005). Subgroup dynamics in internationally distributed teams: Ethnocentrism or cross-national learning? In M. Staw & R. M. Kramer (Ed), *Research in organizational behavior, 26,* 231-263. Greenwich, CT: JAI Press.

Crawford, D. (Ed.). (2003). Special issue on service oriented architectures. *Communications of the ACM*, *46*(10).

Creaser, C. (2010). Open access to research outputs--Institutional policies and researchers' views: Results from two complementary surveys. *New Review of Academic Librarianship*, *16*(1), 4–25. doi:10.1080/13614530903162854

Cushner, K., & Brennan, S. (2007). The value of learning to teach in another culture. In Cushner, K., & Brennan, S. (Eds.), *Intercultural student teaching: A bridge to global competence*. Lanham, MD: Rowman & Littleton Education.

Dafoulas, G., & Macaulay, L. (2001). Investigating cultural differences in virtual software teams. *The Electronic Journal on Information Systems in Developing Countries*, *7*(4), 1–14.

Daft, R. L., & Lengel, R. H. (1986). Organizational information requirements, media richness and structural design. *Management Science*, *32*(5), 554–571. doi:10.1287/mnsc.32.5.554

Daft, R., & Lengel, R. (1984). Information richness: A new approach to managerial behavior. In Straiv, B., & Cummings, L. L. (Eds.), *Research in organizational behavior* (pp. 191–223). Homewood, IL: JAI Press.

Dallimore, E. J., Herenstein, J. H., & Platt, M. B. (2004). Classroom participation and discussion effectiveness: Student-generated strategies. *Communication Education*, *53*(1), 103–115. doi:10.1080/0363452032000135805

Danet, B., Rudenberg-Wright, L., & Rosenbaum-Tamari, Y. (1997). Hmmm... Where's that smoke coming from? Writing, play, and performance on Internet relay chat. *Journal of Computer-Mediated Communication*, *2*(4).

Darling, A. L. (1990). Instructional models. In Daily, J., Friedrich, G., & Vangelisti, A. (Eds.), *Teaching communication: Theory, research, and methods* (pp. 267–278). Hillsdale, NJ: Erlbaum.

Davey, L. (1991). The application of case study evaluations. *Practical Assessment, Research & Evaluation, 2*(9). Retrieved September 18, 2010, from http://PAREonline.net/ getvn. asp?v= 2&n=9

Davies, J., & Graff, M. (2005). Performance in e–learning: Online participation and student grades. *British Journal of Educational Technology*, *36*(4), 657–663. doi:10.1111/j.1467-8535.2005.00542.x

Davis, A., & Khazanchi, D. (2007). Does mutual knowledge affect virtual team performance? Theoretical analysis and anecdotal evidence. *American Journal of Business*, *22*(2), 57–65. Retrieved from http://www.jstor.org. doi:10.1108/19355181200700010

Day, J. M., & Kumar, M. (2010). Using SMS text messaging to create individualized and interactive experiences in large classes: A beer game example. *Decision Sciences Journal of Innovative Education*, *8*(1), 129–136. doi:10.1111/j.1540-4609.2009.00247.x

de Freitas, S. (2008). Emerging trends in serious games and virtual worlds. *Becta: Emerging Technologies for Learning, 3*, 57-72. Retrieved from http://partners.becta.org.uk/ upload-dir/downloads/ page_documents/research /emerging_technologies08 _chapter4.pdf

De Roure, D., Gil, Y., & Hendler, J. A. (Eds.). (2004). Special issue on e-science. *IEEE Intelligent Systems*, *19*(1).

Dede, C. (2008). New horizons: A seismic shift in epistemology. *EDUCAUSE Review*, *43*(3), 80–81.

Dede, C., L'Bahy, T. B., & Whitehouse, P. (2002). Designing and studying learning experiences that use multiple interactive media to bridge distance and time. In Vrasidas, C., & Glass, G. V. (Eds.), *Distance education and distributed learning* (pp. 1–30). London, UK: Eurospan.

Denning, D. E. (2002). Activism, hacktivism, and cyberterrorism: The internet as a tool for influencing foreign policy. In Arquilla, J., & Ronfeldt, D. (Eds.), *Networks and netwars: The future of terror, crime, and militancy* (pp. 239–288). Santa Monica, CA: RAND.

DePew, K. M., & Lettner-Rust, H. (2009). Medaitising power: Distance learning interfaces, classroom epistemology, and the gaze. *Computers and Composition*, *26*(3), 174–189. doi:10.1016/j.compcom.2009.05.002

Deutschmann, M., & Panichi, L. (2009). Talking into empty space? Signaling involvement in a virtual language classroom in Second Life. *Language Awareness*, *18*, 310–328. doi:10.1080/09658410903197306

DeVoe, K. M. (2009). Bursts of information: Microblogging. *The Reference Librarian*, *50*(2), 212–214. doi:10.1080/02763870902762086

Dewald, N., Scholz-Crane, A., Booth, A., & Levine, C. (2000). Information literacy at a distance: Instruction design issues. *Journal of Academic Librarianship*, *26*(1), 33–44. doi:10.1016/S0099-1333(99)00121-4

Dewald, N. H. (1999). Web-based library instruction: What is good pedagogy? *Information Technology and Libraries*, *18*(1), 26–31.

Dey, E. L., Burn, H. E., & Gerdes, D. (2009). Bringing the classroom to the Web: Effects of using new technologies to capture and deliver lectures. *Research in Higher Education*, *50*, 377–393. doi:10.1007/s11162-009-9124-0

Dietz, G. (2010). Unified learning and collaboration: Meeting the needs of education and training in the 21st century [White paper]. Retrieved from http://www.elluminate.com/ Resources/ White_Papers/?id=95

DiMaggio, P., Hargittai, E., Celest, C., & Shafer, S. (2004). Digital inequality: From unequal access to differentiated use. In Neckerman, K. (Ed.), *Social inequality* (pp. 355–400). New York, NY: Russel Sage Foundation.

DiMaggio, P., & Hargittai, E. (2001). *From the digital divide to digital inequality: Studying Internet use as penetration increases*. Retrieved July 2, 2006, from http:// www.webuse. umd.edu/ webshop/ resources/ Dimaggio_ Digital_ Divide.pdf

Dimock, P., Kennedy, E. J., & Dimock, J. P. (2010, April). From just talking to taking a stand: Developing partnerships between the speech classroom and the Minnesota Genocide Intervention Network. In States Advisory Council (Sponsor), *State of Minnesota Advisory Council Report*. Symposium conducted at the meeting of the Central States Communication Association, Cincinnati, OH.

Dixson, M. D. (2010). Creating effective student engagement in online courses: What do students find engaging? *Journal of the Scholarship of Teaching and Learning*, *10*(2), 1–13.

Donath, J. S. (1999). Identity and deception in the virtual community. In Smith, M., & Kollock, P. (Eds.), *Communities in cyberspace* (pp. 29–59). New York, NY: Routledge.

Drago, G. P., Setti, E., Licitra, L., & Liberati, D. (2002). Forecasting the performance status of head and neck cancer patient treatment by an interval arithmetic pruned perception. *IEEE Transactions on Bio-Medical Engineering*, *49*(8), 782–787. doi:10.1109/TBME.2002.800788

Drummond, G. (2008). Success in online education. *Distance Learning*, *5*(4), 43–48.

Duff, P. A. (2002). The discursive co-construction of knowledge, identity, and difference: An ethnography of communication in the high school mainstream. *Applied Linguistics*, *23*(3), 289–322. doi:10.1093/applin/23.3.289

Duke University. (2010). *Mobile Web at Duke University*. Retrieved May 28, 2010, from http://m.duke.edu/

Dumova, T., & Fiordo, R. (2010). Preface. In Dumova, T., & Fiordo, R. (Eds.), *Handbook of research on social interaction technologies and collaboration software: Concepts and trends* (*Vol. 1*, pp. xl–xlvi). Hershey, PA: Information Science Reference.

Duncum, P. (2003). The theories and practices of visual culture in art education. *Arts Education Policy Review*, *105*(2), 19–25. doi:10.1080/10632910309603458

Dunlap, J. C., & Lowenthal, P. R. (2009). Tweeting the night away: Using Twitter to enhance social presence. *Journal of Information Systems Education*, *20*(2), 129–135.

Dunn, P. (1996). More intergrated inter and art. *Art Education*, *49*(6), 6–11. doi:10.2307/3193617

Dutton, W. H., Cheong, P. H., & Park, A. (2004). An ecology of constraints on e-learning in higher education: The case of a virtual learning environment. *Prometheus*, *22*(2), 131–149. doi:10.1080/0810902042000218337

E., M. (2009). Higher ed: A-Twitter or not? [Article]. *University Business, 12*(9), 14-14.

Eckstein, A. D. (2009). *Critical agents of social change: Exploiting media to jump-start radical political ideologies*. Unpublished manuscript on classroom project, University of Missouri-Rolla.Eckstein, J. J. (2010). *Evaluation of sexual violence primary prevention training programs*. Unpublished raw data from collaboration with Connecticut Sexual Assault Crisis Services' Rape Prevention and Education Program Curricula, Connecticut Department of Public Health.

Eckstein, J. J., & Pinto, K. (2010, October). *One man up: Young men resisting society to re-envision selves*. Paper presented at the Women & Society Conference, Poughkeepsie, NY.

Eckstein, J. J., Fried, G., Grosser, D., Miller, B., Robinson, S., & Thakkar, U. … Wetzler, M. (2005, February). *Bugscope*: *Current and future directions*. Project presented at the Imaging Technology Group Forum, Beckman Institute for Advanced Science and Technology, University of Illinois, Champaign, IL.

Edelstein, S., & Edwards, J. (2002). If you build it, they will come: Building learning communities through threaded discussions. *Online Journal of Distance Learning Administration*, *5*(1). Retrieved from http://www.westga.edu/ ~distance/ ojdla.

Elliott, B. (2007). *Assessment 2.0: Assessment in the age of Web 2.0*. Scottish Qualifications Authority. Retrieved June 23, 2010, from http://kn.open.ac.uk/public/ document.cfm?docid=10393

Engebretson, J. (2010). Universities log on to hand-held mobile apps. *Connected Planet*, 36-38.

Engeström, Y. (1987). *Learning by expanding: An activity-theoretical approach to developmental research*. Helsinki, Finland: Orienta-Konsultit Oy.

Engeström, Y. (2001). Expansive learning at work: Toward an activity theoretical reconceptualization. *Journal of Education and Work*, *14*(1), 133–156.

Engeström, Y. (1999). Activity theory and individual and social transformation. In Engeström, Y., Miettinen, R., & Punamaki, R. L. (Eds.), *Perspectives on activity theory* (pp. 19–38). Cambridge, UK: Cambridge University Press.

Ertmer, P., Richardson, J., Lehman, J., Newby, T., Cheng, X., Mong, C., & Sadaf, A. (2010). Peer feedback in a large undergraduate blended course: Perceptions of value and learning. *Journal of Educational Computing Research*, *43*(1), 67–88..doi:10.2190/EC.43.1.e

Ess, C. (2003). Liberal arts and distance education. *Arts and Humanities in Higher Education*, *2*(2), 117–137. doi:10.1177/1474022203002002002

Facebook: Some Implications for Academic Freedom, Intellectual Property, and Education. (2009). *Distance Education Report, 13*(24), 4-6.

Fahy, P., Crawford, G., & Ally, M. (2001). Patterns of interaction in a computer conference transcript. *International Review of Research in Open and Distance Learning*, *2*(1).

Fanderclai, T. L. (1996). Like magic, only real. In Cherny, L., & Weise, E. (Eds.), *Wired women: Gender and new realities in cyberspace* (pp. 224–241). Seattle, WA: Seal Press.

Feenberg, A. (1999). Distance learning: Promise or threat? *Crosstalk*, *7*(1), 12–14.

Feeney, M. (2005). Centralizing information about library services and resources. *Internet Reference Services Quarterly*, *9*(1), 129–146. doi:10.1300/J136v09n01_10

Ferrari-Trecate, G., Muselli, M., Liberati, D., & Morari, M. (2003). A clustering technique for the identification of piecewise affine systems. *Automatica*, *39*(2), 205–217. doi:10.1016/S0005-1098(02)00224-8

Ferris, S. P., & Godar, S. (2006). *Teaching and learning with virtual teams*. Hershey, PA: Information Science.

Fetterman, D. M. (1998). *Ethnography: Step by step*. Thousand Oaks, CA: Sage.

Figlio, D. N., Rush, M., & Yin, L. (2010). *Is it live or is it internet? Experimental estimates of the effects of online instruction on student learning*. NBER Working Paper No. 16089. Cambridge, MA: National Bureau of Economic Research.

Finholt, T., & Sproull, L. (1990). Electronic groups at work. *Organization Science*, *1*(1), 41–64. doi:10.1287/orsc.1.1.41

Fischer, C. S. (1992). *America calling: A social history of the telephone to to 1940*. Berkeley, CA: University of California Press.

Fisher, M., & Coleman, B. (2001-2002). Collaborative online learning in virtual discussions. *Journal of Educational Technology Systems*, *30*(1), 3–17. doi:10.2190/CRDK-VNR8-43DX-V4DB

Fitzgibbon, A., Oldham, E., & Johnston, K. (2008). Are Irish student-teachers prepared to be agents of change in using IT in education? In K. McFerrin et al. (Eds.), *Proceedings of Society for Information Technology and Teacher Education International Conference 2008* (pp. 1397-1404). Chesapeake, VA: AACE.

Fortson, B. L., Scotti, J. R., Chen, Y., Malone, J., & Del Ben, K. S. (2007). Internet use, abuse, and dependence among students at a Southeastern regional university. *Journal of American College Health*, *56*(2), 137–144. doi:10.3200/JACH.56.2.137-146

Foster, I., & Kesselman, C. (Eds.). (2004). *The Grid 2: Blueprint for a new computing infrastructure* (2nd ed.). San Francisco, CA: Morgan Kaufmann.

Foster, I., Kesselman, C., Nick, J., & Tuecke, S. (2002). *The physiology of the Grid: An open Grid services architecture for distributed system integration*. Globus. Retrieved February 15, 2011, from http://www.globus.org/ alliance/ publications/ papers/ ogsa.pdf

(2007a). InFrey, L. R., & Carragee, K. M. (Eds.). Communication activism: *Vol. 1. Communication for social change*. Cresskill, NJ: Hampton Press.

(2007b). InFrey, L. R., & Carragee, K. M. (Eds.). Communication activism: *Vol. 2. Media and performance activism*. Cresskill, NJ: Hampton Press.

Friedman, T. L. (2005). *The world is flat: A brief history of the 21st century*. New York, NY: Farrar, Straus and Giroux.

Friedman, S. L. (2006). Watching twin bracelets in China: The role of spectatorship and identification in an ethnographic analysis of film reception. *Cultural Anthropology*, *21*(4), 603–632. doi:10.1525/can.2006.21.4.603

Fulk, J., Schmitz, J., & Ryu, D. (1995). Cognitive elements in the social construction of technology. *Management Communication Quarterly*, *8*, 259–288. doi:10.1177/0893318995008003001

Gaga, L. (2011, September 2). *Lady Gaga: Bad romance* [Video file]. Retrieved from http://www.youtube.com/ watch?v=qrO4YZeyl0I

Gant, C. (2007). Assessing asynchronous discussion: An exploratory hybrid model. *Online Journal of Distance Learning Administration*, *10*(3). Retrieved from http:// www.westga.edu/ ~distance/ ojdla.

Garatti, S., Bittanti, S., Liberati, D., & Maffezzoli, A. (2007). An unsupervised clustering approach for leukemia classification based on DNA micro-arrays data. *Intelligent Data Analysis*, *11*(2), 175–188.

Garton, L., & Wellman, B. (1995). Social impacts of electronic mail in organizations: A review of the research literature. *Communication Yearbook*, *18*, 434–453.

Garton, L., Haythornthwaite, C., & Wellman, B. (1999). Studying on-line social networks. In Jones, S. (Ed.), *Doing Internet research* (pp. 75–105). Thousand Oaks, CA: Sage.

Gates, W. H. (2007, March). Written testimony of William H. Gates, Chairman, Microsoft Corporation: Before the Committee on Health, Education, Labor and Pensions, U.S. Senate. *Business Week*. Retrieved March 14, 2007, from http://www.businessweek.com/ bwdaily/ dnflash/ content/ mar2007/ db20070307_ 617500.htm

Gay, G. (2002). Preparing for culturally responsive teaching. *Journal of Teacher Education*, *53*(2), 106–116. doi:10.1177/0022487102053002003

George, J., & Martin, K. (2004). Forging the library courseware link: Providing library support to students in an online classroom environment. *College & Research Libraries News*, *65*(10), 594–597, 613.

German, K. (2010, April). *Citizen journalists, viral media, and the call for a new ethics of civic engagement*. Paper presented at the meeting of Central States Communication Association, Cincinnati, OH.

Gilroy, M. (2010). Higher education migrates to YouTube and social networks. [Article]. *Education Digest*, *75*(7), 18–22.

Giovannella, C., & Graf, S. (2010). Challenging technologies, rethinking pedagogy, being design-inspired: The grand challenge of this century. *eLearn Magazine*. Retrieved from http://www.elearnmag.org/ subpage.cfm? section= articles&article= 114-1

Glaser, H. R. (2009). Internet governance in Brazil. *International Telecommunication Union News*, *3*, 18. Retrieved September 19, 2010, from http://www.itu.int/ itunews/ manager/ display.asp? lang= en&year= 2009&issue= 03&ipage= 18&ext=html

Gokhale, A. A. (1995). Collaborative learning enhances critical thinking. *Journal of Technology Education*, *7*, 22–30.

Golan, G. J., & Zaidner, L. (2008). Creative strategies in viral advertising: An application of Taylor's six-segment message strategy wheel. *Journal of Computer-Mediated Communication*, *13*, 959–972. doi:10.1111/j.1083-6101.2008.00426.x

Goodall, B. (2008, March 19). *Ryerson U. Won't expel student over Facebook study group*. Retrieved from http:// chronicle.com/ blogPost/ Ryerson-U-Won-t-Expel/3771

Goode, B. (2002). Voice over Internet Protocol (VoIP). *Proceedings of the IEEE*, *90*(9), 1495–1517. doi:10.1109/ JPROC.2002.802005

Goodfellow, R., & Lamy, M. N. (Eds.). (2009). *Learning cultures in online education*. London, UK: Continuum.

Gordon, T., Holland, J., & Lahelma, E. (2001). Ethnographic research in educational settings. In Atkinson, P., Coffey, A., Delamont, S., Lofland, J., & Lofland, L. (Eds.), *Handbook of ethnography* (pp. 188–203). London, UK: Sage.

Grafstein, A. (2002). A discipline-based approach to information literacy. *Journal of Academic Librarianship*, *28*(4), 197–204. doi:10.1016/S0099-1333(02)00283-5

Gude, O. (2004). Postmodern principles: In search of a 21st century art education. *Art Education, 57*(1), 6–14.

Gully, D. J. (2010, April). *To tweet and blog is to deliberate and engage? Exploring online and offline forms of communication to facilitate increased social capital among young citizens*. Paper presented at the meeting of Central States Communication Association, Cincinnati, OH.

Gupta, A., Mattarelli, E., Seshasai, S., & Broschak, J. (2009). Use of collaborative technologies and knowledge sharing in co-located and distributed teams: Toward the 24-h knowledge factory. *The Journal of Strategic Information Systems, 18*, 147–161. doi:10.1016/j.jsis.2009.07.001

Gurney, L. J., & Wikkes, J. (2008). Creating a library presence in online units. *Australian Academic & Research Libraries, 39*(1), 26–37.

Hajela, S. K. (2005, September). *Role of information and communication technologies in managing globalization at the national and regional levels*. Paper presented at the International Conference on Strengthening Regional Cooperation for Managing Globalization. Moscow, Russia.

Halloran, J., Rogers, Y., & Fitzpatrick, G. (2003). From text to talk: Multiplayer games and VoIP. *Proceedings of Level Up: 1st International Digital Games Research Conference,* (pp. 130-142).

Hammersley, M., & Atkinson, P. (1995). *Ethnography: Principles in practices*. New York, NY: Routledge.

Hanna, B. E., & De Nooy, J. (2004). Negotiating cross-cultural difference in electronic discussion. *Multilingua, 23*, 257–281. doi:10.1515/mult.2004.012

Harboe-Ree, C. (2007). Just advanced librarianship: The role of academic libraries as publishers. *Australian Academic and Research Libraries, 38*(1), 15–25.

Hardesty, L. (1995). Faculty culture and bibliographic instruction: An exploratory analysis. *Library Trends, 44*(2).

Hargadon, S. (2009, December 16). *Educational networking: The important role Web 2.0 will play in education* (White Paper). Retrieved April 24, 2010, from http://www.stevehargadon.com/2009/12/social-networking-in-education.html

Harmon, S. W., & Jones, M. G. (2001). An analysis of situated web-based instruction. *Educational Media International, 38*(4), 271–280. doi:10.1080/09523980110105123

Harrington, S. J., & Floyd, K. S. (2009). *Enhancing engagement and the value of the course to the student through course organization and active learning* [Article]. (pp. 3–5). Online Classroom.

Harris, J. (2002). *Wherefore art thou, telecollaboration?* Eugene, OR: International Society for Technology in Education.

Harter, L. M., Dutta, M. J., & Cole, C. E. (Eds.). (2009). *Communication for social impact: Engaging communication theory, research and pedagogy*. Cresskill, NJ: Hampton Press.

Hashimoto, K., & Lehu, J. (2006). Students International Collaboration Project (SICP): A cross-cultural project using virtual teams to learn communication styles. In Ferris, S. P., & Godar, S. (Eds.), *Teaching and learning with virtual teams* (pp. 221–244). Hershey, PA: Information Science Publishing. doi:10.4018/978-1-59140-708-9.ch010

Hauck, M., & Lewis, T. (2007). The Tridem Project. In O'Dowd, R. (Ed.), *Online intercultural exchange* (pp. 167–192). Clevedon, UK: Multilingual Matters.

Heckman, R., & Annabi, H. (2005). A content analytic comparison of learning processes in online and face-to-face case study discussions. *Journal of Computer-Mediated Communication, 10*(2). Retrieved from http:// jcmc.indiana.edu/ vol10/ issue2/ heckman.html.

Heckman, R., & Annabi, H. (2005). A content analytic comparison of learning processes in online and face-to-face case study discussions. *Journal of Computer-Mediated Communication, 10*(2). Retrieved from http:// jcmc.indiana.edu/ vol10/ issue2/ heckman.html.

Hegna, J., & Johnson, D. (2010). Guidelines for educators using social and educational networking sites. [Article]. *Library Media Connection, 28*(5), 50–51.

Heiberger, G., & Harper, R. (2008). Have you Facebooked Astin lately? Using technology to increase student involvement. *New Directions for Student Services, 124*, 19–35. doi:10.1002/ss.293

Helft, M. (2009, October 9). YouTube: We're bigger than you thought. *The New York Times*. Retrieved from http://bits.blogs.nytimes.com/2009/10/09/youtube-were-bigger-than-you-thought/?hp

Hendley, V. (2000). 30 years of higher education. *AAHE Bulletin, 52*(7), 3–8.

Henri, F. (1992). Computer conferencing and content analysis. In Kaye, A. R. (Ed.), *Collaborative learning through computer conferencing* (pp. 117–136). Heidelberg Germany Springer.

Henry, L. A. (2008). Tertiary level digital divide: A more complex understanding and broadened definition. In P. Kommers & P. Isaías (Eds.), *Proceedings of the IADIS International Conference e-Society 2008* (pp. 24-34). International Association for Development of the Information Society (IADIS).

Herring, S. C. (1999). The rhetorical dynamics of gender harassment on-line. *The Information Society, 15*(3), 151–167..doi:10.1080/019722499128466

Herring, S. C. (2004). Computer-mediated discourse analysis: An approach to researching online behavior. In Barab, S. A., Kling, R., & Grey, J. H. (Eds.), *Designing for virtual communities in the service of learning* (pp. 338–376). Cambridge, MA: Cambridge University Press.

Herrington, J., Oliver, R., & Reeves, T. (2003). Patterns of engagement in authentic online learning environments. *Australian Journal of Educational Technology, 19*(1), 59–71.

Hew, K., Gibbs, M., & Wadley, G. (2004). *Usability and sociability of the Xbox Live VoIP channel*. Paper presented at the Australian Workshop on Interactive Entertainment, Sydney, Australia. Retrieved from http://www.ieconference.org/ie2004/proceedings/051%20hew.pdf

Hewitt, J. (2005). Toward an understanding of how threads die in asynchronous computer conferences. *Journal of the Learning Sciences, 14*(4), 567. doi:10.1207/s15327809jls1404_4

Hewling, A. (2005). Culture in the online class: Using message analysis to look beyond nationality-based frames of reference. *Journal of Computer-Mediated Communication, 11*(1), 337–356. doi:10.1111/j.1083-6101.2006.tb00316.x

Hey, T., & Trefethen, A. (2003). E-science and its implications. *Philosophical Transactions of the Royal Society London A. Mathematical, Physical, and Engineering Science, 361*, 1809–1825.

Heyl, J., & McCarthy, J. (2003, January). *International education and teacher preparation in the US.* Presentation at the National Conference Global Challenges and U.S. Higher Education: National Needs and Policy Implications, Durham, N.C. Duke University.

Hine, C. (2000). *Virtual ethnography*. London, UK: Sage.

Hines, S. S. (2008). How it's done: Examining distance education library instruction and assessment. *Journal of Library Administration, 48*(3), 467–478. doi:10.1080/01930820802289565

Hobbs, R. (2010). *Digital and media literacy: A plan of action*. Whitepaper. New York, NY: The Aspen Institute. Retrieved March 30, 2011, from http://www.knightcomm.org/digital-and-media-literacy-a-plan-of-action

Hoegg, R., Meckel, M., Stanoevska-Slabeva, K., & Martignoni, R. (2006). Overview of business models for Web 2.0 communities. *Proceedings of GeNeMe, 2006*, 23–37.

Hofstede, G. (1991). *Cultures and organizations*. New York, NY: McGraw-Hill.

Hollins, E., & Guzman, M. (2005). Research on preparing teachers for diverse populations. In Cochran-Smith, M., & Zeichner, K. (Eds.), *Studying teacher education: The report of the AERA panel on research in teacher education* (pp. 477–548). Mahwah, NJ: Lawrence Erlbaum Associates.

Holmes, B., & Gardner, J. (2006). *e-Learning: Concepts and practice*. London, UK: SAGE Publications.

Holms, M. (2002). *Web usability & navigation*. New York: McGraw-Hill.

hooks, b. (1994). *Teaching to transgress: Education as the practice of freedom*. New York, NY: Routledge.

hooks, b. (1997). Keeping close to home: Class and education. In A. A. Lunsford & J. J.

Hughes, M., & Daykin, N. (2002). Towards constructivism: Investigating students' perceptions and learning as a result of using an online environment. *Innovations in Education and Teaching International*, *39*(3), 217–224. doi:10.1080/13558000210150036

Hunt, F., & Birks, J. (2004). Best practices in information literacy. *Libraries and the Academy*, *4*(1), 27–39. doi:10.1353/pla.2004.0010

Husarik, S. (2007). The impact of digitalization upon the arts and humanities. *International Journal of the Humanities*, *5*(7), 119–126.

Hutchins, E. (1995). *Cognition in the wild*. Cambridge, MA: MIT Press.

Hutchins, H. M. (2003). Instructional immediacy and the seven principles: Strategies for facilitating online courses. *Online Journal of Distance Learning Administration*, *6*(3). Retrieved from http://www.westga.edu/~distance/ ojdla.

Hymes, D. (1977). Qualitative/quantitative research methodologies in education: A linguistic perspective. *Anthropology & Education Quarterly*, *8*(3), 165–176. doi:10.1525/aeq.1977.8.3.05x1511c

Hymes, D. (1962). The ethnography of speaking. In Gladwin, T., & Sturevant, W. C. (Eds.), *Anthropology and human behavior* (pp. 13–53). Washington, DC: Anthropological Society of Washington.

Hymes, D. (1972). Models of the interaction of language and social life. In Gumperz, J. J., & Hymes, D. (Eds.), *Directions in sociolinguistics: The ethnography of communication* (pp. 35–71). New York, NY: Holt, Rinehart and Winston.

Ingram, A. L., Hathorn, L. G., & Evans, A. (2000). Beyond chat on the internet. *Computers & Education*, *35*(1), 21–35. doi:10.1016/S0360-1315(00)00015-4

Internet World Stats. (2011). *Internet usage statistics: The Internet big picture world Internet users and population stats*. Retrieved September 6, 2011, from http://www.internet worldstats.com/ stats.htm

Islam, R. (2005). *Collaboration for distance learning information literacy instruction (SPEC kit 286)*. Washington, DC: Association of Research Libraries.

Issroff, K., & Scanlon, E. (2002). Using technology in higher education: An activity theory perspective. *Journal of Computer Assisted Learning*, *18*(1), 77–83. doi:10.1046/j.0266-4909.2001.00213.x

Jackson, P. A. (2007). Integrating information literacy into Blackboard: Building campus partnerships for successful student learning. *Journal of Academic Librarianship*, *33*(4), 454–461. doi:10.1016/j.acalib.2007.03.010

Jackson, R., & Pellack, L. J. (2004). Internet subject guides in academic libraries: An analysis of contents, practices, and opinions. *Reference and User Services Quarterly*, *43*(4), 319–327.

Jaggars, S. S., & Bailey, T. (2010). *Effectiveness of fully online courses for college students: Response to a department of education meta-analysis*. New York, NY: Community College Research Center, Teachers College, Columbia University. Retrieved from http://ccrc.tc.columbia.edu/ Publication.asp? UID=796

Jarvenpaa, S. L., Knoll, K., & Leidner, D. E. (1998). Is anybody out there? Antecedents of trust in global virtual teams. *Journal of Management Information Systems*, *14*(4), 29–64.

Jarvenpaa, S. L., & Leidner, D. E. (1999). Communication and trust in global virtual teams. *Organization Science*, *10*, 791–815. doi:10.1287/orsc.10.6.791

Jenkins, C. A. (2004). The virtual classroom as ludic space. In Haythornthwaite, C. A., & Kazmer, M. M. (Eds.), *Learning, culture and community in online education* (pp. 163–176). New York, NY: Peter Lang.

Jensen, C., Farnham, S., Drucker, S., & Kollock, P. (2000). The effect of communication modality on cooperation in online environments. *Proceedings of CHI'00* (pp. 470–477). The Hague, Amsterdam.

Johnson, G., & Rayman, J. R. (2007). E-portfolios: A collaboration between student affairs and faculty. *New Directions for Student Services*, *119*, 17–30. doi:10.1002/ss.246

Johnson, E., & Pitcock, J. (2008). Preparing online instructors: Beyond using the technology. In Orvis, K., & Lassiter, A. (Eds.), *Computer-supported collaborative learning: Best practices and principles for instructors* (pp. 89–113). Hershey, PA: Information Science Publishing. doi:10.4018/978-1-59904-753-9.ch005

Jones, A. R. (2008). Riding the technological rapids with the millennials. [Article]. *Christian Librarian, 51*(2), 64–67.

Jones, S., Johnson-Yale, C., Millermaier, S., & Pérez, F. S. (2009). U.S. college students' Internet use: Race, gender and digital divides. *Journal of Computer-Mediated Communication, 14*, 244–264. doi:10.1111/j.1083-6101.2009.01439.x

Kaleta, R., Garnham, C., & Aycock, A. (2003). *Hybrid courses: Obstacles and solutions for faculty and students.* Paper presented at the 19th Annual Conference on Distance Teaching and Learning, Madison, WI. Retrieved from http://www.uwex.edu/ disted/ conference/ Resource_ library/ proceedings/03_ 72.pdf

Kaptelinin, V., Nardi, B. A., & Macaulay, C. (1999). Methods and tools: The activity checklist: A tool for representing the "space" of context. *Interaction, 6*(4), 27–39. doi:10.1145/306412.306431

Karpova, E., Correia, A. P., & Baran, E. (2009). Learn to use and use to learn: Technology in virtual collaboration experience. *The Internet and Higher Education, 12*(1), 45–52..doi:10.1016/j.iheduc.2008.10.006

Kayworth, T. R., Leidner, D. E., & Mora-Tavarez, M. (2001). *Leadership effectiveness in global virtual teams.* Working Paper, Baylor University.

Keating, E., & Mirus, G. (2003). American Sign Language in virtual space: Interactions between deaf users of computer-mediated video communication and the impact of technology on language practices. *Language in Society, 32*, 693–714. doi:10.1017/S0047404503325047

Keating, E. (2001). The ethnography of communication. In Atkinson, P., Coffey, A., Delamont, S., Lofland, J., & Lofland, L. (Eds.), *Handbook of ethnography* (pp. 285–300). London, UK: Sage.

Kern, R. G., Ware, P., & Warschauer, M. (2004). Crossing frontiers: New directions in online pedagogy and research. *Annual Review of Applied Linguistics, 24*, 243–260. doi:10.1017/S0267190504000091

Kern, R. G., & Warschauer, M. (2000). Theory and practice of network-based language teaching. In Warschauer, M., & Kern, R. (Eds.), *Network-based language teaching: Concepts and practice* (pp. 41–58). Cambridge, UK: Cambridge University Press.

Kiesler, S. (2007). Network nation: Human communication via computer. In Erickson, T., & McDonald, D. W. (Eds.), *HCI Remixed: Reflections on notable HCI papers* (pp. 69–75). Cambridge, MA: MIT Press.

King, K. M. (1994). Leading classroom discussions: Using computers for a new approach. *Teaching Sociology, 22*(2), 174–182. doi:10.2307/1318563

Kling, R. (2000). Learning about information technologies and social change: The contribution of social informatics. *The Information Society, 16*(3), 217–232. doi:10.1080/01972240050133661

Knapp, P. B. (1956). A suggested program of college instruction in the use of the library. *The Library Quarterly, 26*, 224–231. doi:10.1086/618322

Knight, W. B., Keifer-Boyd, K., & Amburgy, P. M. (2005). Visual culture explorations: Un/becoming art educators. *Journal of Social Theory in Art Education, 25*, 255–277.

Knowles, M., Holton, E. F. III, & Swanson, R. A. (2005). *The adult learner: The definitive classic in adult education and human resource development* (6th ed.). Burlington, MA: Elsevier.

Knowles, M. S. (1984). *The adult learner: A neglected species.* Houston, TX: Gulf Pub. Co., Book Division.

Knowles, W., & Kerkman, D. (2007). An investigation of students' attitude and motivation toward online learning. *InSight: A Journal of Scholarly Teaching, 52*, 70-80.

Kock, N. (2005). Media richness or media naturalness? The evolution of our biological communication apparatus and its influence on our behavior toward e-communication tools. *IEEE Transactions on Professional Communication, 48*(2), 117–130. doi:10.1109/TPC.2005.849649

Kolin, P. C. (2010). *Writers at work* (9th ed.). Farmington Hills, MI: Cengage Learning.

Koohang, A. (2004a). Development and validation of an instrument for assessing users' views about the usability of digital libraries. *Issues in Informing Science & Information Technology*, *1*, 55–63.

Koohang, A. (2004b). A study of users' perceptions toward e-learning courseware usability. *International Journal on E-Learning*, *3*(2), 10–17.

Koohang, A., & du Plessis, J. (2004). Architecting usability properties in the e-learning instructional design process. *International Journal on E-Learning*, *3*(3), 38–44.

Koohang, A., & Harman, K. (2005). Usability of digital libraries and their reusable objects in e-learning settings. *Issues in Information Systems*, *6*(2), 272–278.

Koohang, A., et al. (2010). Panel discussion: The current state of blended learning. In *Proceedings of Informing Science & IT Education Conference (InSITE) 2010*. Retrieved May 24, 2010, from http://www.proceedings.informingscience.org /InSITE2010/ InSITE10p705-708Koohang774.pdf

Kraemer, E. W. (2003). Developing the online learning environment: The pros and cons of using WebCT for library instruction. *Information Technology and Libraries*, *22*(2), 87–92.

Kraemer, E. W., Lombardo, S. V., & Lepkowski, F. (2007). The librarian, the machine, or a little of both: A comparative study of three information literacy pedagogies at Oakland University. *College & Research Libraries*, *68*(4), 330–342.

Kramarski, B., & Mizrachi, N. (2006). Online discussion and self–regulated learning: Effects of instructional methods on mathematical literacy. *The Journal of Educational Research*, *99*(4), 218–230. doi:10.3200/JOER.99.4.218-231

Krebs, S. A., Hobman, E. V., & Bordia, P. (2006). Virtual teams and group member dissimilarity: Consequences for the development of trust. *Small Group Research*, *37*(6), 721–741. Retrieved from http://online.sagepub.com. doi:10.1177/1046496406294886

Kress, G. (2005). Gains and losses: New forms of texts, knowledge, and learning. *Computers and Composition*, *22*, 5–22. doi:10.1016/j.compcom.2004.12.004

Kurthen, H. (2008). Transnational online student collaboration: Does it work? In B. E. Dietz & L. H. Ritchey (Eds.), *Scaffolding for student success in learning: Best practices in using instructional strategies* (pp. 202-210). Washington, DC: American Sociological Association (ASA Resource Materials for Teaching).

Kuutti, K. (1996). Activity theory as a potential framework for human-computer interaction. In Nardi, B. A. (Ed.), *Context and consciousness: Activity theory and human-computer interaction* (pp. 17–44). Cambridge, MA: MIT Press.

Kvavik, R., Caruso, J., & Morgan, G. (2004). *ECAR study of students and information technology, 2004: Convenience, connection, and control*. Boulder, CO: EDUCAUSE Center for Applied Research.

Lagace, N., & Chisman, J. (2007). How did we ever manage without the OpenURL? *The Serials Librarian*, *52*(1), 211–222. doi:10.1300/J123v52n01_17

Lai, A. (2002). From classrooms to chatrooms: Virtualizing art education. *Art Education*, *55*(4), 33–39. doi:10.2307/3193966

Lai, A., & Ball, E. L. (2004). Students online as cultured subjects: Prolegmena to researching multicultural arts courses on the Web. *Studies in Art Education*, *46*(1), 20–33.

Lange, P. G. (2008). Publicly private and privately public: Social networking on YouTube. *Journal of Computer-Mediated Communication*, *13*, 361–380. doi:10.1111/j.1083-6101.2007.00400.x

Lange, P. d., Suwardy, T., & Mavondo, F. (2003). Integrating a virtual learning environment into an introductory accounting course: Determinants of student motivation. *Accounting Education*, *12*(1), 1–14. doi:10.1080/0963928032000064567

Lapayese, Y. V. (2003). Toward a critical global citizenship education. *Comparative Education Review*, *47*(4), 493–501. doi:10.1086/379495

Larreamendy-Joerns, J., & Leinhardt, G. (2006). Going the distance with online education. *Review of Educational Research*, *76*, 567–605. doi:10.3102/00346543076004567

Larson, B. E., & Keiper, T. A. (2002). Classroom discussion and threaded electronic discussion: Learning in two arenas. *Contemporary Issues in Technology & Teacher Education, 2*(1). Retrieved from http://www.citejournal.org/ vol2/ iss1/ socialstudies/ article1.cfm.

Lea, M., O'Shea, T., Fung, P., & Spears, R. (1992). "Flaming" in computer-mediated communication: Observations, explanations, implications. In M. Lea (Ed.), *Contexts of computer-mediated communication* (pp. 89-112). New York, NY: Wheatsheaf.

Lee, O. (2002). Cultural differences in e-mail use of virtual teams: A critical social theory perspective. *Cyberpsychology & Behavior, 5*(3), 227–232. doi:10.1089/109493102760147222

Lee, S. J. (2009). Online communication and adolescent social ties: Who benefits more from Internet use? *Journal of Computer-Mediated Communication, 14*, 509–531. doi:10.1111/j.1083-6101.2009.01451.x

Lenhart, A., Madden, M., Macgill, A. R., & Smith, A. (2007). *Teens and social media: The use of social media gains a greater foothold in teen life as they embrace the conversational nature of interactive online media.* Washington, DC: Pew Internet & American Life Project.

Lenhart, A., Purcell, K., Smith, A., & Zickuhr, K. (2010). *Social media and mobile internet use among teens and young adults.* Washington, DC: Pew Internet & American Life Project. Retrieved March 16, 2010, from http:// pewinternet.org/ Reports/ 2010/ Social-Media- and-Young-Adults.aspx

Leontiev, A. N. (1978). *Activity, consciousness, and personality.* Englewood Cliffs, NJ: Prentice-Hall.

Leontiev, A. N. (2005a). The fundamental processes of mental life. *Journal of Russian & East European Psychology, 43*(4), 72–75.

Leontiev, A. N. (2005b). The genesis of activity. *Journal of Russian & East European Psychology, 43*(4), 58–71.

Lessig, L. (2008). *Remix.* Penguin Press, 2008.

Leu, D. J. Jr, Kinzer, C. K., Coiro, J., & Cammack, D. (2004). Toward a theory of new literacies emerging from the Internet and other information and communication technologies. In Ruddell, R. B., & Unrau, N. (Eds.), *Theoretical models and processes of reading* (5th ed., pp. 1568–1611). Newark, DE: International Reading Association.

Lewis, P. R. (2009). *Over 15 billion minutes of VoIP have been delivered in Second Life.* Linden Lab. Retrieved from http://lindenlab.com/pressroom /releases/19_05_09

Liang, X., & Creasy, K. (2004). Classroom assessment in web-based instructional environment: Instructors' experience. *Practical Assessment, Research & Evaluation, 9*(7). Retrieved June 30, 2010, from http://PAREonline.net/ getvn.asp?v=9&n=7

Liberati, D. (2008). Plasmopara viticola germination virulence forecasting via Piece-Wise Affine identification for hybrid systems. *Nonlinear Analysis. Hybrid Systems, 2*, 1217–1221. doi:10.1016/j.nahs.2008.09.014

Liberati, D. (2009). Biomedical applications of Piece-Wise Affine identification for hybrid systems. *Annals of Biomedical Engineering, 37*(9), 1871–1876. doi:10.1007/ s10439-009-9750-x

Ligon, S. (2010). *Digital art revolution.* New York, NY: Watson-Guptill Publications.

Lillard, L. L., Norwood, S., Wise, K., Brooks, J., & Kitts, R. (2009). Embedded librarians: MLS students as apprentice librarians in online courses. *Journal of Library Administration, 49*(1), 11–22. doi:10.1080/01930820802310544

Lima, C. O. (2006). *It's not all about access: A comparative study of global citizenship and ICT Use between Brazilian and American students utilizing a social inclusion framework.* Unpublished doctoral dissertation, University of Connecticut, Storrs.

Linden Research, Inc. (2009). *Second Life wiki.* Retrieved from http://wiki.secondlife.com/ wiki/VoIP

Littlejohn, A., & Higgison, C. (2003). *A guide for teachers: e-learning series No. 3, learning and teaching support network (LTSN).* York, England: Generic Centre.

Liu, X., Magjuka, R., & Lee, S. H. (2008). The effects of cognitive thinking styles, trust, conflict management on online students' learning and virtual team performance. *British Journal of Educational Technology*, *39*(5), 829–846..doi:10.1111/j.1467-8535.2007.00775.x

Liu, X. (2010). Empirical testing of a theoretical extension of the technology acceptance model: An exploratory study of educational wikis. *Communication Education*, *59*, 52–69. doi:10.1080/03634520903431745

Liu, S., Gomez, J., Khan, B., & Yen, C. (2007). Toward a learner-oriented community college online course dropout framework. *International Journal on E-Learning*, *6*(4), 519–542.

Lo, L. S., & Dale, J. M. (2009). Information literacy "learning" via online tutorials: A collaboration between subject specialist and instructional design librarian. [Article]. *Journal of Library & Information Services in Distance Learning*, *3*(3/4), 148–158. doi:10.1080/15332900903375325

Lozar Glenn, J. M. (2007). Mastering the science and the art of teaching online. *Business Education Forum*, *61*(3), 10–15.

Luo, L., & Kemp, J. (2008). Second Life: Exploring the immersive instructional venue for library and information science education. *Journal of Education for Library and Information Science*, *49*(3), 147–166.

Lynch, M. M. (2001). Effective student preparation for online learning. *The Technology Source Archives*. Retrieved from http://technologysource.org/ article/ effective_ student_ preparation_ for_ online_ learning/

Maeroff, G. I. (2003). *A classroom of one: How online learning is changing our schools and colleges*. New York, NY: Palgrave.

Magolis, D. (2008). Connecting with online library research methods course students in Facebook. [Article]. *Pennsylvania Library Association Bulletin*, *63*(6), 8–10.

Maibach, E. W., & Parrott, R. L. (Eds.). (1995). *Designing health messages: Approaches from communication theory and public health practice*. Thousand Oaks, CA: Sage.

Maki, R., & Maki, W. (2007). Online courses. *Handbook of Applied Cognition, 2,* 527-552.

Mann, C., & Stewart, F. (2000). *Internet communication and qualitative research: A handbook for researching online*. London, UK: Sage.

Manning, P. (2008). Barista rants about stupid customers at Starbucks: What imaginary conversations can teach us about real ones. *Language & Communication*, *28*, 101–126. doi:10.1016/j.langcom.2008.02.004

Manovich, L. (2002). *The language of new media*. Cambridge, MA: MIT Press.

Manzo, K. K. (2009). Twitter lessons in 140 characters or less. (Cover story). [Article]. *Education Week*, *29*(8), 1–14.

Markel, S. L. (2001). Technology and education online discussion forums: It's in the response. *Online Journal of Distance Learning Administration*, *4*(2). Retrieved from http://www.westga.edu/ ~distance/ ojdla.

Markgraf, J. S. (2005). Librarian participation in the online classroom. *Internet Reference Services Quarterly*, *9*(1-2), 5–19. doi:10.1300/J136v09n01_02

Marx, L., & Smith, M. R. (Eds.). (1994). *Does technology drive history? The dilemma of technological determinism*. Cambridge, MA: MIT Press.

Mayer, R. E. (2001). *Multimedia learning*. New York, NY: Cambridge University Press.

Mayer, R. E., Dow, G. T., & Mayer, S. A. (2003). Multimedia learning in an interactive self-explaining environment: What works in the design of agent-based microworlds? *Journal of Educational Psychology*, *95*, 806–812. doi:10.1037/0022-0663.95.4.806

Mayer, R. E., Hegarty, M., Mayer, S., & Campbell, J. (2005). When static media promote active learning: Annotated illustrations versus narrated animations in multimedia learning. *Journal of Experimental Psychology. Applied*, *11*, 256–265. doi:10.1037/1076-898X.11.4.256

Mayer, R. E., Sobko, K., & Mautone, P. D. (2003). Social cues in multimedia learning: Role of speaker's voice. *Journal of Educational Psychology*, *95*, 419–425. doi:10.1037/0022-0663.95.2.419

Mayo, S. (2007). Implications for art education in the third millennium: Art technology integration. *Art Education*, *60*(3), 45–51.

Mazer, J. P., Murphy, R. E., & Simonds, C. J. (2009). The effects of teacher self-disclosure via Facebook on teacher credibility. [Article]. *Learning, Media and Technology*, *34*(2), 175–183. doi:10.1080/17439880902923655

McGrath, J. E., & Hollingshead, A. B. (1994). *Groups interacting with technology*. Thousand Oaks, CA: Sage.

McIntosh, P. (2005). Gender perspectives on educating for global citizenship. In Noddings, N. (Ed.), *Educating citizens for global awareness* (pp. 22–39). New York, NY: Teachers College Press.

McKinney, J. P., McKinney, K., Franiuk, R., & Schweitzer, J. (2006). The college classroom as a community: Impact on student attitudes and learning. *College Teaching*, *54*(3), 281–284..doi:10.3200/CTCH.54.3.281-284

McLuckie, J., & Topping, K. J. (2004). Transferable skills for online peer learning. *Assessment & Evaluation in Higher Education*, *29*(5), 563–584. doi:10.1080/0260293041000169144

Merchant, G. (2009). Literacy in virtual worlds. *Journal of Research in Reading*, *32*(1), 38–56. doi:10.1111/j.1467-9817.2008.01380.x

Merryfield, M. M. (2002). The difference a global educator can make. *Educational Leadership*, *60*(2), 18–21.

Meyer, K. A. (2003). Face-to-face versus threaded discussions: The role of time and higher-order thinking. *Journal of Asynchronous Learning Networks*, *7*(3), 55–65.

Meyerson, D., Weick, K. E., & Kramer, R. M. (1996). Swift trust and temporary groups. R. M. Kramer, & T. R. Tyler (Eds), *Trust in organizations: Frontiers of theory and research* (pp. 166-195). Thousand Oaks, CA: Sage Publications.

Michaels, S., O'Connor, C., & Resnick, L. B. (2007). Deliberative discourse idealized and realized: Accountable talk in the classroom and in civic life. *Studies in Philosophy and Education*, *27*(4), 283–297. doi:10.1007/s11217-007-9071-1

Mielke, D. (1999). *Effective teaching in distance education*. ERIC Digest.

Miller, D., & Slater, D. (2001). *The Internet: An ethnographic approach*. Oxford, UK: Berg.

Miller, S., & Redman, S. (2010). Improving web-based student learning through online video demonstrations. *Bulletin of the American Astronomical Society*, *42*, 555–559.

Milliot, J. (2010). Consumers show growing satisfaction with e-books. *Publishers Weekly*, *257*(15), 5.

Milotti, E., Chignola, R., Dalla Pellegrina, C., Del Fabbro, A., Farina. M., & Liberati, D. (2008). VBL: Virtual biophysics lab. *Il Nuovo Cimento, 31 C*(1), 109-118.

Min, S. J. (2007). Online vs. face-to-face deliberation: Effects on civic engagement. *Journal of Computer-Mediated Communication*, *12*, 1369–1387. doi:10.1111/j.1083-6101.2007.00377.x

Mon, L. (2009). Questions and answers in a virtual world: Educators and librarians as information providers in Second Life. *Journal of Virtual Worlds Research*, *2*(1). Retrieved from https://journals.tdl.org/jvwr/article/view/398/452.

Morgan, G. (2003). *Faculty use of course management systems*. Boulder, CO: EDUCAUSE Center for Applied Research.

Mueller, J. (2011). *Resources for teaching social psychology: Topic: Persuasion.* Course Resources on the Web. Retrieved from http://jfmueller.faculty.noctrl.edu/ crow/topicpersuasion.htm

Munck, R. (2005). *Globalization and social exclusion: A transformationalist perspective*. Bloomfield, CT: Kumarian Press, Inc.

Murray, D. E. (1988). The context of oral and written language: A framework for mode and medium switching. *Language in Society*, *17*(3), 351–373. doi:10.1017/S004740450001294X

Murray, S. O. (1993). *Theory groups and the study of language in North America: A social history*. Amsterdam, The Netherlands: John Benjamins Publishing.

Muselli, M., & Liberati, D. (2002). Binary rule generation via Hamming clustering. *IEEE Transactions on Knowledge and Data Engineering*, *14*(6), 1258–1268. doi:10.1109/TKDE.2002.1047766

Nagel, L., Blignaut, A. S., & Cronjé, J. C. (2009). Read-only participants: A case for student communication in online classes. *Interactive Learning Environments*, *17*(1), 37–51. doi:10.1080/10494820701501028

Nakamura, L. (1995). Race in/for cyberspace: Identity tourism and relational passing on the Internet. *Works and Days*, *13*, 181–193.

Nardi, B. A., Whittaker, S., & Bradner, E. (2000). Interaction and outeraction: Instant messaging in action. *Proceedings of the ACM Conference on Computer Supported Cooperative Work (CSCW '00), Philadelphia, PA,* (pp. 79-88).

Nass, C., & Gong, L. (2000). Social aspects of speech interfaces from an evolutionary perspective: Experimental research and design implications. *Communications of the ACM*, *43*(9), 36–43. doi:10.1145/348941.348976

National Association for Media Literacy Education. (2007). *Core principles of media literacy education in the United States*. Retrieved February 2, 2011, from http://namle.net/wp-content/uploads/2009/09/NAMLE-CPMLE-w-questions2.pdf

National Center for Academic Transformation. (2005a). *Institution D: Developmental mathematics*. Retrieved March 10, 2011, from http://www.thencat.org/PlanRes/ProposalExD_DevMath.htm

National Center for Academic Transformation. (2005b). *Welcome to the National Center for Academic Transformation*. Retrieved March 30, 2011, from http://www.thencat.org/R2R/R2R%20PDFs/Six%20Models%20for%20Course%20Redesign.pdf

National Council for the Accreditation of Teacher Education. (2007). *Professional standards for the accreditation of teacher preparation institutions*. Retrieved February 15, 2011, from http://www.ncate.org/public/standards.asp?ch=4

New Media Consortium. (2007). *The 2007 horizon report*. Retrieved from http://www.nmc.org/pdf/ 2007_Horizon_Report.pdf

Newman, H. B., Ellisman, M. H., & Orcutt, J. A. (2003). Data-intensive e-science frontier research. *Communications of the ACM*, *46*(11), 68–77. doi:10.1145/948383.948411

Newman, A. A. (2010, July 16). Old Spice argues that real men smell good. *The New York Times,* B3. Retrieved from http://www.nytimes.com/ 2010/ 07/ 16/ business/ media/ 16adco.html?_ r=2

Ng, K. C., & Murphy, D. (2005). Evaluating interactivity and learning in computer conferencing using content analysis techniques. *Distance Education*, *26*(1), 89–109. doi:10.1080/01587910500081327

Nielsen, J. (1993). *Usability engineering*. Boston, MA: Academic Press.

Nielsen, J. (2000). *Designing web usability*. Indianapolis, IN: New Riders.

Nielsen Company. (2009, June). *How teens use media: A Nielsen report on the myths and realities of teen media trends*. Retrieved January 11, 2010, from http://blog.nielsen.com/ nielsenwire/ reports/ nielsen_ howteensuse-media_ june09.pdf

Nielsen, J. (1989). Usability engineering at a discount. In Salvendy, G., & Smith, M. J. (Eds.), *Designing and using human-computer interfaces and knowledge based systems* (pp. 394–401). Amsterdam: Elsevier Science Publishers.

Nielsen, J. (1999). *Users first: Cheap usability tests*. Retrieved March 11, 2000, from http://www.zdnet.com/devhead/stories/ articles/0,4413, 2224316,00.html

Nielsen, J. (2007, January 2). *Fast, cheap, and good: Yes, you can have it all*. Jakob Nielsen's Alertbox. Retrieved June 24, 2010, from http://www.useit.com/alertbox / fast-methods.html

Nielsen, J., & Molich, R. (1990). Heuristic evaluation of user interfaces. In J. Carrasco & J. Whiteside (Eds.), *Proceedings of the ACM CHI 90 Human Factors in Computing Systems Conference* (pp. 249-256). Seattle, WA: ACM Press.

Norman, D. A. (2002). *The design of everyday things*. New York, NY: Doubleday.

Norris, P. (2001). *Digital divide: Civic engagement, information poverty, and the Internet worldwide*. Cambridge, UK: Cambridge University Press.

O'Brien, A., & Alfano, C. (2009, December). Connecting students globally through video-conference pedagogy. *Journal of Online Teaching and Learning, 5*(4). Retrieved from http://jolt.merlot/org/vol5no4/obrien_1209.htm

O'Hara, K., & Stevens, D. (2006). *Inequality.com: Power, poverty and the digital divide*. Oxford, UK: Oneworld Publications.

O'Brien, J. (1999). Writing in the body: Gender (re)production in online interaction. In Smith, M., & Kollock, P. (Eds.), *Communities in cyberspace* (pp. 76–106). New York, NY: Routledge.

Ogbu, J. U. (1992). Understanding cultural diversity and learning. *Educational Researcher, 21*(8), 5–14.

Oinn, T. (2004). *XScufl language reference*. European Bioinformatics Institute. Retrieved February 20, 2011, from http://www.ebi.ac.uk/~tmo/ mygrid/ XScuflSpecification.html

Old Spice. (2011, September 2). *Old Spice: The man your man could smell like* [Video file]. Retrieved from http:// www.youtube.com/ watch?v= owGykVbfgUE

Olivares, O. J. (2008). Collaborative vs. cooperative learning: The instructor's role in computer supported collaborative learning. In Orvis, K. L., & Lassiter, A. L. R. (Eds.), *Computer-supported collaborative learning: Best practices and principles for instructors* (pp. 20–39). Hershey, PA: Information Science Publishing. doi:10.4018/978-1-59904-753-9.ch002

Oliver, K., & Moore, J. (2008). Faculty recommendations for web tools: Implications for course management systems. *Journal of Computing in Higher Education, 19*(2), 3–24. doi:10.1007/BF03033424

O'Reilly, T. (2005). *What is Web 2.0: Design patterns and business models for the next generation of software*. Retrieved May 11, 2010, from http://oreilly.com/web2/ archive/ what-is-web-20.html

Orvis, K. L., & Lassiter, A. L. (2006). Computer-supported collaborative learning: The role of the instructor. In Ferris, S. P., & Godar, S. (Eds.), *Teaching and learning with virtual teams* (pp. 158–179). Hershey, PA: Information Science Publishing. doi:10.4018/978-1-59140-708-9.ch007

Ostrow, L., & DiMaria-Ghalili, R. (2005). Distance education for graduate nursing: One state school's experience. *The Journal of Nursing Education, 44*(1), 5–10.

Ouzts, K. (2006). Sense of community in online courses. *The Quarterly Review of Distance Education, 7*(3), 285–296.

Overby, A. (2009). The new conversation: Using weblogs for reflective practice in the studio art classroom. *Art Education, 62*(4), 18–24.

Owusu-Ansah, E. K. (2004). Information literacy and higher education: Placing the academic library in the center of a comprehensive solution. *Journal of Academic Librarianship, 30*(1), 3–16. doi:10.1016/j.jal.2003.11.002

Owusu-Ansah, E. K. (2007). Beyond collaboration: Seeking greater scope and centrality for library instruction. *Libraries and the Academy, 7*(4), 415–429. doi:10.1353/ pla.2007.0043

Palfrey, J., & Gasser, U. (2008). *Born digital: Understanding the first generation of digital natives*. New York, NY: Basic Books.

Pankl, E., & Ryan, J. (2008). Information commons and Web 2.0 technologies: Creating rhetorical situations and enacting Habermasian ideals in the academic library. In Kelsey, S., & Amant, K. S. (Eds.), *Handbook of research on computer mediated communication* (*Vol. II*, pp. 845–854). Hershey, PA: Information Science Reference. doi:10.4018/978-1-59904-863-5.ch060

Park, Y. J., & Bonk, C. J. (2007). Synchronous learning experiences: Distance and residential learners' perspectives in a blended graduate course. *Journal of Interactive Online Learning, 6*(3), 245–264.

Park, N., Lee, K. M., & Cheong, P. H. (2008). University instructors' acceptance of electronic courseware: An application of the technology acceptance model. *Journal of Computer-Mediated Communication, 13*, 163–186. doi:10.1111/j.1083-6101.2007.00391.x

Partnership for 21st Century Skills. (2009). *Professional development: A 21st century skills implementation guide*. Retrieved March 1, 2011, from http://p21.org/documents/ p21-stateimp_professional_development.pdf

Partnership for 21st Century Skills (P21). (2004). *A framework for 21st century learning.* Retrieved from http://www.p21.org

Pasteur, E. (2007a, July 12, 2007). *The inferno comes to SecondLife.* Retrieved from http://www.secondlifeinsider.com /2007/07/12/ the-inferno-comes -to-second-life/

Pasteur, E. (2007b, October 24). *Land of Lincoln to come to SL.* Retrieved from http://www.secondlifeinsider.com /2007/10/24/ land-of-lincoln- to-come-to-sl/

Patcha, A., & Scales, G. (2005). Development of an internet based distance learning program at Virginia Tech. *Proceedings of the 6th Conference on Information Technology Education.*

Paul, D. L., & McDaniel, R. R. (2004). A field study of the effect of interpersonal trust on virtual collaborative relationship performance. *Management Information Systems Quarterly, 28*(2), 183–227.

Peltz, C. (2003). Web services orchestration and choreography. *Computer, 36*(10), 46–52. doi:10.1109/MC.2003.1236471

Penbek, S., Yurdakul, D., & Cerit, A. G. (July 2009). *Intercultural communication competence. A study about the intercultural sensitivity of university students based on their education and international experiences.* Paper presented at the European and Mediterranean Conference on Information Systems. Izmir, Turkey.

Peters, T., & Bell, L. (2007). MUVEing toward accessibility. *Computers in Libraries, 27*(4), 34–36.

Peters, L., & Karren, R. J. (2009). An examination of the roles of trust and functional diversity on virtual team performance ratings. *Group & Organization Management, 34*(4), 479–504. Retrieved from http://online.sagepub.com. doi:10.1177/1059601107312170

Philipsen, G. (1975). Speaking "like a man" in Teamsterville: Culture patterns of role enactment in an urban neighborhood. *The Quarterly Journal of Speech, 61*(1), 13–23. doi:10.1080/00335637509383264

Philipsen, G. (1992). *Speaking culturally: Explorations in social communication.* Albany, NY: State University of New York Press.

Philipsen, G., & Carbaugh, D. (1986a). A bibliography of fieldwork in the ethnography of communication. *Language in Society, 15*(2), 387–397. doi:10.1017/S0047404500011829

Philipsen, G., & Coutu, L. M. (2005). The ethnography of speaking. In Fitch, K. L., & Sanders, R. E. (Eds.), *Handbook of language and social interaction* (pp. 355–379). Mahwah, NJ: Lawrence Erlbaum Associates.

Philipsen, G., Coutu, L. M., & Covarrubias, P. (2005). Speech codes theory: Restatement, revisions, and response to criticisms. In Gudykunst, W. (Ed.), *Theorizing about intercultural communication* (pp. 55–68). Thousand Oaks, CA: Sage.

Philipsen, G. (1997). A theory of speech codes. In Philipsen, G., & Albrecht, T. L. (Eds.), *Developing communication theories* (pp. 119–156). New York, NY: State University of New York Press.

Pieper, S., Edwards, E., Haist, B., & Nolan, W. (2010). A survey of effective technologies to assess student learning. In Schreiner, C. (Ed.), *Handbook of research on assessment technologies, methods, and applications in higher education* (pp. 47–64). Hershey, PA: Information Science Reference.

Pinjani, P., & Palvia, P. C. (2007). *Diversity in global virtual teams: Results from a field survey.* Abstract from the International Conference on Advances in Interdisciplinary Statistics and Combinatorics, Greensboro, NC.

Polzer, J. T., Crisp, C. B., Jarvenpaa, S. L., & Kim, J. W. (2006). Extending the faultline model to geographically dispersed teams: How colocated subgroups can impair group functioning. *Academy of Management Journal, 49*(4), 679–692. doi:10.5465/AMJ.2006.22083024

Porter, L., & Golan, G. J. (2006). From subservient chickens to brawny men: A comparison of viral advertising to television advertising. *Journal of Interactive Advertising, 6*(2). Retrieved from http://www.jiad.org/ vol6/ no2/ porter/ index.htm.

Portfolio, M. A. T. C. (2010). *Texas Tech University Department of English.* Retrieved March 1, 2011, from http://www.english.ttu.edu/tcr/MATC/matc_portfolio.asp

Prensky, M. (2000). *Digital game-based learning.* New York, NY: McGraw Hill.

Prensky, M. (2005/2006). Listen to the natives. *Educational Leadership*, *63*(4), 8–13.

Quality Matters. (2010). *Quality Matters program*. Retrieved from http://www.qmprogram.org/ new-website-welcome-page

Rainie, L., Estabrook, L., & Witt, E. (2007). *Information searches that solve problems*. Washington, DC: Pew Internet & American Life Project.

Rainie, L. (2009). *Networked learners* (Keynote address at the 2009 Online Learning Symposium, Michigan Virtual University). Retrieved April 24, 2010, from http://www.pewinternet.org/ Presentations/2009/ 52-Networked-Learners.aspx

Rajaram, G. (2007, May). Viral video success advice from a marketing perspective. In R. LeFurgy (Chair), *Content, consumers, community: The balance of control*. Symposium conducted at the meeting of the Interactive Advertising Bureau Digital Video Forum, New York, NY.

Reynolds, K. C., & Nunn, C. E. (1997). Engaging freshmen in classroom discussion: Interaction and the instructor techniques that encourage it. *Proceedings from Annual Meeting of the Association for the Study of Higher Education*. Albuquerque, NM.

Rice R. E., & Atkin, C. K. (Eds.). (2000). *Public communication campaigns* (3rd ed.). Thousand Oaks, CA: Sage.

Rice R. (2002). *Teaching and learning first-year composition with digital portfolios*. Dissertation. Retrieved February 1, 2011, from http://richrice.com/dissertation.pdf

Richardson, J., & Swan, K. (2003). Examining social presence in online courses in relation to students' perceived learning and satisfaction. *Journal of Asynchronous Learning Networks*, *7*(1), 68–88.

Rideout, V. J., Foehr, U. G., & Roberts, D. F. (2010, January). *Generation M2: Media in the lives of 8-to18-year olds*. Menlo Park, CA: The Henry J. Kaiser Family Foundation. Retrieved from http://www.kff.org/ entmedia/ upload/ 8010.pdf

Ritchie, M. (2009). Chirping about Twitter. [Article]. *Times Educational Supplement*, *4836*, 18–21.

Roberson, K. M. (2010, April). *New media technology and political campaigns: Campaigning through texts, tweets, and Facebook*. Paper presented at the meeting of Central States Communication Association, Cincinnati, OH.

Roberts, G. (2006). So many mobile devices, so little time. *Computers in Libraries*, *26*(4), 29–31.

Roberts, D. F., Foehr, U. G., & Rideout, V. (2005). *Generation M: Media in the lives of 8–18 year-olds*. Washington, DC: The Henry J. Kaiser Family Foundation.

Rockinson-Szapkiw, A. J., & Walker, V. L. (2009). Web 2.0 technologies: Facilitating interaction in an online human services counseling skills course. [Article]. *Journal of Technology in Human Services*, *27*(3), 175–193. doi:10.1080/15228830903093031

Rodrigues, D. B., & Sedo, D. R. (2008). Experiencing information literacy in Second Life. *Canadian Journal of Library and Information Practice and Research*, *3*(1). Retrieved from http://journal.lib.uoguelph.ca/ index.php/ perj/article /view/426/861.

Rodriguez, J., Ortiz, I., & Dvorsky, E. (2006). Introducing evolution using online activities in a non-major biology course. *Journal of College Science Teaching*, *35*(6), 31–35.

Rogers, E. M. (1983). *Diffusion of innovations* (3rd ed.). New York, NY: The Free Press.

Roland, C. (2010). Preparing art teachers to teach in a new media landscape. *Art Education*, *63*(1), 17–24.

Rourke, L., & Anderson, T. (2002). Using peer teams to lead online discussions. *Journal of Interactive Media in Education*, *1*(1), 1–21. Retrieved from http://www-jime.open.ac.uk/ 2002/ 1/ rourke- anderson- 02-1.pdf.

Rovai, A. P. (2000). Building and sustaining community in asynchronous learning networks. *The Internet and Higher Education*, *3*(4), 285–297. doi:10.1016/S1096-7516(01)00037-9

Rovai, A. P. (2002). Sense of community, perceived cognitive learning, and persistence in asynchronous learning networks. *The Internet and Higher Education*, *5*(4), 319–332..doi:10.1016/S1096-7516(02)00130-6

Rovai, A. P. (2007). Facilitating online discussions effectively. *The Internet and Higher Education*, *10*(1), 77–88. doi:10.1016/j.iheduc.2006.10.001

Rovai, A. P., & Barnum, K. T. (2003). On-line course effectiveness: An analysis of student interactions and perceptions of learning. *Journal of Distance Education, 18*(1), 57–73.

Rovai, A. P., & Jordan, H. (2004). Blended learning and sense of community: A comparative analysis with traditional and fully online graduate courses. *International Review of Research in Open and Distance Learning, 5*(2). Retrieved from http://www.irrodl.org/ index.php/ irrodl/ article/ view/192.

Rovai, A. P., & Wighting, M. J. (2005). Feelings of alienation and community among higher education students in a virtual classroom. *The Internet and Higher Education, 8*(2), 97–110. doi:10.1016/j.iheduc.2005.03.001

Rüschoff, B., & Ritter, M. (2001). Technology-enhanced language learning: Construction of knowledge and template-based learning in the foreign language classroom. *Computer Assisted Language Learning, 14*(3-4), 219–232. doi:10.1076/call.14.3.219.5789

Russo, T., & Benson, S. (2005). Learning with invisible others: Perceptions of online presence and their relationship to cognitive and affective learning. *Journal of Educational Technology & Society, 8*(1), 54–62.

Ruszkiewicz (Eds.), *The presence of others: Readings for critical thinking and writing* (pp. 85-96). New York, NY: St. Martin's Press.

Rutkowski, A. F., Vogel, D., Van Genuichten, M., & Saunders, C. (2008). Communication in virtual teams: Ten years of experience in education. *IEEE Transactions on Professional Communication, 51*(3), 302–312. doi:10.1109/TPC.2008.2001252

Ryan, J., Miller, R., & Porter, M. (in press). Library services in virtual worlds: An examination of the potential for library services in immersive environments. *Journal of Information Technology Education.*

Ryckman, B., Long, V., & Yedinak, E. (2009). Establishing a baseline: History, evolution and evaluation of Grand Valley State University's off-campus library services. *Journal of Library Administration, 49*(1), 75–88. doi:10.1080/01930820802310890

Salam, U. (2009). *Students' participation in virtual learning environments. Unpublished Ph.D thesis.* Melbourne, Australia: Monash University.

Salaway, G., & Caruso, J. B. (2007). *The ECAR study of undergraduate students and Information Technology.* 2007: EDUCAUSE Center for Applied Research.

Sallnäs, E.-L. (2005). Effects of communication mode on social presence, virtual presence, and performance in collaborative virtual environments. *Presence (Cambridge, Mass.), 14*(4), 434–439. doi:10.1162/105474605774785253

Salmon, G. (2009). The future for (second) life and learning. *British Journal of Educational Technology, 40*(3), 526–538. doi:10.1111/j.1467-8535.2009.00967.x

Salt, B., Atkins, C., & Blackall, L. (2008, October). *Engaging with Second Life: Real education in a virtual world.* Retrieved from http://slenz.files.wordpress.com /2008/12/slliteraturereviewa1.pdf

Salter, D., Pang, M. Y. C., & Sharma, P. (2009). Active tasks to change the use of class time within an outcomes based approach to curriculum design. *The Journal of University Teaching & Learning Practice, 6*(1), 27–38.

New Literacies Research Team. (2007). New literacies, new challenges, and new opportunities. In Sampson, M. B., Szabo, S., Falk-Ross, F., Foote, M. M., & Linder, P. E. (Eds.), *Multiple literacies in the 21st century: The twenty-eighth yearbook of the College Reading Association* (pp. 31–50). Logan, UT: College Reading Association.

Sartorio, A., De Nicolao, G., & Liberati, D. (2002). An improved computational method to assess pituitary responsiveness to secretagogue stimuli. *European Journal of Endocrinology, 147*(3), 323–332. doi:10.1530/eje.0.1470323

Sauers, D., & Walker, R. C. (2004). A comparison of traditional and technology–assisted instructional methods in the business communication classroom. *Business Communication Quarterly, 67*(4), 430–442. doi:10.1177/1080569904271030

Saunders, L. (2007). Regional accreditation organizations' treatment of information literacy: Definitions, collaboration, and assessment. *Journal of Academic Librarianship, 33*(3), 317–326. doi:10.1016/j.acalib.2007.01.009

Saville-Troike, M. (1982). *The ethnography of communication: An introduction*. Baltimore, MD: University Park Press.

Schoenecker, T. S., Martell, K. D., & Michlitsch, J. F. (1997). Diversity, performance, and satisfaction in student group projects: An empirical study. *Research in Higher Education, 38*(4), 479–495. doi:10.1023/A:1024966627400

Scholl, J., McCarthy, J., & Harr, R. (2006). *A comparison of chat and audio in media rich environments*. Paper presented at the CSCW, Banff, Alberta, Canada.

Schonfeld, R. (2007). Getting from here to there, safely: Library strategic planning for the transition away from print journals. *The Serials Librarian, 52*(1), 183–189. doi:10.1300/J123v52n01_15

Schrader, P. G., & Lawless, K. A. (2009). The hidden literacies of massively multiplayer online games. In Pullen, D. L., Gitsaki, C., & Baguley, M. (Eds.), *Technoliteracy, discourse and social practice: Frameworks and applications in the digital age* (pp. 200–219). Hershey, PA: IGI Global. doi:10.4018/978-1-60566-842-0.ch012

Schrire, S. (2004). Interaction and cognition in asynchronous computer conferencing. *Instructional Science, 32*, 475–502. doi:10.1007/s11251-004-2518-7

Schroeder-Moreno, M. S. (2010). Enhancing active and interactive learning online: Lessons learned from an online introductory agroecology course. *NACTA Journal, 54*(1), 21–30.

Schwartz, H. L. (2009). Facebook: The new classroom commons? [Article]. *The Chronicle of Higher Education, 56*(6), B12–B13.

Second Life. (2010). *Private region pricing*. Retrieved May 21, 2010, from http://secondlife.com/land /private-pricing.php

Selfe, C. L., & Selfe, R. J. Jr. (1994). The politics of the interface: Power and its exercise in electronic contact zones. *College Composition and Communication, 45*(4), 480–504. doi:10.2307/358761

Selwyn, N. (2009). Faceworking: Exploring students' education-related use of Facebook. *Learning, Media and Technology, 34*(2), 157–174. doi:10.1080/17439880902923622

Semchuk, S., & Tien, L. (2004). Telling story! Voice in photography: An online visual art studies program evaluation. *International Review of Research in Open and Distance Learning, 5*(3). Retrieved from http://www.irrodl.org/ index.php/ irrodl/ article/ view/ 206.

Shachaf, P. (2008). Cultural diversity and information and communication technology impacts on global virtual teams: An exploratory study. *Information & Management, 45*(2), 131–142..doi:10.1016/j.im.2007.12.003

Shachaf, P., & Hara, H. (2007). Behavioral complexity theory of media selection: A proposed theory for global virtual teams. *Journal of Information Science, 33*(1), 63–74. doi:10.1177/0165551506068145

Shackel, B. (1991). Usability – Context, framework, design and evaluation. In Shackel, B., & Richardson, S. (Eds.), *Human factors for informatics usability* (pp. 21–38). Cambridge, UK: Cambridge University Press.

Shenton, A. K. (2008). The information seeking problems of English high schoolers responding to academic information need. *Library Review, 57*(4), 276–288. doi:10.1108/00242530810868715

Sherblom, J. C. (2010). The computer-mediated communication (CMC) classroom: A challenge of medium, presence, interaction, identity, and relationship. *Communication Education, 59*, 497–523. doi:10.1080/03634523.2010.486440

Shneiderman, B., & Plaisant, C. (2005). *Designing the user interface: Strategies for effective human-computer interaction*. Boston, MA: Pearson/Addison Wesley.

Short, J., Williams, E., & Christie, B. (1976). *The social psychology of telecommunications*. New York, NY: Wiley.

Silver, S. L., & Nickel, L. T. (2007). Are online tutorials effective? A comparison of online and classroom library instruction methods. *Research Strategies, 20*, 389–396. doi:10.1016/j.resstr.2006.12.012

Simkins, S., & Maier, M. (2004). Using just-in-time teaching techniques in the Principles of Economics course. *Social Science Computer Review, 22*(4), 444–456. doi:10.1177/0894439304268643

Simonson, M., Smaldino, S., Albright, M., & Zvacek, S. (2000). *Teaching and learning at a distance: Foundations of distance education*. Upper Saddle River, NJ: Merrill.

Sims, R. (2003). Promises of interactivity: Aligning learner perceptions and expectations with strategies for flexible and online learning. *Distance Education, 24*(1). doi:10.1080/01587910303050

Sitzmann, T., Kraiger, K., Stewart, D., & Wisher, R. (2006). The comparative effectiveness of web-based and classroom instruction: A meta-analysis. *Personnel Psychology, 59*(3), 623–664..doi:10.1111/j.1744-6570.2006.00049.x

Sivunen, A. (2006). Strengthening identification with the team in virtual teams: The leaders' perspective. *Group Decision and Negotiation, 15*(4), 345–366..doi:10.1007/s10726-006-9046-6

Slebodnik, M., & Riehle, C. F. (2009). Creating online tutorials at your libraries: Software choices and practical implications. *Reference and User Services Quarterly, 49*(1), 33–37, 51.

Sloan Consortium. (2010). *Learning on demand: Online education in the United States, 2009* (pp. 1–29). Babson Survey Research Group.

Smillie, D. (2009, November 25). The business of Lady GaGa. *Forbes Magazine*. Retrieved from http://www.forbes.com

Smith, S. S. (2006). *Web-based instruction: A guide for libraries* (2nd ed.). Chicago, IL: American Library Association.

Smith, R. G. (1980). The contract net protocol: High-level communication and control in a distributed problem solver. *IEEE Transactions on Computers, 29*(12), 1104–1113. doi:10.1109/TC.1980.1675516

Smith, K. A. (1994). Cooperation in the college classroom. In *What works: building effective collaborative learning experiences* (pp. 65–70). National Center on Postsecondary Teaching, Learning and Assessment.

Smith, V. (2001). Ethnographies of work and the work of ethnographers. In Atkinson, P., Coffey, A., Delamont, S., Lofland, J., & Lofland, L. (Eds.), *Handbook of ethnography* (pp. 220–233). London, UK: Sage.

Snyder, C., Logue, S., & Preece, B. (1996). *Role of libraries in distance education (SPEC Flyer 216)*. Washington, DC: Association of Research Libraries.

Sodt, J. M., & Summey, T. P. (2009). Beyond the library's walls: Using library 2.0 tools to reach out to all users. *Journal of Library Administration, 49*(1), 97–109. doi:10.1080/01930820802312854

Sproull, L., & Kiesler, S. (1991). *Connections: New ways of working in the networked organization*. Cambridge, MA: MIT Press.

St.Amant, K. (2002). Integrating intercultural online learning experiences into the computer classroom. *Technical Communication Quarterly, 11*(3), 289–315. doi:10.1207/s15427625tcq1103_4

Staley, S. M. (2007). Academic subject guides: A case study of use at San Jose State University. *College & Research Libraries, 68*(2), 119–139.

Staples, D. S., & Zhao, L. (2006). The effects of cultural diversity in virtual teams versus face-to-face teams. *Group Decision and Negotiation, 15*(4), 389–406..doi:10.1007/s10726-006-9042-x

Starke-Meyerring, D., & Andrews, D. (2006). Building a shared virtual learning culture: An international classroom partnership. *Business Communication Quarterly, 69*(1), 25–49. doi:10.1177/1080569905285543

Starke-Meyerring, D., Duin, A. H., Palvetzian, T., & Wilson, M. (2008). Enabling and sustaining globally networked learning environments. In Starke-Meyerring, D., & Wilson, M. (Eds.), *Designing globally networked learning environments*. Rotterdam, The Netherlands: Sense Publishers.

Starke-Meyerring, D., & Wilson, M. (2008). Learning environments for a globally networked world: Emerging visions. In Starke-Meyerring, D., & Wilson, M. (Eds.), *Designing globally networked learning environments*. Rotterdam, The Netherlands: Sense Publishers.

Stefani, L., Mason, R., & Pegler, C. (2007). *The educational potential of e-Portfolios: Supporting personal development and reflective learning*. New York, NY: Routledge.

Steinkuehler, C. (2007). Massively multiplayer online game as a constellation of literacy practices. *E-learning, 4*(3). Retrieved from http://website.education.wisc.edu /steinkuehler/papers/ Steinkuehler_eLearn.pdf. doi:10.2304/elea.2007.4.3.297

Steinkuehler, C. A., Black, R. W., & Clinton, K. A. (2005). Researching literacy as a tool, place, and way of being. *Reading Research Quarterly*, *40*(1), 95–100.

Steinkuehler, C. A. (2004). Learning in massively multiplayer online games. In Y. B. Kafai, W. A. Sandoval, N. Enyedy, A. S. Nixon & F. Herrera (Eds.), *Proceedings of the Sixth International Conference of the Learning Sciences: Embracing Diversity in the Learning Sciences* (pp. 521-528). Mahwah, NJ: Lawrence Erlbaum Associates. Retrieved from http://website.education.wisc.edu/steinkuehler/papers/ SteinkuehlerICLS2004.pdf

Sterne, J. (1999). Thinking the Internet: Cultural studies versus the millenium. In Jones, S. (Ed.), *Doing Internet research: Critical issues and methods for examining the Net* (pp. 257–288). Thousand Oaks, CA: Sage.

Stewart, M. (n.d.). Teaching art at a distance. *eLearn Magazine.* Retrieved from http://www.elearnmag.org/subpage.cfm? section= case_studies& article=39-1

Stoerger, S. (2010). In a virtual classroom, who has a "voice"? A discourse analysis of student-instructor interactions in two Second Life-based courses. In Park, J., & Abels, E. (Eds.), *Interpersonal relations and social patterns in communication technologies: Discourse norms, language structures and cultural variable* (pp. 237–255). Hershey, PA: IGI Global. doi:10.4018/978-1-61520-827-2.ch013

Stone, A. R. (1995). *The war of desire and technology at the close of the mechanical age.* Cambridge, MA: The MIT Press.

Storchi, R., Biella, G., Liberati, D., & Baselli, G. (2009). Extraction and characterization of essential discharge patterns from multisite recordings of spiking ongoing activity. *PLoS ONE*, *4*(1), e4299. doi:10.1371/journal.pone.0004299

Strain, J. (1987). The role of the faculty member in distance education. *American Journal of Distance Education*, *1*(2), 61–65. doi:10.1080/08923648709526585

Sull, E. C. (2008). *Teaching online with Errol* (pp. 33–38). Online Classroom.

Swidler, A. (2001). *Talk of love: How culture matters.* Chicago, IL: University of Chicago Press.

Symantec Corporation. (2009). *Norton online living report 09.* Mountain View, CA: Author. Retrieved from http://www.symantec.com/ about/ news/ release/ article.jsp? prid=20090317_02

Tam, M. (2000). Constructivism, instructional design and technology: Implication for transforming distant learning. *Journal of Educational Technology & Society*, *3*(2).

Taylor, M. C., & Saarinen, E. (1994). *Imagologies: Media philosophy.* New York, NY: Routledge.

Tempelman-Kluit, N. (2006). Multimedia learning theories and online instruction. *College & Research Libraries*, *67*(4), 364–369.

Teng, T. L., & Taveras, M. (2004-2005). Combining live video and audio broadcasting, synchronous chat, and asynchronous open forum discussions in distance education. *Journal of Educational Technology Systems*, *33*(2), 121–129. doi:10.2190/XNPJ-5MQ6-WETU-D18D

Tesone, D. V., & Ricci, P. (2008). Student perceptions of web-based instruction: A comparative analysis. *MERLOT Journal of Online Learning and Teaching*, *4*(3), 317–324.

Thames, G. (2009). *Twitter as an educational tool* (*Vol. 22*, pp. 235–235). Wiley-Blackwell.

Thomas, M. (2002). Learning within incoherent structures: The space of online discussion forums. *Journal of Computer Assisted Learning*, *18*, 351–366. doi:10.1046/j.0266-4909.2002.03800.x

Thomsett-Scott, B., & May, F. (2009). How may we help you? Online education faculty tell what they need from libraries and librarians. *Journal of Library Administration*, *49*, 111–135. doi:10.1080/01930820802312888

Torres, C. A. (1998). Democracy, education, and multiculturalism: Dilemmas of citizenship in a global world. *Comparative Education Review*, *42*(4), 421–447. doi:10.1086/447522

Travica, B. (2005). Virtual organization and electronic commerce. *ACM SIGMIS Database*, *36*(3), 45–68. doi:10.1145/1080390.1080395

Traxler, (2009). Mobile subcultures. In S. Wheeler (Ed.), *Connected minds, emerging cultures: Cybercultures in online learning* (pp. 17-28). Charlotte, NC: Information Age Publishing.

Turkle, S. (1995). *Life on the screen: Identity in the age of the Internet*. New York, NY: Simon & Schuster.

Twigg, C. A. (2003). Improving quality and reducing costs: Designs for effective learning. *Change*, *35*, 22–29. doi:10.1080/00091380309604107

Twitter. (2010). *Twitter is the best way to discover what's new in your world*. Retreived November 23, 2010, from http://twitter.com/about

U.S. Department of Education, National Center for Education Statistics. (2009). *Digest of Education Statistics, 2008* (NCES 2009-020), Table 188. Retrieved from http:// nces.ed.gov/ fastfacts/ display.asp?id=98

U.S. Department of Education, Office of Planning, Evaluation, and Policy Development. (2009). *Evaluation of evidence-based practices in online learning: A meta-analysis and review of online learning studies*. Washington, D.C. Retrieved from http://www2.ed.gov/ rschstat/ eval/ tech/ evidence- based- practices/ finalreport.pdf

US Census Bureau. (2009). Newsroom: Census bureau reports nearly 6 in 10 advanced degree holders age 25-29 are women. Retrieved October 14, 2010, from http:// www.census.gov/ newsroom/ releases/ archives/ education/ cb10-55.html

Valdes, R. (2001). How VoIP works. *HowStuffWorks*. Retrieved from http://communication. howstuffworks. com/ ip-telephony.htm

Van der Aalst, W., & Van Hee, K. M. (2002). *Workflow management: Models, methods, and systems*. Cambridge, MA: MIT Press.

VanSlyke Turk, J., & Martin, E. F. (2004). Opportunities and challenges for intercultural virtual teams: Human factors in project Connect and their implications for using virtual teams in public relations. *Proceedings of the 7th Annual International Public Relations Conference*. Miami, FL. Retrieved from http://instituteforpr.org

Varsity Outreach. (2010). *Facebook page directory for colleges & universities*. Retrieved May 14, 2010, from http://www.varsityoutreach.com/ pagedirectory.aspx

Verduin, J. R., & Clark, T. A. (1991). *Distance education: The foundations of effective practice* (1st ed.). San Francisco, CA: Jossey-Bass.

Vonderwell, S. (2003). An examination of asynchronous communication experiences and perspectives of students in an online course: A case study. *The Internet and Higher Education*, *6*(1), 77–90. doi:10.1016/S1096-7516(02)00164-1

Vrasidas, C., & McIsaac, M. S. (2000). Principles of pedagogy and evaluation for web-based learning. *Educational Media International*, *37*(2), 105–111. doi:10.1080/095239800410405

Wadley, G., Gibbs, M., Hew, K., & Graham, C. (2003, November). Computer supported cooperative play, "third places" and online videogames. *Proceedings of the Thirteenth Australian Conference on Computer Human Interaction (OzChi 03), Brisbane, Australia*, (pp. 238-241).

Wales, T., & Robertson, P. (2008). Captivating open university students with online literature search tutorials created using screen capture software. *Program: Electronic Library and Information Systems*, *42*(4), 365–381. doi:10.1108/00330330810912052

Walther, J. (1994). Anticipated ongoing interaction versus channel effects on relational communication in computer-mediated interaction. *Human Communication Research*, *20*, 473–501. doi:10.1111/j.1468-2958.1994.tb00332.x

Walther, J. (1995). Relational aspects of computer-mediated communication: Experimental observations over time. *Organization Science*, *6*(2), 186–203. doi:10.1287/ orsc.6.2.186

Walther, J., & Parks, M. (2002). Cues filtered out, cues filtered in: Computer-mediated communication and relationships. In Knapp, M., & Daly, J. (Eds.), *Handbook of interpersonal communication* (3rd ed.). Thousand Oaks, CA: Sage.

Walther, J. (1999). *Visual cues and computer mediated communication: Don't look before you leap*. Paper presented at the International Communication Association, San Francisco, CA.

Walther, J. B., Bunz, U., & Bazarova, N. N. (2005). The rules of virtual groups. *Proceeding of the 38th Hawaii International Conference on System Sciences* (pp. 1-10). Waikoloa, HI.

Wang, R. (2006). The lasting impact of a library credit course. *Portal: Libraries and the Academy*, *6*(1), 79–92. doi:10.1353/pla.2006.0013

Warburton, S. (2009). Second Life in higher education: Assessing the potential for and the barriers to deploying virtual worlds in learning and teaching. *British Journal of Educational Technology*, *40*(3), 414–426. doi:10.1111/j.1467-8535.2009.00952.x

Warschauer, M. (2000). Online learning in second language classrooms: An ethnographic study. In Warschauer, M., & Kern, R. (Eds.), *Network-based language teaching: Concepts and practice* (pp. 41–58). Cambridge, UK: Cambridge University Press.

Waschull, S. B. (2001). The online delivery of psychology courses: Attribution, performance, and evaluation. *Teaching of Psychology*, *28*, 143–147. doi:10.1207/S15328023TOP2802_15

Washburn, A. (2008). Finding the library in Blackboard: An assessment of library integration. *MERLOT Journal of Online Learning and Teaching*, *4*(3), 301–316.

Welker, C. B. (2002). The paradigm of viral communication. *Information Services & Use*, *22*, 3–8.

Wellin, C., & Fine, G. A. (2001). Ethnography as work: Career socialization, settings and problems. In Atkinson, P., Coffey, A., Delamont, S., Lofland, J., & Lofland, L. (Eds.), *Handbook of ethnography* (pp. 323–338). London, UK: Sage.

Wendler, C., Bridgeman, B., Cline, F., Millett, C., Rock, J., Bell, N., & McAllister, P. (2010). *The path forward: The future of graduate education in the United States*. Princeton, NJ: Educational Testing Service.

West, R., Waddoups, G., & Graham, C. (2007). Understanding the experiences of instructors as they adopt a course management system. *Educational Technology Research and Development*, *55*(1), 1–26. doi:10.1007/s11423-006-9018-1

Wheeler, S. (2009). *Connected minds, emerging cultures: Cybercultures in online learning*. Charlotte, NC: Information Age Publishing.

Wick, N. (1997). Speech community in the virtual world: The case of one listserv. Unpublished Doctoral dissertation. University of Washington.

Wikipedia. (2007). *Voice over IP*. Retrieved from http://en.wikipedia.org/ wiki/ Voice_over_IP

Williams, D., Caplan, S., & Xiong, L. (2007). Can you hear me now? The impact of VoIP in an online gaming community. *Human Communication Research*, *33*, 427–449. doi:10.1111/j.1468-2958.2007.00306.x

Willson, M. (2010). Technology, networks and communities. *Information Communication and Society*, *13*, 747–764. doi:10.1080/13691180903271572

Witmer, D. F. (1998). Introduction to computer-mediated communication: A master syllabus for teaching communication technology. *Communication Education*, *47*, 162–173. doi:10.1080/03634529809379120

Wojciechowska, I. (2010, July 16). News: Continuing debate over online education. *Inside Higher Ed*. Retrieved from http://www.insidehighered.com/ news/ 2010/ 07/ 16/ online

Wolcott, H. F. (1999). *Ethnography: A way of seeing*. Walnut Creek, CA: AltaMira Press.

Wooldridge, M. J. (2002). *An introduction to multiagent systems*. New York, NY: Wiley.

Xia, J. (2010). A longitudinal study of scholars' attitudes and behaviors toward open-access journal publishing. *Journal of the American Society for Information Science and Technology*, *61*(3), 615–624.

Yakimovicz, A. D., & Murphy, K. L. (1995). Constructivism and collaboration on the Internet: Case study of a graduate class experience. *Computers & Education*, *24*(3), 203–209. doi:10.1016/0360-1315(95)00015-E

Yang, S. (2009). Information literacy online tutorials: An introduction to rationale and technological tools in tutorial creations. *The Electronic Library*, *27*(4), 684–693. doi:10.1108/02640470910979624

Yildiz, S. (2009). Social presence in the web-based classroom: Implications for intercultural communication. *Journal of Studies in International Education*, *13*, 46–65. doi:10.1177/1028315308317654

Yoon, S. W., & Johnson, S. D. (2008). Phases and patterns of group development in virtual learning teams. *Education Technology Development*, *56*, 595–618..doi:10.1007/s11423-007-9078-x

York, A. C., & Vance, J. M. (2009). Taking library instruction into the online classroom: Best practices for embedded librarians. *Journal of Library Administration*, *49*(1), 197–209. doi:10.1080/01930820802312995

Young, J. R. (2008). Forget e-mail: New messaging service has students and professors atwitter. [Article]. *The Chronicle of Higher Education*, *54*(25), A15–A15.

Young, J. R. (2010). Teaching with Twitter. [Article]. *Education Digest*, *75*(7), 9–12.

Young, J. R. (2008, January 25). YouTube professor scholars as online video stars. *The Chronicle of Higher Education*, *54*, A19.

YouTube.com. (2010). *Fact sheet*. Retrieved from http:// www.youtube.com/ t/ fact-sheet

YouTube.com. (2011). *Press room: Statistics*. Retrieved from http://www.youtube.com/ t/ press_statistics

Yunus, A., Kasa, Z., Asmuni, A., Samah, B., Napis, S., & Yusoff, M. (2006). Use of webcasting technology in teaching higher education. *International Education Journal*, *7*(7), 916–923.

Zakaria, E., Chin, L. C., & Daud, M. Y. (2010). The effects of cooperative learning on students' mathematics achievement and attitude towards mathematics. *Journal of the Social Sciences*, *6*(2), 272–275. doi:10.3844/ jssp.2010.272.275

Zellmer-Bruhn, M. E., Maloney, M. M., Bhappy, A. D., & Savaldor, R. (2008). When and how do differences matter? An exploration of perceived similarity in teams. *Organizational Behavior and Human Decision Processes*, *107*(1), 41–59..doi:10.1016/j.obhdp.2008.01.004

Zhang, L., Watson, E. M., & Banfield, L. (2007). The efficacy of computer-assisted instruction versus face-to-face instruction in academic libraries: A systematic review. *Journal of Academic Librarianship*, *33*(4), 478–484. doi:10.1016/j.acalib.2007.03.006

Zhao, Y., & Bryant, F. (2006). Can teacher technology integration training alone lead to high levels of technology integration? A qualitative look at teachers' technology integration after state mandated technology training. *Electronic Journal for the Integration of Technology in Education*, *5*, 62–63.

Zhao, Y., Lei, J., Yan, B., Lai, C., & Tan, H. S. (2005). What makes the difference? A practical analysis of research on the effectiveness of distance education. *Teachers College Record*, *107*(8), 1836–1884. doi:10.1111/j.1467-9620.2005.00544.x

Zhu, E. (2006). Interaction and cognitive engagement: An analysis of four asynchronous online discussions. *Instructional Science*, *34*, 451–480. doi:10.1007/s11251-006-0004-0

Zhu, E. (2008). Breaking down barriers to the use of technology for teaching in higher education. In Robertson, D. R. (Ed.), *To improve the academy* (*Vol. 26*). San Francisco, CA: Jossey-Bass.

Zimmerman, E. (2009). Reconceptualizing the role of creativity in art education theory and practice. *Studies in Art Education*, *50*(4), 382–399.

Zurita, G., & Nussbaum, M. (2007). A conceptual framework based on activity theory for mobile CSCL. *British Journal of Educational Technology*, *38*(2), 211–235. doi:10.1111/j.1467-8535.2006.00580.x

About the Contributors

Sigrid Kelsey is a Librarian at Louisiana State University. She is editor of *Catholic Library World*, and has co-edited several books including *Best Practices for Corporate Libraries* (Libraries Unlimited, 2011), and *Handbook of Research on Computer Mediated Communication* (Information Science Reference, 2008). She a recipient of the *Baton Rouge Business Report*'s Forty under 40 award (2010), the Association of College and Research Libraries (ACRL) College Libraries Section (CLS) ProQuest Innovation in College Librarianship Award (2010), the Louisiana Library Association (LLA) Anthony H. Benoit Mid-Career Award (2010), and the ACRL-Louisiana Scholar Librarian of the Year Award (2009).

Kirk St.Amant is an Associate Professor of Technical and Professional Communication and of International Studies at East Carolina University.

* * *

Tricia Amato is an independent scholar and adjunct instructor in art history, graphic design, and humanities. She has taught traditional and online courses since 2007 and is interested in the possibilities and problems associated with online education, especially in relation to the interactivity of the online platform. She has presented at conferences in the US and abroad on a variety of topics and is an active researcher. Tricia received her MA from the University of Arizona and was the recipient of several awards and scholarships. In addition to online teaching, Tricia's research interests center around women as makers of medieval art.

Lilianna Aniola-Jedrzejek is a Senior Lecturer in the Department of Foreign Languages at Poznan University of Technology, Poznan, Poland, teaching in the faculty of Physics and Mechanical Engineering. She has also served as the Coordinator for International Programmes in the Department of Research and International Cooperation. She has created English for Special Purposes programs at her University and recently has developed a Teaching & Learning Centre for the entire university. She also has significant experience with international projects, and has coordinated or contributed to multiple European Union online learning projects, including the EU Lifelong Learning Programme–LaProf, the EU-financed Engineering Era project, and three EU Leonardo da Vinci programmes, partnering with educational organisations from Germany, Lithuania, Estonia, Finland, and Great Britain.

Diane Boehm has been Director of the University Writing Program at Saginaw Valley State University in Michigan, USA, since 1995. She founded and directs the SVSU Writing Center, which conducts well over 4,000 individual tutorial sessions each year. She is also the director of the Saginaw Bay Writing

Project, one of over 200 federally-funded National Writing Project sites in the US that work to develop teachers of writing at all levels. She teaches writing courses, including both first-year composition and upper level professional writing courses. In these courses, her students build a wiki with students from Poznan University of Technology, Poland, using the Sakai course management system, which includes collaboration tools such as discussion forums, virtual chat, wikis, resource folders, and the like. One of her primary research interests is cross-cultural collaboration using electronic media.

Tatyana Dumova (PhD, Bowling Green State University) is an Associate Professor of Digital Media in the School of Communication at Point Park University (Pittsburgh, Pennsylvania). Her research focuses on the social implications of information and communication technologies and the role of technology in teaching and learning. She has presented and published her research nationally and internationally.

Gaelle Picherit-Duthler (PhD, University of Kentucky) is currently Associate Professor at Zayed University in the United Arab Emirates. She teaches undergraduate and graduate courses in public relations, organizational communication, computer-mediated communication, and research methods. She has written and presented several papers on issues of international public relations, electronic communication, organizational diversity, virtual teams, and organizational culture at regional and international conventions. Her most recent work appears in Public *Relations Journal, Communication Research Reports* as well as a book chapter on Virtual and Collaborative Teams.

Jessica J. Eckstein is currently an Assistant Professor of Communication at Western Connecticut State University (WCSU), where she focuses on teaching and researching interpersonal relationships. Leading classrooms since 2001, she has taught interpersonal relations, ethical responsibilities, gender communication, research methods, nonverbal communication, abusive relationships, and many other communication courses. She is the recipient of grants and awards pertaining to training, technology, research, and teaching. Dr. Eckstein's primary passion is working with university-level students and making things "matter" (in an un-cheesy way) in their daily lives.

Mariela Gunn is an Assistant Professor and Digital Services Librarian at Kresge Library, Oakland University. She holds an MS in Information Studies from the University of Texas at Austin (2002), where she developed a strong interest in how technology interplays with pedagogy in the areas of information literacy and instructional design. Gunn also earned an MA in English from Kansas State University in 2000, which instilled in her a high respect for pedagogically grounded approaches to computer-mediated technologies well before her decision to enter the field of librarianship. Her current research agenda includes information literacy, instructional design, content management, web development, digital tools for research, and open access publishing.

Tabitha Hart is a doctoral candidate in the Department of Communication at the University of Washington. Some of her areas of interest and specialization are the ethnography of communication; speech codes theory and cultural communication; online research methods; intercultural communication, particularly in online settings; customer insight; customer service, particularly international and intercultural customer service; and organizational communication. Her regional foci include the United States, China, Germany, and India. For more information, see her blog at http://tabithahart.net/.

Laurie A. Henry has a PhD in Educational Psychology with emphases in the areas of Cognition & Instruction and Literacy & Technology from the University of Connecticut. She also holds a Master of Arts degree in Education and a Bachelor of Science degree in Special Education from the same institution. Dr. Henry is an Assistant Professor of Early Adolescent Literacy at the University of Kentucky and Co-Director of the P20 College & Career Readiness Lab. Prior to this appointment, she was a member of the New Literacies Research Team and helped develop a new instructional model for Internet-based literacies known as Internet Reciprocal Teaching. Dr. Henry has served on the ReadWriteThink.org advisory board since 2003, and her research interests focus on 21st century literacy skills, instructional uses of technology, issues related to the digital divide, and global/cultural awareness.

Peter Hesseldenz is the Academic Liaison for Business and Economics for the University of Kentucky Libraries. He received a Master's of Library Science degree from the University of Kentucky. He also holds a Master's of Arts degree in English from the University of Kentucky and a Bachelor of Arts degree in English from the University of Wisconsin, Madison. He currently lives in Lexington, Kentucky with his wife and two sons.

Elizabeth W. Kraemer is a librarian and Associate Professor at Oakland University's Kresge Library, where she serves as the library's Coordinator of Information Literacy. She has published a number of articles on library instruction, including the co-authored "The Librarian, the Machine, or a Little of Both: A Comparative Study of Three Information Literacy Pedagogies at Oakland University," which was recognized as a finalist for the 2010 McGraw-Hill—Magna Publications Scholarly Work on Teaching and Learning Award. Kraemer holds a Master's in Library in Information Science from Wayne State University and a Bachelor of Arts degree from Western Michigan University. She is a member of the American Library Association and the Association of College and Research Libraries, and is a past President of ALA's New Members Round Table.

Diego Liberati has a PhD in Electronic and Biomedical Engineering from the Milano Institute of Technology. He is Director of Research, Italian National Research Council and author of 50 papers in ISI journals and editor and author of books and chapters. He is also Secretary of the Biomedical Engineering Society of the Italian Electronic Engineering Association (and Milano prize laureate in 1987); he has chaired Scientific Committees for Conferences and Grants. Dr. Liberati has been a visiting scientist at Rockefeller University, New York University, University of California, and the International Computer Science Institute and has directed joint projects granted by both private and public institutions and mentored dozens of pupils toward and beyond their doctorate.

Clarisse Olivieri de Lima has a PhD in Educational Psychology from the University of Connecticut (2006), a Master's in Brazilian Education (1997) and a Technologist in Data Processing (1994), both from the Pontifical Catholic University of Rio de Janeiro – Puc-Rio, Brazil. She worked as the webmaster for the international virtual organization Kidlink; she was coordinator and instructional designer responsible for the development and implementation of the virtual campus for the Gama Filho University in Rio de Janeiro; she worked as a research assistant for the GlobalEd project and coordinated the Technology Assessment program of State of Connecticut's teachers at the University of Connecticut; she also worked as a technology coordinator at a private K-12 school in Rio de Janeiro. Clarisse has large experience in education, particularly in educational technology, acting mainly in the areas of global citizenship education, distance education, international development, and ICT integration to education.

Heath Martin is Head of Collection Management at the University of Kentucky Libraries. He also serves as Academic Liaison to the Department of Philosophy and the Division of French and Italian Studies. He received a Master's of Library Science degree from Queens College and a Master of Arts in English from Illinois State University. His publications and presentations have focused on the evolution of selection, assessment, and delivery of library resources in higher education. He has previously held faculty appointments in library collection development and acquisitions at Western Carolina University and the Maritime College of the State University of New York.

Rich Rice is Associate Professor of English at Texas Tech University where he specializes in Technical Communication and Rhetoric (see http://richrice.com). He directs the Multiple Literacy Lab, which explores intersections between new media composing and teaching, research, and service. He also teaches online and face-to-face courses in new media and rhetoric, grant writing, multimodal composition, and technical communication. Recent work related to CMC includes topics on ePortfolios, photo essays, remediation and film, and the rhetoric of family. His current projects include an edited collection on ePortfolio rhetoric, a preservice teacher best practice video database, and a new media-rich first-year composition reader focusing on intercultural communication. The reader is called *Writing for Life*.

Laura M. Rusnak, an Ohio native, has a BFA in Illustration from the Cleveland Institute of Art, a MFA in Computer Art from Syracuse University, and is currently working on a PhD in Mass Communication at Ohio University. Her research and teaching interests include Art/Tech studio practices: communication design (online and print), net.art, dynamic web scripting, digital imaging and alternative print processes, and interactive installation along with Art/Media studies: history of new media, internet studies, visual culture, electronic rhetoric, gender/sexuality in media, and women's and gender studies. Laura currently teaches visual arts in the web design and interactive media department for the Art Institute of Pittsburgh's online division.

Jenna Ryan is a Science and Engineering Librarian at Boston University. She earned her MLIS and MA in English from the University of South Carolina in Columbia in 2005 and has undergraduate degrees in English and Marine Science from the same institution. She has been published in the *Electronic Journal of Academic and Special Librarianship*, *The Handbook of Research on Computer Mediated Communication*, and has a forthcoming publication in the *Journal of Information Technology Education*. Her research interests include virtual reference, social networking in the classroom, emerging technologies, and bibliographic instruction.

Urai Salam, BEd (Indonesia), MCALL (Melbourne University, Australia), PhD (Monash University, Australia), is a Lecturer at the English Education and Training, Faculty of Education, Tanjungpura University, Indonesia. He has just started his career in Educational Technology Research. Most recently, he has made an initiative to introduce technology with limited supports in his faculty; even though the faculty does not provide computer facilities and Internet connection for students, he makes efforts to integrate the technology into his classroom processes. As well as lecturing in English Education, Urai

Salam conducts research in educational technology, particularly in teaching and learning English. His research interests extend to computer literacies and their role in effective learning and teaching.

Sharon Stoerger is an instructional design consult in the Learning Technology Center at the University of Wisconsin-Milwaukee. She is also an adjunct Instructor at Mount Mary College, Milwaukee, WI. Her research interests include computer-mediated communication, communities of practice, and social media. More specifically, she has been investigating virtual worlds, their use in formal and informal educational settings, and the ways in which visitors to these online spaces communicate with each other. Much of her recent work examines continuing education courses in Second Life (SL) through an ethnographic approach, including participant observation and informal interviews. In addition, she has been using discourse analysis methods to better understand the interactions between students and instructors in virtual environments, as well as the learning potential of these worlds.

Kathy Tally is an educational technologist at the University of Kansas Medical Center, online faculty for the University of Central Missouri, and a Master Reviewer for Quality Matters, an online quality assurance organization. Her research interests include student learning styles, increasing rigor, relevance, and relationships in the online learning environment, and online teaching pedagogy surrounding best practices and quality assurance. Tally is a graduate of the University of Central Missouri and Central Missouri State University.

René Tanner is a Life Sciences Librarian at Arizona State University. Previously, she was a Reference Librarian at Montana State University. She has experience in the online classroom both as a student and an instructor. She earned her Master's degree in Information Resources and Library Science in 2007 and took the majority of her courses online. This experience gave her insight, from a student's perspective, regarding classroom community building and best practices to motivate and engage students in an online or blended course. René has taught library instruction, electronic library research skills, and a seminar course in philosophy, all of which included a blended learning component.

Maura Valentino is the Coordinator of Digital Initiatives at the University of Oklahoma. In this role she manages the digital collections as well as the Institutional Repository. She also writes and produces a weekly podcast and an occasional vodcast. She graduated from Syracuse University with an MSLIS and a Certificate of Digital Studies. Before graduating, she traveled the east coast teaching Microsoft certified courses in a variety of programming languages. She also taught high school students in standardized test taking skills. It is from these experiences that she developed her interest in education theory and has begun to pursue another graduate degree in that area.

Kristin Whitehair is a Biomedical Librarian at the University of Kansas Medical Center A.R. Dykes Libraries. As a professional Librarian, Whitehair has served community college, undergraduate, and graduate students, staff, practicing health care clinicians, and faculty members. Her research interests include providing library services to distance students, meeting the information needs of non-traditional students, and analyzing data to better inform library decision-making. Whitehair is a graduate of Kansas State University and Louisiana State University.

Yun Xia (PhD, Southern Illinois University, 2002) is currently an Associate Professor in the Department of Communication and Journalism at Rider University (USA). At Rider, he teaches print media and digital media design courses. His research interests are on the social impact of computer-mediated communication, educational applications of communication technologies, visual intelligence of graphic communications in new media, and semiotic analysis of communication signs in new media. His current studies include the exploration of logic forms in analog and digital communication and technology use in different cultures. His works have appeared in journals such as *Human Communication*, *China Media Research*, and *The American Journal of Semiotics*.

Index

V

W

Y

CPSIA information car be obtained at www.ICGtesting.com
Printed in the USA
BVOW021626241011
273893BV00008BC/4/P